Beyond the Apparent Banality of the Mathematics Classroom

Edited by

COLETTE LABORDE, MARIE-JEANNE PERRIN-GLORIAN & ANNA SIERPINSKA

This book was reprinted from *Educational Studies in Mathematics*, Volume 59, Nos. 1–3, 2005.

Springer

Library of Congress Cataloging-in-Publication Data

A. C.I.P. Catalogue record for this book is available from the Library of Congress

ISBN 0-387-25353-x

Published by Springer,
P.O. Box 990, 3311 GX Dordrecht, The Netherlands.

Sold and distributed in the U.S.A. and Canada
by Springer,
101 Philip Drive, Norwell, MA 02061, U.S.A.

In all other countries, sold and distributed
by Springer,
P.O. Box 990, 3311 GX Dordrecht, The Netherlands.

Printed on acid-free paper

Printed in the Netherlands

TABLE OF CONTENTS

BEYOND THE APPARENT BANALITY OF THE MATHEMATICS CLASSROOM

Preface

This book is addressed to researchers in mathematics education. It presents examples of application of a restricted number of theories in the design and/or study of mathematics teaching situations. It shows how these theories work in the practice of research. Without taking into account this 'second order' perspective, the articles may appear as telling banal stories and drawing banal conclusions. The theories reveal that the everyday banality of the mathematics classroom situations is fraught with deep, unresolved didactic problems. They show that these problems will never be solved if we continue to consider these situations as banal.

One example of such apparent banality is the importance of memory in the learning of mathematics, discussed in (Flückiger, this volume). At first sight, it is difficult to see why one would have to observe 50 classroom sessions and analyze them, using, in a subtle way, two sophisticated theories such as the theory of didactic situations (Brousseau, 1997) and the theory of conceptual fields (Vergnaud, 2002), to arrive at this statement. It is hard to understand why student's memory is qualified as 'didactic' and endowed with the status of a theoretical concept which calls for a definition. In what sense is this 'pupil's didactic memory' different from the trivial phenomenon of 'recalling what previously happened in the classroom' and why is this concept 'essential in the understanding of didactic phenomena? To see beyond the banality, it is necessary to abandon the psychological-cognitive perspective and look at the classroom situation as an integral but dynamic system evolving in time. It is also necessary to look at this system not from a social or cultural point of view but from the point of view of a didactician whose objective is to engage students in autonomous mathematical thinking and independent validation of its results. Mathematical validation is based on noticing logical and semantic relations within a system of statements whose meanings must be stable and independent from the individuals who made them and from the time of their production. Thus, if statements — made by students who verbalize the results of their work orally or in writing over one or more classroom sessions — are to be seen by these students as belonging to a common system of knowledge and compared with respect to agreed upon rules of mathematical consistency, they must be *re-called from* their "historical" context of utterance *to* the "logical" here and now of the mathematical debate. For a *mathematical* debate to take place in the classroom, then, students should not consider the work of verbalization of their mathematical work as an isolated, individual school task to be evaluated by the teacher and then quickly forgotten, but as a mere phase in a collective project of common knowledge construction. The class must work as a system, to produce a system of knowledge and this implies the necessity of "record keeping" or memory. Students must be given access to the work of other students and it is the teacher's responsibility to make it possible. This implies, among others, that learning may be very difficult to obtain in situations, not unusual in some places, where students get a new teacher every month or so and the student population is not stable.

The research papers in this collection have at least the following aspects in common:

- The unit of analysis is not a mathematical concept to be taught, students' understanding of this concept, a teacher's project of teaching this concept and choice of didactic means, the institutional and cultural constraints of this choice, the interactions amongst the students and the teacher in the classroom, but all these things at once and more, experienced by the teacher and students as an integral whole.
- This unit of analysis is modeled using a combination of a small and inter-related set of theories, mainly the theory of didactic situations (Brousseau, 1997), and the anthropological theory of didactics (Chevallard, 1992).
- Sharp distinction between the perspectives of the researcher and the teacher in describing and interpreting classroom events.
- Meticulous attention to epistemological and didactic analyses of the mathematical tasks and classroom settings in which they were proposed to students.
- More or less explicit concern with the extreme difficulty, in the ordinary mathematics classroom, of endowing students with a sense of personal intellectual agency and mathematical means of exercising this agency over the directions and results of their mathematical work. In particular – the concern with the difficulty of obtaining that students' reasonings be based on their mathematical knowledge and motivated by intellectual needs such as reduction of uncertainty, relevance, consistency or completeness, and not by opportunistic reasons such as their interpretation of the teacher's expectations. Some papers show how, despite teachers' best intentions and use of innovative approaches (often popularized by research), students' intellectual agency could not be obtained in an

observed classroom situation; others describe the conditions under which it was achieved.

- Abstraction from issues related to such aspects as affect in the teaching and learning of mathematics, multilingualism and multiculturalism of mathematics classes or political decisions regarding the place of mathematics in general compulsory education. The sources of the above mentioned difficulty are sought not in those affective, cultural or political factors, but more in the epistemological features and didactic treatment of mathematical tasks, relative to students' mathematical knowledge and customary behavior.

All this may be regarded as a narrow mathematical-didactical perspective, and criticized for this reason. But the theoretical frameworks used by the authors do not, in principle, rule out taking into account any aspect of the mathematics teaching and learning processes that the researcher finds important or interesting. The aspects chosen by the authors in this volume were a result of conscious decisions, which must be respected as long as the chosen perspective provides plausible and cogent explanations of the observed didactic phenomena. For example, the study, by Brousseau & Gibel of a seemingly interesting lesson based on an investigative activity or "problem situation", shows that, given the students' mathematical knowledge, the characteristics of the mathematical task alone could explain the failure of the teaching situation to provide students with any kind of control over the mathematical validity of their and their peers' solutions, and therefore with an opportunity to develop their mathematical reasoning skills.

Contributors to this volume have so far published mostly (but not solely) in French and Spanish. This English language publication contributes to broadening the readership of their work and opening up a vast domain of theoretical and empirical studies to international debate and, hopefully, further applications and theoretical developments.

The collection of papers that constitutes this book first appeared in 2005 as a special issue of *Educational Studies in Mathematics* (volume 59), under the guest-editorship of Colette Laborde and Marie-Jeanne Perrin-Glorian, and myself in the role of managing editor. In view of the size of the collected material (over 300 pages), the coherence and demonstrated usefulness of the theoretical frameworks, and the fact that these frameworks are not yet well known or understood among the English language readership, the publisher and the editors decided that it is worthwhile to publish a "book spin-off" of the journal special issue.

I finish this preface with a few words of warning as well as encouragement, for readers not familiar with the theoretical perspectives used in the volume. The book is not an "easy read". But, as the persevering reader will see, it is well worth the effort. The reader is generously rewarded with examples of concrete applications of theories that many have found difficult to understand, a multitude of research avenues to pursue, and a coherent set of analytical tools for studying the phenomena of teaching and learning in their full complexity.

REFERENCES

Brousseau, G.: 1997, *Theory of Didactic Situations in Mathematics. Didactique des Mathématiques 1970–1990*, Kluwer Academic Publishers, Dortrecht.

Vergnaud, G.: 2002, 'Towards a cognitive theory of practice', in A. Sierpinska and J. Kilpatrick (eds.), *Mathematics Education as a Research Domain: A Search for Identity. An ICMI Study*, Kluwer Academic Publishers, Dortrecht, pp. 227–240.

Chevallard, Y.: 1992, 'Fundamental concepts in didactics: Perspectives provided by an anthropological approach', in R. Douady and A. Mercier (eds.), *Research in Didactique of Mathematics, Selected Papers*, extra issue of *Recherches en Didactique des Mathématiques*, La Pensée Sauvage, Grenoble, pp. 131–167.

Anna Sierpinska
Montreal
March 15, 2005

COLETTE LABORDE and MARIE-JEANNE PERRIN-GLORIAN

INTRODUCTION TEACHING SITUATIONS AS OBJECT OF RESEARCH: EMPIRICAL STUDIES WITHIN THEORETICAL PERSPECTIVES

ABSTRACT. This volume gathers contributions that share the same double concern: to focus on teaching situations in classrooms, especially the work of the teacher, and to be strongly anchored in original[1] theoretical frameworks allowing to take the classroom situation as unit of analysis. The contributions are not a representative sample of all research sharing this focus worldwide. The theoretical frameworks are grounded mainly (but not solely) in the theory of didactic situations (Brousseau, 1997) and the anthropological theory of didactics (Chevallard, 1992, 1999). There are 11 articles altogether, 9 of which present research works within the chosen theme and focus. The other two are commentary papers offering a reflection on studies of classroom situations from the point of view of other theoretical viewpoints.

KEY WORDS: teaching situations, teacher's activity, classroom situation, theory of didactic situations, anthropological theory of didactics, intertwining of theoretical frameworks and empirical data, dynamics of the teaching/learning process, knowledge progress in class, long term studies, ordinary teaching, time management

1. The classroom teaching situation as unit of analysis

Now that students' learning processes of specific mathematical notions are better known, research in mathematics education may turn to dealing with the complexity of the mathematics classroom.

The classroom is a place where knowledge is transmitted through various processes, in particular through situations that contextualize knowledge and through interactions about this knowledge amongst people (teacher and students) who act within and on these situations. At the same time, teaching in the classroom is part of a broader social project, which aims at educating future adult citizens according to various cultural, social and professional expectations. Thus situated at an intermediate position between the global educational system and the microlevel of individual learning processes, the classroom teaching situation constitutes a pertinent *unit of analysis* for didactic research in mathematics, that is, research into the ternary didactic relationship which binds teachers, students and mathematical knowledge.

[1] "Original" in the sense of having been developed specifically for research in mathematics education and not borrowed from other domains such as psychology, sociology, etc.

Educational Studies in Mathematics (2005) 59: 1–12
DOI: 10.1007/s10649-005-5761-1

The classroom can be considered as a complex didactic system, where one can observe the interplay between teaching and learning as partly shaped by the school institution, which assigns the syllabi and imposes time constraints, but also as not completely determined by the institution. As a result, the study of the classroom offers the researcher an opportunity to gauge the boundaries of the freedom that is left with regard to choices about the knowledge to be taught and the ways of organizing the students' learning.

While the subsystem reduced to one individual learner excludes, in principle, the social dimension, the classroom teaching situation is essentially social in various respects. It reflects the social and cultural education project; it is the place of social interrelations between the teacher and students shaped by the difference of position of the two kinds of actors with respect to knowledge and giving rise to *sociomathematical norms* (Yackel and Cobb, 1996) or to a *didactic contract* (Brousseau, 1989, 1997). It also allows social interactions among students that can be used as a *milieu* (in the sense of the theory of didactical situations) by the teacher to foster learning processes. The size of the classroom teaching situation as a unit of analysis seems to be appropriate for the study of didactic phenomena to grasp the multifaceted complexity of the interrelations between the teaching and learning processes in school.

Taking the classroom situation as a unit of analysis requires the study of the *interrelations* between three main components of the teaching process: the mathematical content to be taught and learned, the management of the various time dimensions, and the activity of the teacher who prepares and manages the class so as to ensure the progress of students' knowledge as well as his or her own teaching experience.

1.1. *The mathematical content*

The mathematical content is itself subject to questioning as regards the way it is introduced, presented, transformed into tasks by the teacher or understood by students. All the papers in the present volume take into account the specific mathematical content in studying classroom situations and develop an analysis which is shaped, to some extent, by that mathematical content. The analysis of the evolution of the memory of the pupil (Flückiger, this volume) is carried out from the perspective of teaching and learning long division in primary school. The social interactions among students as a didactic means to organize the transition from arithmetic to algebra (Sadovsky and Sessa, this volume) are analyzed through the notions of variable and dependency between variables. Robert and Rogalski (this volume) analyze how a teacher uses the rather narrow space available when faced with teaching the use of absolute value in grade 10. The content to be taught may be the core of the paper in that it is subject both to institutional constraints (for example, in the form of a curriculum) and to the choices of the teacher within these constraints. This is exactly the case of the paper by Barbé et al. (this volume), where authors analyze how a teacher adapted his approach to teaching the notion of limit of functions to cope with the existing disjunction between the algebra of limits on the one hand and the topology of limits on the other, institutionally imposed by the organization of the contents to be taught.

1.2. *The issue of time in classroom teaching situations*

The issue of time underlies all studies presented in the papers, whether as one of the dimensions to be taken into account in analyzing the progress of the class over time with respect to knowledge or as the central issue addressed in the paper (Assude, this volume). Taking teaching situations as the object of analysis leads quite naturally to considering time as an important aspect of the teaching process: indeed teaching consists in helping students to construct knowledge which is new relative to what they already know. The management, by the teacher, of this process, called, in the anthropological theory (Chevallard, 1985; Chevallard and Mercier, 1987; Sensevy et al., this volume), progress of didactic time or chronogenesis – is an explicit object of analysis in several papers (Hersant and Perrin-Glorian, Sensevy et al., Robert and Rogalski). But it is also implicit in the study of the evolution of students' solution procedures from arithmetic to algebra in Sadovsky and Sessa (this volume), and in the observation of absence of progress over time in the students' reasoning in a seemingly open "problem-situation" studied by Brousseau and Gibel (this volume).

1.3. *The role of the teacher*

The role of the teacher necessarily becomes central as soon as the classroom situation is taken as the object of study. All the papers address this question by analyzing, for example

- the segmentation of the content to be taught and the organization of the tasks by the teacher as in the papers by Robert and Rogalski, Assude, and Barbé et al.;
- how the teacher is organizing an interplay between the didactic contract and the milieu in order to let students progress in the solving process of a problem situation as in the papers by Hersant and Perrin–Glorian, Sensevy et al., and Sadovsky and Sessa;

– or how the teacher learns or does not learn from the classroom situation and the students' solving procedures as in Margolinas et al. (this volume).

Through classroom situations, various levels of phenomena can be studied.

– *At a micro level*: The solving processes of a particular problem by students in the classroom and its management by the teacher that allows the students to advance in solving the problem.
– *At a meso level*: The teaching of a mathematical theme at a specific level of schooling (several classrooms can be observed in several sessions).
– *At a macro level*: Study of the teaching of a mathematical theme through analyzing the curriculum and time constraints.

Almost all the papers in this volume deal with interactions between two levels: the micro and the meso levels for most of them, the meso and macro levels in the papers by Barbé et al. and Brousseau & Gibel, while Sensevy et al. deals with the micro and macro levels.

2. Theoretical Frameworks

One way of studying the complexity of the mathematics classroom is to use a variety of theoretical frameworks borrowed or adapted from other sciences such as psychology, sociology or epistemology, and analyze each such aspect almost independently from the others. Another is to develop comprehensive theoretical frameworks specific to the study of the mathematics classroom, to model the behaviors of the students and the teacher with respect to the mathematical knowledge to be taught and learnt, while taking into account the situated and institutional character of learning and teaching processes. Papers in this volume illustrate the latter approach.

The specific theoretical frameworks evoked and further developed in the papers presented here have originated, for the most part, in two theories: the theory of didactic situations (Brousseau, 1997); the anthropological theory of didactics and, in particular, the theory of practice or praxeology (Chevallard, 1992, 1999). The theory of conceptual fields (Vergnaud, 1991) has been taken into account in one of the papers, as well. It is important to mention that new developments and extensions of these theories, elaborated over the past ten years or so, conflict neither with the first versions of the theories, nor among each other. Indeed, in the papers, the theories are often combined to offer a deeper and richer understanding of the complexity of the classroom situations.

At the beginning of the development of the theory of didactic situations, research aimed mainly at identifying those features of learning situations (i.e. in the form of problems to be solved by students) that were quasi-independent from the teacher, allowing an almost autonomous construction of knowledge by students (i.e. the so-called *adidactic situations*). The situations were defined in terms of conditions relative to the economy of the functioning of knowledge: knowledge called for by the situation was supposed to make possible an efficient solution to the given problem. These aspects of the theory are presented briefly at the beginning of Brousseau and Gibel's paper. Adidactic situations are designed with a didactic intention but because they are experienced by students as devoid of any teaching intention, they are called *adidactic*. To solve the problem, students must try to seek reasons inherent to mathematical knowledge and not external to mathematics (such as satisfying what they believe to be the teacher's expectations). In such situations, students do not immediately find an efficient solving strategy and the features of the situation must be carefully chosen to allow the evolution of their strategies. This can be modeled as a system of interactions between the student and the situation.

The concept of *milieu* models the elements of the material or intellectual reality, on which the student acts and which may impinge on his/her actions and thought operations. The system of interactions between the student and the milieu is both a consequence and a source of knowledge. When the student acts upon the milieu, he or she receives information and feedback that can destabilize his/her previous knowledge. The equilibrium of the system characterizes a state of knowledge. The destabilized system can lead to the learning of new knowledge. The *objective milieu* is independent of the teacher and of the students. The paper by Sadovsky and Sessa (this volume) is strongly based on this notion of milieu, focusing on a specific milieu of social interactions organized between students. The written judgment of students on other students' solutions is a means used to open up the range of arithmetic solutions to a problem situated at the borderline between arithmetic and algebra; the enlargement of the scope of solutions may lead to considering a systematic variation of the solutions and thus adopting an algebraic point of view.

The design of the milieu is critical for giving the students full responsibility with regard to knowledge. On the other hand, the different positions of the students and the teacher with respect to knowledge shape the interactions between them (this is called the *didactic contract*). The teacher can play on these positions to prompt students' solving strategies that no longer originate fully from mathematical reasons but also from didactical reasons. Examples of such teacher's actions may be found in several papers in this volume, for instance Sensevy et al. and Hersant and Perrin-Glorian.

Some aspects of the theory of didactic situations will be illuminated by their use in the papers Brousseau and Gibel, Sensevy et al., Hersant and Perrin-Glorian, Sadovsky and Sessa, Flückiger, and Margolinas et al. New developments of the theory are used and discussed in most of these papers.

The anthropological theory of didactics focuses, on the one hand, on the organization of mathematics themselves as a human activity involving semiotic instruments and their actual organization within various institutions (*mathematical organizations*), and on the other, on the complex processes carried out by the teacher for organizing interactions between knowledge and students (*didactic organizations*). From the early 1980s, Chevallard (1985) pointed out the constraints bearing on the organization of knowledge in school and the difference between these constraints and those leading to the production of knowledge by mathematicians (didactic transposition). Later, this author (Chevallard, 1992) developed a theoretical framework based on the fundamental notions of institution (in a broad sense: the whole educational system as well as the sixth grade in a country or some particular classroom may be considered as institutions), and of institutional and personal relationships to knowledge. The different scales for institutions allow the researcher to take into account and to study the different expectations concerning the same piece of knowledge and the ways to address the same problem through school levels or in different parts of the school system, i.e. changes in the institutional relationship to knowledge. This framework also allows one to observe the agreement (or not) between the personal and institutional relationships to knowledge and to make a connection between the notion of didactic contract at a microdidactic level and the notion of didactic transposition at a macrodidactic level. The latest developments of the theory, referred to in this volume by Barbé et al., help characterize the mathematical organization actually taught as well as the didactic organization designed by the teacher.

As most papers in this volume summarize the main elements of the theoretical frameworks needed for their study, these elements will not be extensively presented in this introduction. We prefer to focus on the ways they are used in the papers, and, in particular, on showing how several theoretical frameworks may be intertwined. This use of the theories is indeed a critical feature of several papers and it reflects present advances in research on mathematics education.

3. THE FOCUS

3.1. *Towards an analysis of ordinary classroom situations based on the concepts of contract and milieu*

Until ten years ago, the notions of milieu and contract were used mainly as tools for the design of learning situations. Since then, however, partly under the influence of the anthropological theory of didactics, their field of application started to change. By allowing to grasp the responsibilities of students and the teacher with regard to knowledge, they became tools for analyzing the activity of teachers and students in ordinary classroom situations taking into account two main elements of classroom dynamics: time and teacher.

The paper by Brousseau and Gibel (this volume) addresses the issue of a teacher using an open problem situation that does not offer an adidactic milieu for the students' actual knowledge. The students could not enter the problem as it was intended. Since the situation did not present an adidactic nature for the students, the teacher had to use rhetorical means to support the learning. This can serve as a prototypical example of analysis of situations in which what is expected in terms of learning does not occur: in this case the open nature of the problem and the large scope of solving strategies could let one believe there were good conditions for supporting an evolution of the arguments of the students. The paper by Sensevy et al. shows how milieu and contract may be under the control of the teacher and how the teacher, by changing the milieu, is jointly introducing a new rule of action, a new contract to move the didactic time forward. The paper by Hersant and Perrin-Glorian also makes an extensive use of the concepts of milieu and contract in order to analyze the management of the classroom by the teacher. The notions of contract and milieu are unfolded and structured, in particular, by means of a model of layers of the milieu proposed by Brousseau (1989) and adapted by Margolinas (1995). It illustrates very well how the teacher's actions and decisions in everyday conditions in a long-term teaching sequence can be interpreted in terms of milieu and contract: preparation of a milieu, managing breaks of contract or relying on contract in the absence of feedback provided by the milieu. The paper by Sadovsky and Sessa (this volume) uses the notions of milieu and contract to analyze the role of social interactions in the classroom and compares this analysis to studies using other frameworks. The notion of milieu is extended to the learning potential of the teacher in the paper by Margolinas et al. (this volume), which describes how the teacher may acquire "observational didactical knowledge" enabling him/her to interpret unexpected students' strategies in a problem solving activity even if the classroom situation was not designed "to teach the teacher"!

3.2. *Institutional aspects framing the teacher's work analyzed from an anthropological perspective or combining several frameworks*

Anthropological theory contributes, among others, to the analysis of (1) the role of time and its use by the teacher and (2) the various relationships to knowledge of the different actors in the class. Assude's paper (this volume) refers mainly to the first aspect of this theoretical framework and Barbé et al. (this volume) – mainly to the second one. The progress of the teaching is analyzed as a change of positions of students and teacher with regard to knowledge (*topogenesis*) and at the same time as the evolution of knowledge over time (*chronogenesis*) (cf. the paper of Sensevy et al.). Several papers combine concepts coming from both theories (of didactic situations and the anthropological theory) in order to analyze the techniques used by the teacher to move the class forward. The paper by Sensevy et al. expresses it in a very eloquent way by creating the word *mesogenesis* (inspired by chronogenesis and topogenesis) to describe a change of the milieu operated by the teacher. In the same vein, the paper by Flückiger, based on a long-term study of teaching, combines the design of a specific milieu and contract with the journal writing of students and an analysis of the teacher's activity for guiding the individual memory of the students and using it for the progress of knowledge. This paper is also based on a third theoretical framework, the theory of conceptual fields, and identifies invariants in the students' schemes to pinpoint topogenetic shifts and chronogenetic changes in the classroom. The paper by Robert and Rogalski combines didactic and psychological theoretical frameworks in a study of teaching practice in ordinary classroom situations. It carries out a micro didactic analysis of the mathematical tasks given by a teacher (in terms of cognitive and epistemological dimensions) and an ergonomic analysis of the teacher as a professional who must involve the students in the situation. All these papers give evidence of the benefit of using two theoretical frameworks for interpreting the teachers' practice. Often this dual view transforms what could be called a 2D picture of the teacher's activity into a 3D picture and is a good way of grasping the complexity of this activity. Two ways of combining theories are illustrated in this volume: crossing two perspectives on the same object of study or linking concepts coming from different theories.

4. METHODOLOGY

The cornerstone of the constitution of this volume lies in the intertwining of theory and empirical research. As it is clear from the preceding section, questions addressed in the papers are formulated from the perspective of one or more theoretical frameworks but in all papers they are tackled by means of empirical investigations. The papers analyze empirical data obtained in different ways and this analysis is grounded in theoretical foundations. There is a mutual benefit in such approach for both the understanding of teaching phenomena and the robustness of theories; as Goldin wrote (2003, pp. 197–198):

> We need theoretical frameworks that are neither ideological nor fashion-driven. They should be such as to allow their constructs to be subject to validation. Their claims should be, in principle, open to objective evaluation, and subject to confirmation or falsification through empirical evidence.

Empirical data are obtained in the papers through two different means: teaching sequences designed by the researcher with the intention to play on didactic variables in order to allow construction of knowledge by the students (didactic engineering) (Sadovsky and Sessa, this volume; Flückiger, this volume) or analysis of ordinary classroom situations. In the latter case, the classroom sessions have been chosen by the researchers for the following reasons.

- The researchers' interest in the mathematical content and the problems faced by the teachers in subdividing the knowledge to be learned, in order to make links with the students' prior knowledge, or designing tasks (Robert and Rogalski, Hersant and Perrin-Glorian, Margolinas et al., Barbé et al., Assude).
- The researchers' interest in the situations with which the students are faced as in the papers by Brousseau and Gibel – an open problem of the kind that can now be found in primary school textbooks – and Sensevy et al. where the observed situation comes from a well-known experimental teaching process ("Race to 20") designed from the perspective of the theory of didactical situations.

5. THE STRUCTURE

The contents of this volume can be structured according to the aspects of the classroom complexity studied in the articles. The first three papers focus mainly on didactic situations, including their impact on students' behavior and learning or on the teacher's decisions. The next four papers focus on the teacher's management of the classroom situation by analyzing the tension between the progress of his/her teaching project and students' productions, i.e. the time management. Among these papers, the last two address the way the teacher may learn some professional knowledge while managing the classroom situations. The subsequent two papers analyze, from different theoretical perspectives, teacher practice as a professional practice shaped by various institutional constraints, from the macrodidactic level of organizing the curriculum for a given mathematical notion to the microdidactic level of classroom management.

6. RESULTS AND PERSPECTIVES

Classroom situations have been the object of study in other research works grounded in different theoretical backgrounds, of course. This special issue is not intended to be exhaustive. It brings together papers that share common theoretical frameworks in order to build a coherent whole and to allow possible interrelations between various papers. Nevertheless, it is worth noticing that other research trends also focus on the teaching/learning situation as a whole; for example, the projects built around long-term teaching sequences based on the theoretical concepts of *field of experience* and *processes of semiotic mediation*, as presented in the special issue of ESM 39/1.3 (Boero, 1999) or in (Mariotti, 2002). The meaning of mathematical signs and symbols as it develops in the interactive social processes of teaching and learning in the classroom has been analyzed by Steinbring from an epistemological perspective (1998, cf. also his commentary paper in this volume). The role of the teacher in the construction of a shared meaning in the mathematics classroom has also been analyzed in other research (see, for example, Yackel, 2001 or Voigt, 1985). The notion of socio-mathematical norms developed in these studies overlaps with the notion of didactic contract.

A global result certainly coming from the set of studies presented in this volume deals with the dynamics of the teaching/learning process with respect to knowledge progress in class. While it is widely recognized that the relationship to knowledge is variable for students, the studies bring in a new perspective by showing how knowledge taught in the classroom is also changing over time through the teachers' decisions and the interactions between the teacher and the students. Now, as Liping Ma's (1999) study of Chinese and US primary teachers showed, the consideration of mathematical content inside teaching practices is of great importance to the study of these practices and their effects on students' learning.

Several facets of the dynamics have been analyzed by the papers but they all show the importance of several dialectics.

- The interaction between the global and the local level of the teaching/learning processes (or the micro and the macro levels); in particular that it is very important for the teacher to play on the interactions between the local and the global levels for the progress of this dynamic.
- The trajectory of the class with respect to knowledge between, on the one hand, the constraints coming from the teaching system, from knowledge to be taught and from students' knowledge, and, on the other, the teacher's choices. In other words – between the determinants of teaching and the freedom of action of the teacher.

Some of them also express conditions on didactic situations (in terms of mathematics organization as well as of didactic organization) that are needed in order to allow these dialectics to take place between the three poles of the didactic relationship (teacher, students, knowledge).

The very technical nature of the job of the teacher emerges from several papers. The teacher is in charge of moving between local and global levels, as mentioned above. But also in managing a time capital and moving forward the didactic time, the teacher must elaborate and refine strategies. S/he has to plan a cognitive route for the students and must be able to implement it in the reality of the class when interacting with the students, and to adapt it when incidents occur. The theoretical tools used in the papers allow one not only to speak about the techniques of the teacher but also to analyze their functioning. One of the novel aspects brought forth by the papers is to show that the teacher may learn how to improve time management, how to interpret the students' strategies and take them into account.

We believe that the papers presented in this volume will be of interest for the research community as well as contribute to enriching the resources for teacher education through the tools of analysis it provides for tackling the complexity of the role of the teacher in the classroom.

REFERENCES

Boero, P. (ed.): 1999, 'Special issue: Teaching and learning mathematics in context', *Educational Studies in Mathematics* 39, Kluwer Academic Publishers, Dortrecht.

Brousseau, G.: 1989, 'Le contrat didactique: le milieu', *Recherches en Didactique des Mathématiques* 9(3), 309–336.

Brousseau, G.: 1997, *Theory of Didactical Situations in Mathematics. Didactique des Mathématiques 1970–1990*, Kluwer Academic Publishers, Dordrecht, 336 pp.

Chevallard, Y.: 1985, *La Transposition Didactique. Du savoir savant au savoir enseigné*, La Pensée Sauvage, Grenoble.

Chevallard, Y.: 1992, 'Concepts fondamentaux de la didactique: Perspectives apportées par une approche anthropologique', *Recherches en Didactique des Mathématiques* 12(1), 73–111. Translated as 'Fundamental concepts in didactics: Perspectives provided by an anthropological approach', in R. Douady and A. Mercier (eds.), *Research in Didactique of Mathematics*, Selected Papers, extra issue of *Recherches en didactique des mathématiques*, La Pensée sauvage, Grenoble, pp. 131–167.

Chevallard, Y.: 1999, 'Pratiques enseignantes en théorie anthropologique', *Recherches en Didactique des Mathématiques* 19(2), 221–266.

Chevallard, Y. and Mercier, A: 1987, *Sur la formation historique du temps didactique*, Publication de l'IREM d'Aix-Marseille, no. 8, Marseille.

Goldin, G.: 2003, 'Developing complex understandings: On the relation of mathematics education research to mathematics', in R. Even and D.L. Ball (eds.), *Connecting Research, Practice and Theory in the Development and Study of Mathematics Education, Educational Studies in Mathematics, Special Issue*, Vol. 54(2–3), pp. 171–202.

Ma, L.: 1999, *Knowing and Teaching Elementary Mathematics*, Lawrence Erlbaum Associates Publishers, Mahwah, New Jersey.

Margolinas, C.: 1995, 'La structuration du milieu et ses apports dans l'analyse a posteriori des situations', in Margolinas (ed.), *Les débats en didactique des mathématiques*, La Pensée Sauvage, Grenoble, pp. 89–102.

Mariotti, M.-A.: 2002, 'The influences of technological advances on students' mathematical learning', in L. English (ed.), *Handbook of International Research in Mathematics Education*, Lawrence Erlbaum, Mahwah, New Jersey, pp. 695–723.

Steinbring, H.: 1998, 'Elements of epistemological knowledge for mathematics teachers', *Journal of Mathematics Teacher Education* 1(2), 157–189.

Vergnaud, G.: 1991, 'La théorie des champs conceptuels', *Recherches en Didactique des Mathématiques* 10(2–3), 133–169.

Voigt, J.: 1985, 'Patterns and routines in classroom interaction', *Recherches en Didactique des Mathématiques* 6(1), 69–118.

Yackel, E. and Cobb, P.: 1996, 'Sociomathematical norms, argumentation, and autonomy in mathematics', *Journal for Research in Mathematics Education* 22, 390–408.

Yackel, E.: 2001, 'Explanation, justification and argumentation in mathematics classrooms', in M. van den Heuwel-Panhuizen (ed.), *Proceedings of the 25th Conference of the International Group for the Psychology of Mathematics Education*, Vol. 1, Freudenthal Institute, Utrecht University, Utrecht, pp. 9–24.

COLETTE LABORDE
IUFM of Grenoble & University Joseph Fourier, Grenoble

MARIE-JEANNE PERRIN-GLORIAN
IUFM Nord-Pas-de-Calais
& Equipe DIDIREM, University Paris 7, Paris

GUY BROUSSEAU and PATRICK GIBEL

DIDACTICAL HANDLING OF STUDENTS' REASONING PROCESSES IN PROBLEM SOLVING SITUATIONS

ABSTRACT. In this paper, we analyze an investigative situation proposed to a class of 5th graders in a primary school. The situation is based on the following task: In a sale with group rates on a sliding scale, the students must find the lowest possible purchase price for a given number of tickets. A study of students' arguments made it possible to identify a large number of rhetorical forms. However, it turned out that one of the intrinsic features of the situation restricted the teacher's possibilities of making didactical use of the students' forms of reasoning and led him to try to support students' learning with "didactical reasons" rather than with "reasons for knowing".

RÉSUMÉ. L'article analyse une situation de recherche proposée dans une classe de 5ième année de primaire. Dans une vente par lots à tarif dégressif, les élèves doivent minimiser le prix d'achat pour une quantité donnée. L'étude des arguments des uns et des autres fait apparaître de nombreuses formes rhétoriques, mais une propriété intrinsèque de la situation va limiter les possibilités du professeur dans l'utilisation didactique des raisonnements des élèves et va dissocier les raisons de savoir et les raisons didactiques utilisées.

KEY WORDS: didactics, mathematics education, "didactique" of mathematics, didactical situation, devolution, situation of autonomous learning, observation in classroom, reasoning, argument, proof, teaching-learning process, primary school, secondary school, problem solving, word problems, story problems, open problems, linear optimization, theory of situations, *a priori* analysis, *a posteriori* analysis

1. Introduction

The study presented in this paper is a part[1] of an ongoing research on the role of the different forms of reasoning in the didactical relation,[2] in mathematics, at the primary school level.

We start by explaining what we mean by "reasoning" (Section 2). The term is widely used by teachers of all subjects and by researchers, with a variety of meanings. Conversely, many other terms have been used to name different kinds of reasoning. Unfortunately, each usage of the term is linked with a theoretical approach or practice which determines its meaning and makes this usage inappropriate within other approaches. Therefore, we had to directly define the object and the methodology of our study before classifying the different forms of reasoning we were concerned with. Moreover, we define and classify the forms of reasoning

Educational Studies in Mathematics (2005) 59: 13–58
DOI: 10.1007/s10649-005-2532-y

according to their functions in the didactical relation, which has not been done hitherto, at least not systematically. We apologize to the reader for this new presentation of a well-known term. A comparison of this presentation with the existing definitions and classifications is outside the scope of this article.

In mathematics, the teaching of reasoning used to be conceived of as a presentation of model proofs, which then had to be faithfully reproduced by the students. But for today's teachers, as well as for psychologists, reasoning as a mental activity is not a simple recitation of a memorized proof. Whence the idea that it is necessary to confront students with "problems", where it would be "natural" for them to engage in reasoning. If model proofs are still presented to students, they are meant to serve as "model reasoning" which the students could then use in producing their own original forms of reasoning. But there is always the risk of reducing problem solving to an application of recipes and algorithms, which eliminates the possibility of actual reasoning. The risk increases when, to prevent their students from failing, teachers try to teach solving problems in a way which strips these problems of their nature of being problems requiring live mathematical thinking. Using various formal procedures, teachers then try to create more open problems, called "problem situations". We will characterize these in Section 3.

The implementation, however, of these problem situations is beset with a number of difficulties. For example, the student is subject to a greater uncertainty with regard to very heterogeneous questions, while the teacher has to analyze, evaluate and make quick decisions regarding unpredictable student behaviors, which may also be hard to explain or use. Assessment of students' learning becomes more complex. What could be regarded as evidence of the advantages versus the disadvantages of this type of practice for various populations of teachers?

As mentioned above, our study belongs to a larger study whose aim is to determine the main features of all kinds of [teaching] situations and their bearing on the kinds of reasoning which appear in the course of lessons where these situations are being used. In this article, we will confine ourselves to a clinical analysis of a lesson based on the implementation of a problem situation in arithmetic. The problem situation and its development in class will be generally outlined in Section 4. In Section 5, we will identify several forms of reasoning which appeared in class during students' investigation [in small groups] and subsequent whole class presentations and discussions.

In Section 6, we will address the following questions: Did the proposed problem situation favor students' production of forms of reasoning? What is the value of these forms of reasoning? Are they linked with useful learning?

Which didactical decisions of the teacher strongly determine the presence, the meaning and the actual possibilities of processing and using students' forms of reasoning?

We will deal with these questions from the perspective of the Theory of Didactical Situations (TDS), its concepts and its methods. Other contributions will be used as well, if necessary. TDS appeared at the beginning of 1970s and since then has been developed by many researchers from a variety of countries. It has been used and presented also in English language publications. The initiative of Educational Studies in Mathematics (ESM) to present in a Special Issue some studies focused on the teaching situation as a unit of analysis has given us an opportunity to present some of the concrete reasons for the construction and use of this theory to those of ESM readers who have not had a chance to become acquainted with the empirical, theoretical and teaching design studies which led to the development of TDS. We will do our best to restrict the specialized terminology of the theory to those that are indispensable for understanding our particular study, and we will try to justify their use in each case. We hope that these "didactical" precautions will not prevent readers more familiar with our theoretical approach from appreciating our work.

2. Reasoning in the classroom

2.1. *Constructing a model of a subject's reasoning: The notion of "situation"*

The word "reasoning" refers to a domain which is not restricted to that of formal, logical or mathematical forms of reasoning. This is why we decided to start from a rather broad definition, proposed by Oléron (1977; 9), who said that *a reasoning is an ordered set of statements, which are purposefully linked, combined or opposed to each other respecting certain constraints that can be made explicit.*

Let us consider a student stating: "If A then B, by Thales theorem". This statement has the form of a reasoning in the sense of the above definition. But the student will not be credited for this reasoning if he[3] only repeats, upon the teacher's request, a theorem that has been established and written up on the board. On the other hand, the teacher may accept as a reasoning a student's statement of the form, "If A then B" which does not contain a justification, if he regards this justification as obvious (e.g., reference to a common algebraic identity such as $(a + b) = a^2 + 2ab + b^2$). The teacher may even find it important that the student knows what to say and what to omit in his reasoning. Even in the case of a simple action by the student (e.g. drawing a certain straight line in a given figure) the teacher may infer a reasoning, correct or not, which led the student to undertake this action. In this case, only the statement B is observable.

The teacher may interpret the same sentence uttered by different students differently. In particular, the same sentence uttered by a teacher and a student, may be given different interpretations.

Therefore, to be able to claim that a given observable behavior is a sign of a reasoning whose elements are, for the most part, implicit, it is necessary to go beyond the formal definition and examine the conditions in which a "presumed reasoning" can be considered as an "actual reasoning".

Quite often, the teacher interprets students' statements more according to their usefulness for the overall course of the lesson than according to the student's [presumed] initial intentions. It is different for the observer of the lesson (the researcher), who has to justify how a presumed reasoning, of which only a part is explicit or otherwise signaled, can be attributed to the author of this explicit part. The observer has to show that:

- The subject would be able to formulate the presumed reasoning, because he knows or is somehow aware of the rule or fact expressed in the premise A of the reasoning.
- The reasoning is useful (for example, it reduces the level of uncertainty in case a choice has to be made between several possible premises), but its usefulness is intellectual, under the control of the subject's judgment and will, and not based on a cause-effect relationship.
- The reasoning is motivated by an advantage that it affords the subject, by bringing about a positive (from the subject's point of view) change in his environment.
- The reasoning is motivated by "objective" and specific reasons, such as relevance, coherence, adequacy, appropriateness, which justify this particular reasoning (and not any other), as opposed to opportunistic reasons such as conforming to the teacher's expectations. If the student infers B from his understanding of the teacher's expectations, he engages in a reasoning very different from one which is grounded only in his knowledge of mathematics and the premise A.

Thus, in brief, the observer has to show that the reasoning attributed to the subject is intentional, purposeful and useful from the subject's point of view, with respect to his mathematical knowledge.

Thus, amongst all the circumstances in which a reasoning is produced, only some – the one which are necessary – can serve to determine and justify it. These circumstances are not arbitrary. They constitute a coherent set, which we have called "the situation". The situation is only a part of the "context" or the environment in which the actions of the student or the teacher take place, and it includes, among other things, a question to which the reasoning is an answer. The situation can be reduced neither to the action of the subject nor to the knowledge which motivates it, but it is the set of circumstances that create a rational relationship between the two. The situation can explain why a false reasoning has been produced by pointing to causes other than a mistake or inadequacy of the subject's knowledge.

This point of view is a little different from the one that teachers commonly share (for good reasons), that the only really usable forms of reasoning are those that are completely correct. Seldom if ever does a false reasoning become an object of study [in class].

The objective of TDS is to study and construct theoretical models of situations in the sense described above. It is an instrument for the construction of minimal explanations of newly observed facts that would be compatible with already established knowledge.

2.2. *Actual forms of reasoning*

The forms of reasoning studied in this paper will be, essentially, those that can be modeled by inferences of the form, "If the condition A is satisfied, so is (or will be) the condition B". But we need a supplement to this definition because we want to be able to

- distinguish actual reasoning from recitations;
- include reasoning manifested by actions and not only by declarations;
- consider metamathematical and didactical statements as well as mathematical ones;
- distinguish the meaning of the same reasoning according to whether it has been produced by a student or a teacher.

We define, therefore, a *reasoning* as a relation *R* between two elements A and B such that,

- A denotes a condition or an observed fact, which could be contingent upon particular circumstances;
- B is a consequence, a decision or a predicted fact;
- *R* is a relation, a rule, or, generally, something considered as known and accepted. The relation *R* leads the acting subject (the reasoning "agent"), in the case of condition A being satisfied or fact A taking place, to make the decision B, to predict B or to state that B is true.

An *actual* reasoning contains, moreover,

- an agent E (student or teacher) who uses the relation *R*;
- a project, determined by a situation S, which requires the use of this relation.

We can say that, to carry out a project determined by a situation S, the subject uses the relation *R* which allows him to infer B from A. This project can be acknowledged and made explicit by the agent, or it can be attributed to him by the observer on the basis of some evidence.

This definition will still need to be completed for us to be able to distinguish between the students', the teacher's and the observer's reasons and to be able to formally discuss the modalities of our analysis.

2.3. *First classification of forms of reasoning according to their function and type of situation*

As implied in the previous section, a reasoning is characterized by the role it plays in a situation, i.e. by its function in this situation. This function may be to decide about something, to inform, to convince, or to explain. The function of a reasoning varies according to the type of situation in which it takes place; on whether it is a situation of action, formulation, validation or other (Brousseau, 1997: 8–18).

Accordingly, we may expect to be able to distinguish several "levels" of more or less degenerate forms of inferences that are adapted to the different types of situations.

Reasoning of level 3 (N3) is defined as complete formal reasoning based on a sequence of correctly connected inferences, with explicit reference to the elements of the situation or of knowledge considered as shared by the class. It is not postulated that this reasoning be correct. Reasoning of this level is characteristic of situations of validation.

Reasoning of level 2 (*N*2) is defined as reasoning that is incomplete from the formal point of view, but with gaps that can be considered as implicitly filled by the actions of the subject in a situation where a complete formulation would not be justified. Reasoning of this type appears in situations of formulation. It plays a more important role in situations of communication[4] (formulation to a real interlocutor).

Reasoning of level 1 (*N*1) is defined as reasoning that is not formulated as such but can be attributed to the subject based on his actions, and construed as a model of this action (called "an implicit model of action[5]" or "theorem-in-act"[6]).

2.4. *Didactical functions of reasoning according to types of situations*

At any given moment of a lesson, depending on the participants' intentions, there are a large number of more or less overlapping situations. But we are only interested in those that emerge from and influence the collective process, and on which the teacher wants to capitalize to advance the work of the class.

2.4.1. *Reasoning as a solution to a classical mathematical problem*
The teacher presents the students with the *text* of a classical *problem*. This text normally presents a so-called "objective milieu" or situation.

The student regards as "objective milieu" the collection of objects and relations that depend neither on his actions and knowledge (mathematical or meta-mathematical) nor on those of the teacher. The objective milieu is mobilized in a situation of action. It may be real or imaginary. If it is real, the student acts on it and observes the consequences of his actions. If it is imaginary, the student must imagine the functioning of the milieu and how it is transformed under hypothetical actions on it. In either case, the student is an agent operating in function of his Implicit Model of Action (IMA).

The objective milieu can be a situation by itself, in which case we call it "objective situation" (e.g. a story problem, a geometric construction). The student is expected to take it as such, even if it is a made-up situation.

The problem calls for solutions and/or proofs whose validity is assumed to be independent from the didactical circumstances in which the problem is given. The standard solution, i.e. a solution that could be produced by the teacher and is expected of the student, has the form of a sequence of inferences (and calculations), which is correctly connected, i.e. conform to rules of logic. The teacher calls this *the* solution or *the* correct reasoning associated with the problem. We will call it the "standard solution" in this paper. Each step of the reasoning is supplemented (if necessary for understanding) with standard logical and mathematical justifications, whose validity and relevance appear to the student and the teacher as well as the observer, to be independent of the situation.

2.4.2. *The student's actual reasoning in solving a classical problem*
However, a student's actual reasoning is the product of a mental activity which may be different from the standard solution, and it is a response to a situation containing, but not confined to, the formulation of the problem. The student does all sorts of things to find the expected solution but he doesn't have to give an account of all this process in the final product. Therefore the observer's, just like the teacher's, interpretation of students' solutions must take into account a much larger and more complex system, if he wants to be able to challenge them or explain why such and such forms of reasoning, correct or not, have been produced. Therefore, to be able to discuss students' solutions in class, the teacher must assume, at least implicitly, that students are working under assumptions about reality that are more open than those of the "objective situation", stated explicitly in the text of the problem alone. These assumptions about reality may be at the source of students' tactical, strategic or ergonomic justifications about the

validity and the adequacy of the choice of inferences and their connections, which are not part of the standard solution.

The teacher has several options in justifying the standard solution for the students, while discrediting the solutions he considers incorrect.

One option is to justify the construction of the solution of the problem by

- bringing into play the knowledge students had been taught and that they are supposed to have learned;
- taking into account information about the "objective situation" as given in the text of the problem.

Another is to use an original reasoning which is, nevertheless, logically reducible to the information given in the problem and the presumed students' knowledge, as in the previous case. In doing this, the teacher, more or less consciously, lays a wager on students' heuristic abilities (which he wins with respect to some students and loses with others).

A third option is to refer to conditions which are not included in the presumed students' knowledge and which cannot be logically deduced from the text of the problem. This option is rarely taken by the teacher in the case of classical problems; it is more likely to occur in the so-called "open" problems. In this case, the students alone cannot construct the standard solution and the teacher must intervene at some point to bring it forth. Moreover, the teacher cannot make the solution appear to the students as a "reasoned" consequence of a combination of the conditions given in the text of the problem with the presumed students' knowledge.

In the first two options, the conditions of the objective situation are sufficient for explaining and justifying all students' productions; the [expected] solution can therefore be communicated to the whole class. The reasoning is produced by the student as a reasoned action, based on the conditions which define the objective situation: using the rule *R*, the student justifies that, given the premise A, the conclusion or decision B appears as a necessary condition of the situation S. In this case, the reasoning appears as a "reason for knowing", by which we mean that the reasoning makes it possible to justify the validity of an element of knowledge by reference to its logical connections with other elements of this kind of knowledge, in other words, by means of internal reasons, specific of this knowledge. In the third option, the student can accept the solution only upon his trust in the teacher's authority; there can be no autonomous learning in this situation.

2.4.3. *Reasoning as a cause and a means of learning autonomously*

In the first two options, the reasoning can be produced by the students for the purpose of solving the problem without the teacher's intervention, support or help:

- as a way for one or more students to make their decisions in autonomous "situations of action";
- as a somewhat formal support for clarifying a piece of information in a simple communication;
- as a way to convince peers of the validity of a statement, or, more generally, to justify a statement.

A new reasoning is learned when it is promoted from being just a particular means of solving a given problem to a "universal" means of solving all problems of a certain type, and becomes integrated as such with the subject's knowledge. In an autonomous situation, the reasoning is based on induction, but this induction is supported by a chain of inferences that can be made explicit.

In the third case, the autonomous learning cannot back up this integration, which can only result from [more or less direct] teaching.

Connecting new knowledge ("cognition") to knowledge already acquired or to known circumstances is a way to remember it. The more familiar the supporting knowledge is, the better remembered will be the connection and the easier and more faithful will be the recollection ("recognition") of the new knowledge. However, as the amount of new knowledge to be learned increases, it becomes harder and harder to keep in mind the growing number of independent circumstantial connections, and there is a considerable risk of confusion. Rational connections create an organization of knowledge which is much more economical. Knowledge is only an organization and reasoning provides a systematic means of connecting facts so that they don't have to be learned separately. Therefore, students' use of reasoning can be strongly enhanced if motivated by the necessity to learn a large number of apparently isolated facts.

2.4.4. *Reasoning as a means of teaching*

Consequently, teachers demonstrate reasoning underlying the knowledge they teach to promote its learning, and to reduce the effort of teaching. However, if the students themselves cannot produce this reasoning, they increase the memory load instead of reducing it. However, understanding is not always a sufficient condition of learning. Therefore, teachers sometimes resort to "didactical reasons" by establishing artificial links between different pieces of knowledge, unrelated to the scientific meaning of this knowledge: review, mnemonic devices, and metaphors, metonymies, analogies, which we call "the rhetorical means of didactics". These didactical reasons, which cannot be justified by a logical reasoning, are completely unrelated with the "reasons for knowing" which are specific of the knowledge in question. However, didactical reasons are quite often an object of teaching and can be considered, by both the teacher and by the observer, as "causes of learning". They are associated with a whole didactical culture, and equally needed by the teacher and the students.

In this case, a reasoning formulated by the teacher is either

- an explicit object of teaching demonstrated in the phase of institutionalization or given as a reference, but unrelated to the conditions which define the objective situation (provided there is one),
- or a support for learning and remembering the statement taught (i.e. as something like a "legitimate" mnemonic device),
- or else a rhetorical argument used as didactical means for helping the student to understand the statement.

In these conditions, a reasoning produced by a student is addressed mainly to the teacher, and its purpose is

- to justify an action or an answer, or
- to satisfy the teacher's explicit or implicit request, where, formally, the reasoning is considered to be an object of teaching, independently from its relationship with the student's action (recitation, quotation, etc.), and, more precisely, independently of the situation the student had been confronted with.

2.5. *Autonomous learning and devolution of situations*

To be ready to take the risk of responding in conditions of uncertainty is part of the student's "job" and characteristic of any didactical situation. Most children undertake risky jobs quite naturally, unlike professionals, who would normally refuse to undertake a job and make promises about the results of their work if they didn't know beforehand how to complete it successfully. A professional cannot take the risk of accepting a task except to the degree that he possesses the means to limit the risks and consequences of his/her/its possible failure. The necessary means of control are the known and accepted definitions of tasks, techniques, technologies and theories. The professor "imposes" tasks but communicates only a part of the means to do and to control them. The pupil must combine and complete the means. The missing part is the object of the teaching.

In each of these types of means of control, reasoning plays roles that are different, but not independent. For example, in the achievement of tasks, reasoning relieves memory of keeping track of the order of stages, permits the anticipation of failures, etc.

In the aim of promoting students' practice and learning of reasoning the teacher must propose problems (whose solutions require knowledge that has not been institutionalized in class yet) to be solved completely autonomously by the students. The part of the teacher's job which consists in getting the students to accept the risk of not knowing how to solve a problem is called "devolution".

Devolution is ethically acceptable if

a) it is realistic to assume that students will be able to solve the problem on their own;
b) the problem can be solved by a simple use of the given information, the knowledge already taught and correct reasoning, and
c) this reasoning can be made explicit by the teacher, at least, and understood by the students at the time of the presentation and discussion of solutions in class.[7]

In case the answer is not related to the information given in the problem by intelligible reasoning, the student remains dependent on the teacher's judgment and good will, and consequently will have difficulty in reinvesting his knowledge in a non-didactical environment.

Teachers have recourse, necessarily and often advantageously, to situations which are not solvable by students on their own, in the framework of various didactical strategies and stages. But students' activity can be guaranteed only on the basis of steps accomplished in an autonomous situation.

2.5.1. *Translation of causes of learning into reasons for knowing*

When a student has learned a fact or developed a behavior by connecting it rationally to his previous knowledge, we say that his learning is motivated by a *reason*; otherwise, we say that it is an effect of a *cause*. When teachers, intentionally or not, use various rhetorical didactical means to get the students to learn certain things, they have to face the problem of connecting these things by culturally acceptable relations. The role of reasoning is then to translate or convert the causes of learning into reasons for knowing. This is an important function of reasoning in teaching. Sometimes, the "translation" consists only in a substitution of reasons for causes; sometimes, it is an explanation, but sometimes the translation becomes a real rational transformation.

2.6. *Conclusion*

In a didactical analysis of a lesson, it is necessary to distinguish several situations:

- the mathematical situation (the objective situation) the student is faced with and has to act upon;
- the situation of autonomous learning (what is it about in terms of the didactical objectives of the lesson); and
- the didactical situation (the way the teacher conducts the lesson: his interventions, the arguments he uses).

Some teachers believe that it is important to induce the students to learn in situations of autonomous action rather than to teach them some knowledge in a formal way first, and then let them apply it in problems where the use of this knowledge is artificially described. They try to simulate "natural" processes of production and use of knowledge. But we know that such a radical constructivist approach has its own, equally radical, limitations. Theoretically, there is no way for knowledge developed in an autonomous situation in class to have the same properties as culturally developed knowledge. Its learning must be supported by specific didactical actions.

3. Problem situations

Mathematics and mathematics education have produced a huge number of problems that require all sorts of reasoning. But direct teaching of solutions and later of problem solving methods tends to close what is supposed, by the very definition of "problem", to stay open. "Didactical engineering" (i.e. an approach to teaching design based on the theory of didactical situations) developed a number of complex situations for teaching specific knowledge, by defining the conditions that must be satisfied in order to make a situation open for the students, that is, endowed with an uncertainty (for the students, but not for the teachers), which justifies the use of reasoning.

The designed situations have sometimes been based on classical problems, but these classical situations were set up in such a way as to engage students with causes and reasons of solution other than just mathematical validity. There were also other ideas with similar intent; for example,

- creating situations open not only for the students but also for the teacher (who would not know the solution),
- building situations based on classical problems, such as word problems, story problems or embedded context problems (where a mathematical relation is "dressed up" with elements of some extra-mathematical context), with insufficient data, or, on the contrary, superfluous data; without questions or with absurd questions; with a debate of the solutions and with students working in small groups ...

At the same time, there appeared, under the name of "problem situations", a global didactical activity, with intertwined phases of direct didactical intervention and autonomous or quasi-autonomous learning. This term may be understood as containing all the other situations. But there is a tendency to use it in a way that reduces the possibilities of monitoring the didactical properties of each phase. In a problem situation there may occur a large number of sometimes unexpected situations of (presumably) autonomous learning. But the effectiveness of these phases of apparently autonomous learning depends, in principle, on the actual features of the aspects of the situation that are left to the students.

In our research, we have been interested in knowing to what extent situations – like the problem situations described above – that cannot, theoretically, be devolved to the students, make it nevertheless possible for the students to produce forms of reasoning.

3.1. *Presentation of problem situations*

3.1.1. *The origin of problem situations*

The use of problem situations, that is of open problems, has expanded greatly in classrooms (at least in France) in recent years. The reasons given are well known:

- to provide the students with "models" of situations of research or of the natural functioning of knowledge,
- to stimulate autonomous work and enhance students' motivation;
- to combat the formal teaching of algorithms which are then applicable to a limited set of conventional exercises;
- to make the students engage in reasoning rather than just in performing calculations.

Problem situations are expected to create favorable conditions for all kinds of mathematical activities that are difficult to obtain in the more classical situations: problem posing, information search, verification of the plausibility and relevance of the information, organization of a series of activities . . .

The classical problem and its solution are, this way, embedded in a "context" with a complex relationship to the knowledge that is the object of teaching.

The actual conditions created in the problem situations can be very diverse: autonomous interactions with a material milieu (e.g. geometric figures) or a system (e.g. a software system); autonomous interactions with other students (in small or bigger groups); didactical interactions with the teacher . . . .

A problem situation is a conglomerate of conditions based on simpler elements, which correspond to the various types of situations proposed in the theory of didactical situations: situations of action, communication, validation, institutionalization or devolution. But these component situations may appear unexpectedly, as a result of an unanticipated initiative of a student or of the teacher. This potentially unpredictable character of problem situations contributes to giving the teacher the impression of great freedom, artlessness, and naturalness. This impression may contribute to the success of the problem situation approaches. Knowledge is produced in an apparently spontaneous, unprepared way.

3.1.2. *Characteristics, conditions, results of problem situations and questions underlying the present study*

Problem situations can be used in a very flexible way. They allow and stimulate the teachers to choose from a broad range of didactical possibilities, from a radical devolution, to guided investigation or teacher presentation of an imaginary Socratic teacher-student dialogue (maïeutique).

On the other hand, one can question the outcomes of the use of problem situations. In spite of the variety of didactical methods associated with this use, are there any common features? What are the students actually doing? What are the advantages, but also risks or drawbacks, of these situations for the students and the teachers? What assessment tools do they make available to the teacher? One can also ask more specific questions regarding the expected advantages: is there an increase, qualitative and quantitative, of the produced forms of reasoning?

3.2. *The context of the observed lesson*

3.2.1. *The teacher, the students and teacher training*

The teacher prepared the "ski passes" problem to show student teachers, doing their training course at school, the following advantages of the associated problem situation:

1. The richness, the originality and the variety of forms of reasoning produced by students confronted with the problem situation. (This way, he tries to highlight the students' actual ability to produce forms of reasoning in new situations).
2. The students' ability to articulate their reasoning when they come to present their solutions during the plenary discussion phase.
3. The arguments and logical reasoning the teacher uses when dealing with students' reasoning.

4. The means he uses to provoke the students to engage with their peers' reasoning, and discuss their validity and relevance with respect to the solutions.

His aim is, therefore, to make the trainee teachers aware of the advantages of problem situations for learning and reasoning.

3.2.2. *Presentation and rationale of the teacher's didactical strategies*
This teacher often leaves his students in charge of the situations of action, based on pre-experimented teaching projects published in teacher journals. He justifies his choice of this type of situation by the fact that they allow the students

- to develop one or more procedures, based on what they already know,
- to test their procedures,
- to become aware of the decisions underlying their reasoning.

Indeed, these situations have the following features:

1. The student has sufficient knowledge to develop the basic strategies.[8]
2. Knowledge necessary to develop the strategies for solving the problem is not too far from what the students already know.
3. The students are able to determine the validity of their solutions by themselves.
4. The students can make several attempts.

The phase of whole class presentation and discussion of solutions then allows the students to reflect on their procedures and analyze the decisions underlying their reasoning. Our study shows that students manage to use their reasoning, produced in the situation of action, as arguments making it possible to justify or reject the solutions.

The intentions of the teacher in managing the observed lesson are, on the one hand, not to intervene in the phase of investigation, so that the development of this phase resembles that of a situation of action, and on the other – to intervene minimally during the phase of the plenary presentation of solutions and thus stimulate the students to debate the validity of the underlying reasoning.

4. THE OBSERVED LESSON

4.1. *The components of the situation*

The lesson took place in a 5th grade mathematics class.

4.1.1. *The problem and the objective situation*

The teacher starts by handing out the following problem:

> A one-day ski trip to the resort of Gourette is being organized next Saturday for students from the Oloron area. For this exceptional event, the local city council has decided to pay for the ski passes for the day. The resort of Gourette offers the following group rates:
>
> 216 passes: 1,275 F
>
> 36 passes: 325 F
>
> 6 passes: 85 F
>
> 979 children have signed up for the trip but when the morning of departure comes 12 children do not turn up because they are sick, of course. The council accountant says to himself "Too bad for these kids, but never mind, it'll work out less expensive for us this way".
>
> What do you think?

The "objective situation" is the situation presented in the problem; the student is expected to deal with it without questioning the status of reality of what is thus presented to him as "objective".

4.1.2. *The planned phases of the lesson*

The development of the lesson, chosen by the teacher, follows a plan that has become quite common in France:

- the research activity is presented by the teacher (phase 1);
- students read the problem (phase 2);
- the teacher provides additional information, if necessary; for example - explains the terms used in the formulation of the problem (phase 3),
- students work on the problem individually for about 10 minutes (phase 4)
- students are divided into small groups (phase 5)
- students work in small groups, and prepare a written report; this phase (phase 6), lasts about 25 minutes;
- whole class presentation and discussion of the reports, with each group going to the board in turn to present their results (phase 7).

4.1.3. *A mathematical analysis of the problem*

In the presented wording, the problem is not a problem in the usual sense of the word in that it does not contain an explicit question. This kind of formulation is sometimes used to encourage students to spontaneously pose questions and conjectures, rather than content themselves with answers to already asked or even taught questions, or with standard solutions to habitual problems.

However, not asking an explicit question after presenting a situation is legitimate only if:

- several questions that are *interesting* for the students are reasonably suggested in the description of the objective situation;
- the teacher is prepared to intervene in the course of the determination of the problem, and therefore of its solution, which means that the students themselves no longer obtain this solution autonomously.

Although the declared objective of this method is to allow the students to ask original and personal questions, in fact, students soon become more dependent on what the teacher does and says than they would be if the problem were given in a more conventional form.

In the case under discussion, students have to understand that the information provided in the formulation of the problem can be used to calculate the total expenditures in two different cases (depending on the number of the excursion participants, with and without the 12 sick students), that these two amounts may be unequal and that the aim is to establish if they are equal or not. The only interesting question is, "Is the expense for 979 students higher than for 967 students, as assumed by the accountant?"

The proposed mathematical situation is, in essence, one of optimization and linear programming: evaluation of the prices of the required passes amounts to calculating the lowest possible value of a linear function, corresponding to the cost of expenditure for a given number of children N. If the functional is denoted by $J(n_1, n_2, n_3)$ where n_1 denotes the number of 216-pass packages bought, n_2 – the number of 36-pass packages, and n_3 – the number of 6-pass packages, the corresponding expenditure can be written as

$$J(n_1, n_2, n_3) = 1275n_1 + 325n_2 + 85n_3$$

the constraint being that

$$216n_1 + 36n_2 + 6n_3 \geq N$$

where N denotes the number of children; i.e. the number of passes purchased must be no less than the number of students N.

The aim is to calculate the minimum value of the functional $J(n_1, n_2, n_3)$ for a given N relative to the above inequality constraint, and therefore to solve the following problem of linear optimization with a constraint:

$$\begin{cases} \underset{n_1,n_2,n_3}{\text{Min}} \quad J(n_1, n_2, n_3) = \underset{n_1,n_2,n_3}{\text{Min}} \ (1275 \times n_1 + 325 \times n_2 + 85 \times n_3) \\ \qquad\quad N - n_1 \times 216 - n_2 \times 36 - n_3 \times 6 \leq 0 \end{cases}$$

The solution of this problem makes sense only for positive integer values of the variables n_i. The recognition of this problem as a problem of optimization allows us to identify the possible alternatives to the usual conceptions of the students and to the knowledge to which they will more or less explicitly refer to in the resolution process.

The above mathematical analysis of the situation points to the influence of several didactical variables. For each of them, the teacher chooses values that increase its complexity: the number of the parameters (8) and the unknowns (3) is very high and their verbal names are neither simple nor familiar to the students. Moreover, the values of the parameters do not make the understanding and the solution any easier. Let us first consider the numbers of passes included in each of the group rates. The choice of values 1, 10, 100 would result in working with decimal notation; the choice of values 6, 36, 216, leads to working, in fact, in base 6, which makes the task considerably more complex, especially given that the logic of group rates excludes the possibility of buying the passes per unit. The number of children who turn up for the trip (967) is chosen to be very large, so that the use of three types of group rates can be "justified". This number is divisible neither by 216, nor by 36, or even by 6, which further complicates the task. The total number of children (979) justifies the remark of the accountant, but doubles the number of calculations to do.

4.1.4. *Analysis of the research situation: The possible and expected students' actions*

The solution that the teacher expects the students to produce is of an arithmetic nature and, in principle, within the reach of children in this grade, but it is still very complex. If we assume that each stage of this solution of the problem constitutes a "module", then the solution can be seen as composed of five such modules, each of which requires several operations:

Module 1: Calculation of the number N_p of children that turn up for the trip; N_t denotes the total number of children.

Module 2: Comparison of the three group rates in the aim of ordering them.

Module 3: Research of an organization or a way of conceiving of the purchase of N tickets, where $N \geq N_p$, combining the different rates so that the expense is minimal.

Module 4: Research of an organization or a way of conceiving of the purchase of N tickets, where $N \geq N_t$, combining the different rates so that the expense is minimal.

Module 5: Calculation of how much has been saved by the accountant: calculation of the difference.

The complexity of the Modules 3 (presented in the Annex) and 4 is particularly high compared to problems commonly used in autonomous situations. One can presume that the Modules 1 and 2, which are much simpler, are aimed at allowing the weaker students to participate in solving the problem at least to some extent.

4.1.5. *Situations of autonomous learning*

The above-described situation of group rates is not a situation of autonomous learning because of the theoretical impossibility of students producing a solution to the given problem without external help and by means of a simple play with legitimate reasoning. They have neither the time nor even the possibility of adapting to the situation. None of the justifications based on a reference to the milieu will be valid.

Whatever the teacher or the student will pretend to justify, will not be really justified because the situation does not offer this possibility. The students have no resources other than their knowledge and their imagination, because they will receive no information and no "objective" control. Only their memory, or the teacher's interventions will allow them to realize which hypotheses, methods or conclusions are valid or not. Moreover, since the proposed situation does not provide feedback in response to the actions of the students, they will be unable to judge of the value of their production during the whole class presentation and discussion of solutions.

Therefore, since the situation cannot function in an autonomous fashion, everything will depend on the teacher, his choices, his didactical decisions and, finally, on his didactical rhetoric, in case the students do not possess a correct representation of the "objective situation" which serves as a milieu for their action.

4.1.6. *Analysis of the didactical situations (phase by phase)*

The complexity of the objective situation shown above reveals the complexity of the didactical situation, and permits the anticipation of the teacher's difficulties in managing those phases of the lesson during which he will have to intervene. In the conditions made explicit above, the teacher has many responsibilities, which he cannot share with the students. Thus, the teacher will have to intervene,

1. To present the activity to the students (phase 1).
2. To explain, if necessary, certain terms (phase 3).
3. To institutionalize the question (phase 3) so that the problem to be investigated by the students is completely determined (phase 3).
4. To manage the phase of whole class presentation and discussion of students' solutions (phase 7). In view of his didactical strategy, he will have to bring forth a plan of resolution on the basis of the bits of

solution that he hopes to find in the work of the small groups. During this phase, he will have to guess the intentions and the occasionally erroneous interpretations of the situation, and correct them at the opportune moment.

The time assigned to the realization of the task will leave little room for the identification and study of the necessary knowledge, for the teaching of the solution, and, indeed, for any really autonomous exercise.

In conclusion, we expect to see, in the students' solutions, many questions and many elements of "reasoning" related to the conception of the various components of the solution, or for the organization of the solution, but none of them partaking in the process of teaching.

4.2. *How the lesson developed*

4.2.1. *The research activity and the written traces of it*

In the observed lesson, the research activity was based on the research and formulation of the question, which completely determines the problem (in the classical sense of the term). But the students were not able to perceive what is at stake (mathematically) in the problem situation and it is the teacher himself who formulated the question: "When, do you think, is the ski trip more expensive: when there are 979 students or when there are 967 students?"

The written traces of the students' research contain some calculations. In Section 5, we will try to determine if the students have produced some forms of reasoning and if so, of what type.

4.2.2. *The phase of whole class presentation and comparison of students' solutions*

Our theoretical, *a priori*, analysis of the problem situation led us to expect a failure of the teacher's plan: The management of the didactical phase of the lesson (phase 7) appeared all the more delicate that the reduction of the complexity was essentially in the hands of the teacher; it depended on his choices, his decisions and his "opportune" interventions.

But upon viewing the video recording of the lesson (which we haven't seen before the theoretical analysis), we had to admit that the teacher managed to conduct his class without being challenged with any major difficulties.

4.2.3. *The surprise of the observers*

An external observer, and in particular the student teachers who viewed the video recording, could see nothing unusual in the lesson. This divergence with our theoretical analysis led us to examine in more detail the knowledge and especially the forms of reasoning that appeared during this lesson. Who produced these forms of reasoning? What were they used for? How did the teacher use them? What type of arguments did the teacher use in managing the reasoning? Did the students use reasoning as means of justification? What did the students learn?

5. The Observed Forms of Reasoning, Their Function and Use

5.1. *Forms of reasoning in students' written productions*

Tables IA and IB display the forms of reasoning underlying the written productions of students working in small groups. The forms of reasoning are analyzed according to the different modules, defined in Section 4, which represent the "standard solution", expected by the teacher. The corresponding levels of reasoning, defined in Section 2, are labeled N_1, N_2 and N_3, respectively. For each group, the table describes the procedures used by the group, relative to the corresponding modules. For each written production, we have indicated the associated IMA.

In summary, it appears that the majority of students used the model of a classical commercial situation: the buyer purchases the quantity of passes he wants, at a constant rate. Therefore, for them, the prices for different quantities could only represent prices proposed by different traders. The different quantities could appear, to these students, as a didactical device with the purpose of making them calculate the different unit prices. Even if some students had had practical experience with sales with group rates on a sliding scale, they lacked the vocabulary to speak about the hypothetical unit price, for a given group rate.

In the model of sales with group rates on a sliding scale, almost all forms of reasoning imported from the classical model are contradicted. For example, the actual price of an object is not the total price divided by the number of objects; the number of purchased objects may be not the number one really needs, etc. Therefore, the questions posed by the teacher during the whole class discussion and comparison of solutions cannot result in clarifying the difficulties.

The analysis of the different forms of reasoning which appear in the students' solutions shows that what is really at stake in the problem situation, namely the problem of minimizing the expense, has not been grasped by the majority of students.

In this lesson, it is clear that the devolution of the situation did not work; the students were not able to take responsibility of the proposed situation. Indeed, in the phase of whole class discussion and comparison of solutions, it appears that:

TABLE IA
Different forms of reasoning in the written productions (Groups 1–4)

Module	Group 1: Xavier, Sylvain, Yannick	Group 2: Marine, Deborah	Group 3: Alexandre (alone)	Group 4: Julien (alone)
1	Reasoning N_1 Implicit	Reasoning N_2	Reasoning N_2	Reasoning N_2
2				
3		Reasoning N_2 Calculation, with rounding up, of the number of group passes, purchased at the middle rate (by groups of 36).	Reasoning N_3 Calculation, with rounding up, of the number of group passes purchased at the lowest rate (by groups of 216). Procedure 3.3, Annex 2	
4	Reasoning N_2 Calculation of the hypothetical unit price of a pass, at the highest rate. The total expense is obtained by means of the use of proportionality. The price of a pass is obtained by dividing the price of the group pass for 6 by 6.		The result is derived from the previous module (the purchase is identical to the one obtained in module 3).	Reasoning N_2 Calculation of the unit price for each group rate, by dividing the group rate by the number of passes in the group. The calculation of the total expense for each of the unit prices is obtained by the use of proportionality.

5	Reasoning N_3 Calculation of the savings by computing the price of two group passes of 6.	Reasoning N_2 Calculation of the savings by computing the price of two group passes of 6.	There are no savings.	
The implicit model of action	Calculation of the hypothetical unit price at the highest rate. The total expense is obtained by the use of proportionality. The price of a pass is obtained by dividing the price of 6 passes by 6.	Students use the model of wholesale commerce. There is no attempt at minimizing the expense.	The student uses the model of wholesale commerce. There is no attempt at minimizing the expense.	The student uses the classical model of retail commerce for each mode of sale. He uses the model of proportionality.

TABLE IB
Different forms of reasoning in the written productions (Groups 5–8)

Module	Group 5: Stephanie-Audrey	Group 6: Fabrice-Thomas	Group 7	Group 8: Caroline
1	Reasoning N_1 Implicit	Reasoning N_2	Reasoning N_2	Reasoning N_2
2	Reasoning N_3			
3	Reasoning N_2 Calculation, with rounding down, of the number of group passes at the lowest rate (by groups of 216), followed by the calculation, with rounding up, of the remaining number of group passes at the medium rate (by groups of 36). Procedure 3.3 (Annex 3)	Looking for the number of groups of 216 (with rounding down), of 36 (rounding down) and of 6 (rounding up) necessary for 967 students. Calculation of the hypothetical unit price of a pass at the highest rate. The total expense is obtained by the use of proportionality. The price of a pass is obtained by dividing the price of the group pass for 6 by 6.	(1) Calculation, with rounding down, of the number of group passes at the highest rate (by groups of 6); the remainder is 1. (2) Calculation, with rounding down, of the number of group passes at the medium rate (by groups of 36), followed by the calculation, without rounding down, for the remaining students, the number of group passes at the highest rate (by groups of 6); the remainder is 1. (3) Comparison.	Calculation of the hypothetical unit price of a pass at the highest rate. The total expense is obtained by the use of proportionality. The price of a pass is obtained by dividing the price of the group pass for 6 by 6.

4	The result is derived from the one obtained in Module 3.			
5	Reasoning N_3			
Implicit model of action	The students use the model of wholesale commerce and manage to carry out a combination of the proposed rates.	The students use the model of retail commerce to calculate the total expense, although they have reasoned in terms of group rates.	The students use the model of wholesale commerce. They try different combinations aimed at minimizing the expense.	Calculation of the hypothetical unit price of a pass at the highest rate. The total expense is obtained by the use of proportionality. The price of a pass is obtained by dividing the price of the group pass for 6 by 6.

- The students do not possess the necessary knowledge to conceive of the basic strategies.
- The students cannot obtain, as feedback to their actions, the information necessary for the solution of the problem.
- There is not enough time for the students to produce a solution, because of the complexity of the problem.
- The students have no means to judge, by themselves, of the validity of their solutions.

Moreover,

- The formulation of the problem does not sufficiently determine the objective situation and therefore leads many students to construct mistaken implicit models.

5.2. *Analysis of an episode of interactions during the whole class discussion phase*

For this paper, we have chosen to present an analysis, in terms of the theory of didactical situations in mathematics, of an excerpt from the transcript of phase 7, i.e. the whole class discussion and comparison of students' solutions phase.

The episode focuses on interactions related to one student's work. This student, Julien, chose to work alone. His written work is presented in Figure 1.

Our analysis of this episode is presented in Table II. The first column of the table contains the code of the intervention, where the first

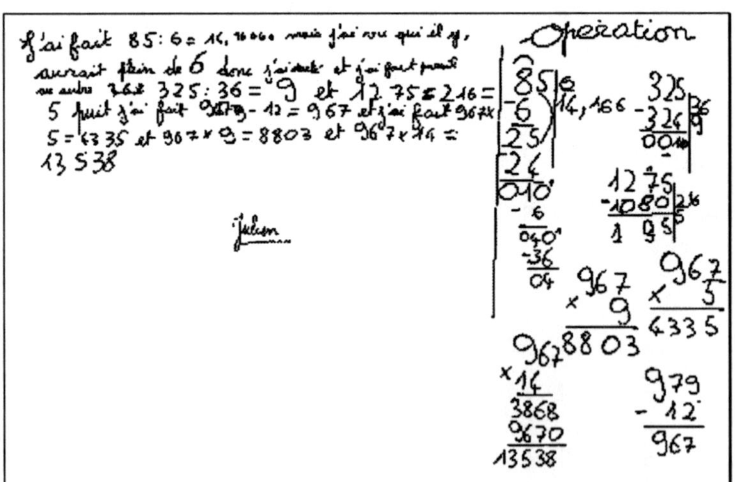

Figure 1. Julien's written work.

TABLE II
Transcript and analysis of some interactions

No. Min.	Transcript	Comments	Analysis	Nature and function of the intervention
4.1 12′35	Julien: Okay, I started by doing. . . (1) I divided 6 into 85 (2) and I got 14,166; I took the 14 and then I saw. . . (3) I did the same with 325, in short, I did the same, (4) I did the same with all three operations.	Julien comes to present his work. He describes his calculation, without defining or naming the variable that he calculates.		(1) Direct description of an action (calculation) (2) Formulation of a result (3) Indirect reference to an action: by analogy (4) Organization of the calculation Strategic or organizational reasoning, local and expressed orally.
4.2	Teacher: (1) the three proposals, (2) the group passes	The teacher reformulates a part of the student's statement to introduce a vocabulary.	The teacher wants to establish a link between the performed calculations and the objective situation.	(1) Correction of the terminology (2) Suggestion of a terminology and giving a name to a result.
4.3	Julien: (1) 325 divided by 36 and 1275 divided by 216, (2) and then I did. . .	Julien continues to describe his calculations		(1) Direct description of an action (2) Organization of calculation. Strategic or organizational reasoning, local and expressed orally.

(*Continued on next page*)

TABLE II
(Continued)

No. Min.	Transcript	Comments	Analysis	Nature and function of the intervention
4.4	Teacher: (1) [Your] first conclusion after these calculations? [to the whole class] Have you heard the operations he had done? (2) What is the price of a pass, relative to each of the three proposed conditions, right?	The teacher asks Julien what he got from the calculations he performed. He intervenes to provide an interpretation of the calculations. He points to the nature of the results as the "price of a pass relative to each of the three conditions".	The teacher gives an interpretation of each of the calculations performed by Julien. His didactical intention is to construct Julien's calculations as a support for introducing the stages of reasoning underlying the Module 2.	(1) Giving a statement the status of a "conclusion" in the development of a reasoning. Invitation to comment on Julien's results and to position them relative to an action. (2) Use of rhetorical didactical means: Element of a local explicit reasoning of the teacher who aims at representing the calculations as part of Module 2.
4.5	Julien: Yeah!			Agreement, approval.

4.6	Teacher: (1) Okay, first conclusion after that?	The teacher questions Julien on what he gets from his calculations.	.	(1) Request to make an inference. The teacher waits for the student to continue his reasoning and articulate a conclusion.
4.7	Julien: And then I did...	No answer; Julien seems to want to continue to describe his calculations.		
4.8	Teacher: No, your first conclusion after that? When you were done with these calculation, what did you think to yourself?	The teacher reiterates his question.	The teacher makes a second attempt, with the same aim as in 4.4. But the formulation is more precise.	Recall of what is a conclusion; invitation to comment on a result.
4.9	Another student: Which one was less expensive.	A student puts into words the question that the teacher has previously asked in an implicit manner.	A student points to Julien what he could get from his calculations, namely a comparison of prices.	Question on an order relation. Project formulation.
4.10	Julien: Yeah! Which one was less expensive... But, no, I couldn't see...			But "which one" does not denote a well-determined object. A passive explanation. Impossibility to realize a project.

(*Continued on next page*)

TABLE II
(*Continued*)

No. Min.	Transcript	Comments	Analysis	Nature and function of the intervention
4.11	A student: But yes, you can see!	A student points out to Julien that he has all the necessary information.	The student pushes Julien to produce a reasoning, by pointing out to him that he has all the necessary elements to conclude (i.e. to compare the prices).	Possibility of realizing a project.
4.12	Julien: (1) Yes, it was 1275, (2) because a pass cost 5 F, (3) more or less and then, (4) so then I tried, in short, I did 979 less 12, I got 967 and then I multiplied 967 by all the results of the divisions.	Julien gives the expected answer and continues to describe his calculations.	Julien articulates the conclusion, expected in the Module 2. He goes back immediately to his initial reasoning, in describing his calculations.	(1) Implicit conclusion, (2) Explanation, (3) Estimation, (4) Direct description of a sequence of actions and organization. Strategic or organizational reasoning, local and expressed orally.
4.13	Teacher: To find what?	The teacher questions Julien on the aim of his calculations.		Project; request to name a result. Request for an explanation.

4.14	Julien: To find the price of how much it was going to cost.	Julien points to the aim of his calculation: to calculate the total expense (for the students who were present at the trip).	Julien indicates the purpose or project he has in mind: for each group rate, to calculate the total expense.	Naming the result. Articulation of the purpose of his procedure.
4.15	Teacher: Yes, the price... to find which one was the least expensive.		The teacher starts from the formulation of the student and transforms it. Julien stated that his aim is to calculate the total expense for each of the three cases. But the teacher focuses on the comparison of the group rates. The IMA developed by Julien does not conform to the teacher's expectations. The teacher will establish that Julien's calculations are useless for the comparison of the rates.	Rhetorical didactical means: Element of a local explicit reasoning of the teacher, which aimes at re-positioning the calculations in the perspective of Module 2. Recall of the necessity to subordinate a result to the main task.

(*Continued on next page*)

TABLE II
(*Continued*)

No. Min.	Transcript	Comments	Analysis	Nature and function of the intervention
			Didactical intention: reject the calculations by making them appear as useless, redundant, with respect to the previously established conclusion (Module 2).	
4.16	Julien: Yes.			Accord
4.17	Teacher: And you did the three calculations?	The teacher wishes to make Julien aware of the fact that the calculations were not necessary, that a reasoning could help to avoid doing calculations.		Effectiveness of an action.
4.18	Julien: Yes.			
4.19	Teacher: It was necessary?			Call for a judgement of the relevance or adequacy of a calculation.
4.20	Julien: Well. . . yeah. . .			Agreement
4.21	Another student: To see which one was the least expensive.			Subordination recall, as in 4.15.

4.22	Teacher: You didn't know it before?	The teacher wants to incite Julien to reflect on his reasons for doing the calculations.	The intention would be: "could you know it beforehand, without doing the calculations?" It is, therefore, a call for a direct reasoning.	Call for the anticipation of the role of a result in the resolution of a problem. Call for a formulation of a direct local reasoning.
4.23	Julien: Yes, I knew it. . . but. . .	The student cannot distinguish between his opinion and the justification required by the teacher.		
4.24	Teacher: Okay then, so what is the result?	The teacher re-asks Julien to formulate his conclusion.		
4.25	Julien: So I saw which one was the least expensive, and then. . .			Validity status: subjective certitude
4.26	Teacher: And what did you get?			
4.27	Julien: In my opinion, it's 967 times 5 equals 4335.			Statement of a subjective value. Formulation of a result.

(*Continued on next page*)

TABLE II
(Continued)

No. Min.	Transcript	Comments	Analysis	Nature and function of the intervention
4.28	Teacher: Hum! Why 4300? 5 times 900?	The teacher questions Julien on the order of magnitude of the result of the multiplication.		Challenging the result (but not its relevance). Hint at the nature and position of the mistake: the 3. Hint at the cause of the mistake. Request of justification.
4.29	Julien: 36 000... er... 3600.			Mistakes in mental calculation, table of multiplication and rule of zeros.
4.30	A student: 5000...4500.			Hint that 9 times 5 is calculated as 10 times 5 minus 5: scaffolding.
4.31	Julien: Yes, 4500.			Agreement
4.32	Teacher: And 5 times 967, 4300... There is a little mistake here!	The teacher contradicts the result stated by Julien. He picks up the mistake but does not correct it.		
4.33	Julien: Yes, 4500.			

4.34	Teacher: Anyway, the least expensive would be the price of a pass sold in packages of 216 multiplied by the number of children, 967, how do you think?	The teacher reformulates and summarizes the procedure of Julien. He wants to put this procedure up for debate, he wishes to test it. He is trying to "reposition" the problem: he uses the conditional mode.	The teacher faces the obstacle of change of model: the expression, "sold in packages of 216" doesn't have, for the students, the meaning that would be necessary for understanding the question. The appeal to the students is doomed to failure.	Re-interpretation of Julien's calculation. Request to comment which suggests that the calculation is not satisfactory. Putting the solution up for debate. Repositioning of the reasoning in the perspective of the debate over the solution of the problem. Repetition of the "guilty" data. Global reasoning. Use of rhetoric didactical means.
4.35	Many students: inaudible			
4.36	Teacher: This is probably true, but what...?			Conditional reasoning: suppose (what we are saying is true). Call for a re-examination of the method of calculation. Request for justifications.

(*Continued on next page*)

TABLE II
(*Continued*)

No. Min.	Transcript	Comments	Analysis	Nature and function of the intervention
4.37	Melanie: Under the condition that...			*Identification of a condition*
4.38	Teacher: Yes, Melanie?	Teacher's hopeful encouragement.		
4.39	Melanie: You calculated the others to know if it was... Why have you then taken this one? Have you tried the others?	This is a formal objection; the division is not understood.	But Melanie believes that the justification should deal with the choice of the values and not with the method of calculation.	Request to justify. Request to explain.
4.40	Julien: I did 85 because it was the first.			
4.41	Teacher: Yes.			
4.42	Julien: I don't know...			
4.43	Teacher: You understand that this was the least expensive but you haven't tried the others.	Unjustified intervention.	The teacher would like to get the students to pass from the retail model to the wholesale model with rates on a sliding scale but doesn't know how to make it happen.	

4.44	Julien: Yes, I have tried everything, yes, I have tried 85 divided by 6 and the others, and from the three, I have seen that it was the 1275 divided by 216 that worked.			Explanation aimed at the justification of his procedure. Statement about having exhausted the possible calculations (from the perspective of retail sales model).
4.45	Teacher: 1275 divided by 216 and then multiplied by 979, okay… and Alexandre, what would you say about all that, in relation to your proposal?	The teacher reformulates the procedure and asks Alexandre (who correctly interpreted the information in the problem) to comment on Julien's implicit model.	Call upon a student to participate in the debate.	Rhetoric didactical means: extraction of the doubtful part of a reasoning.
4.46	Alexandre: I am saying that it's true…		Alexandre stays in the frame of the retail sales model.	Statement (in the frame of an implicit model).

(*Continued on next page*)

TABLE II
(*Continued*)

No. Min.	Transcript	Comments	Analysis	Nature and function of the intervention
4.47	Teacher: Okay, so we buy, we go to the cash of the ski station, and we ask for 967 passes at the price of 1275 divided by 216, is that the idea?	Third formulation of Julien's solution by the teacher with the aim of launching a debate.		Confrontation with a milieu assumed to be sufficiently familiar to "impose" some contradictions.
4.48	The students: Yes.			Agreement
4.49	Teacher: There we go, again. . . Okay. . .	The teacher gives up, temporarily, and Julien returns to his seat.	The failure of the attempt is proved.	The milieu provides nothing of the expected feedback.

number (4) indicates that Julien's "small group" (composed of him alone) was the fourth to present its results. For some interventions, the timing is shown (since the beginning of phase 7). The second column contains the transcript, and the third some comments on the intervention. In the fourth column we analyze the nature and function of the intervention with regard to the locutor's intended project. The fifth column aims at articulating the function of the intervention.

5.3. *Discussion*

The forms of reasoning which appear in the written work of Julien (Figure 1) are level 2 reasoning: the calculations are neither justified nor explained. However, the analysis of the IMA (Table I) allows us to identify the implicit mathematical model and Julien's representation of the objective situation. His model is that of the classical commercial situation, based on selling the passes per unit, corresponding to the mathematical model of proportionality.

The transcript (Table II) shows that, in phase 7, Julien describes his calculations without providing the class with more explanations on why he did them. This is why his project is not accessible to the class, which makes it necessary for the teacher to intervene. By proceeding this way, he presents the teacher with the opportunity to interpret his calculations in a way which does not necessarily correspond to his (Julien's) initial project. The teacher seizes this opportunity; using rhetorical didactical means, in the sense defined in Section 2, he manages to divert Julien's initial project to the benefit of his own, which is to develop the reasoning underlying Module 2 of the standard solution.

Moreover, our analysis shows that the teacher tries, several times, to engage a discussion on the validity of the presented procedures, or, more precisely, on the validity of the decisions underlying students' reasoning. However, all his attempts fail. Although he chooses to take in for questioning students whose representations of the situation conform to his expectations, in the hope that these students' contributions will help invalidating the incorrect representations, these students do not try to undermine the incorrect decisions underlying the presented work (see, e.g., the case of Alexandre, lines 4.45–4.46 in Table II).

6. Conclusions and Conjectures

6.1. *Students' reasoning*

The object of our analysis was the influence of certain features of the situation proposed to the students on the elaboration of the different forms of reasoning, their use and the possibilities of their processing available to the teacher during the whole class presentation and discussion of the solutions phase.

This analysis (see Table I) shows that the forms of reasoning elaborated by the students were few, that they were not very complex in terms of the number of calculations and the number of stages involved, and were mostly of level 2.

Moreover, the identification of the implicit models of action – level 1 reasoning underlying the students' work – was not particularly difficult for the observers and the teacher. However, this identification of students' models required an *a priori*, theoretical, analysis of the behaviors, the difficulties and the procedures likely to appear in the different phases of the lesson.[9]

A detailed analysis of the whole class presentation and discussion of students' solutions (Table II is a representative part of this analysis) aims at identifying the different types of arguments used by the teacher in taking into account and processing the students' reasoning. This analysis implies that the teacher has no means for an effective processing of the produced reasoning, i.e. he cannot use logical reasoning directly related to the objective situation in arguing with the students' solutions. In fact, he fails in all his attempts to take into account the reasoning produced by some of the students and get the rest of the class to share and discuss them.

This brings us to the first conjecture: the factor which constraints the teacher's possibilities of taking into account, articulating and processing students' reasoning is not so much the complexity of this reasoning but another feature which is related to the very nature of the situation proposed to the students.

6.2. *The effect of the lesson on students' behavior and learning*

6.2.1. *The effect of the lesson on the validity of the reasoning and students' conviction*

In the complete analysis of the transcript there is a lot of evidence that the students, having produced a reasoning based on a representation conforming to the teacher's expectations, have not become aware of the conditions which define the objective milieu. Indeed, in phase 7, they are unable to formulate the reasons that led them to elaborate these forms of reasoning, or even to react to the reasoning of their classmates when these are based on erroneous representations of the objective situation.

This can be partly explained by the fact that the situation does not provide the students with the possibility of testing their decisions: the objective milieu does not respond with any feedback to the students' actions. Therefore the students have no means to validate or reject their reasoning and therefore to reflect on the decisions underlying their implicit models of action or their representations of the objective milieu.

6.2.2. *The effect on the actions, language and opinions of the students*
The students, unable to gauge the validity of their work, cannot use the reasoning they have produced as arguments in a debate. The debate amongst peers wished for by the teacher is out of the students' reach.

6.3. *The effect on the didactical process*

6.3.1. *The devolution*
Decisions underlying the elaboration of each of the models are closely linked with the students' representations of the objective milieu. But this situation is not happening in real time and the students have to imagine the rules governing its functioning. Since the objective milieu is not clearly defined, this leads the students to construct different representations of the situation and therefore also different implicit models of action. Thus, the objective situation cannot be devolved to the students, i.e. the students cannot challenge the retail sales model adopted by the majority, or even calculate the results of the different possible choices.

6.3.2. *Didactical corrections*
The complete analysis of the transcript shows that the teacher cannot bring the students to articulate the reasons underlying their implicit models of action. To avoid a block, related to the fact that the students do not understand the decisions made by their peers, the teacher is forced to use rhetorical didactical means (Table II). These means make it possible for the teacher to divert the initial project of a student to the benefit of his own, i.e. the establishment of certain modules of the standard solution. However, the real reasons that justify the elaboration of the module are not there for the students to see; the reasons which underlie and justify the connections between the data given in formulation of the problem situation are hidden.

6.3.3. *Assessment*
The objective situation cannot be devolved to the students. Therefore the teacher cannot assess the students' ability to mobilize and use their knowledge and produce the expected reasoning.

6.3.4. *Institutionalization of learning*
The teacher cannot bring forth the modules of the standard solution as a reasoned consequence of putting together the information given in the objective situation and the presumed knowledge of the students. Therefore, on the one hand, he is forced to use rhetorical didactical means in responding to students' reasoning, and on the other, he cannot institutionalize the knowledge inherent in students' calculations (such as multiplication, Euclidean division, decimal division), because he cannot extract it from the situation proposed to the students.

The complete analysis of the transcript shows the impossibility, for the teacher, of bringing out and sharing the different organizations (i.e. the different means of conceiving of the purchase of passes according to the wholesale model) that conform to his expectations and appear in some students' work.

Consequently, another source of the teacher's difficulty in elaborating the expected solution with the students is that his project is not visible for the class.

Moreover, one of the consequences of the teacher's practice, more precisely, of the use of rhetorical means, is that, in phase 7, the students do not have the opportunity to reflect on and revise, perhaps, the reasonings they have produced. Nor have they become aware of the project of learning or even of the way to elaborate the expected solution.

Therefore, this problem situation has not allowed the students to make any progress in their practice of reasoning.

No new mathematical knowledge suitable for institutionalization appears in this situation. Besides, the teacher had reserved no time for "extracting" what could be remembered from this lesson. On the other hand, the situation does open up an opportunity to study the different models of sales: original ones and those commonly used at school.

7. Final conclusions

The study shows that although the students, faced with a problem situation elaborated and conducted by the teacher, have certainly produced forms of reasoning, they have not made much progress in their practice of reasoning. Indeed, they have not reflected back on their reasoning, on its validity, relevance or adequacy because the teacher was not able to process it. He could not respond to this reasoning by logical arguments based on the objective situation; he was forced to use rhetorical means.

Now, it is not the complexity of the students' reasoning that forced the teacher to use this type of means but the fact that the problem situation could not be devolved to the students. This implies that it is not the teacher's management of the whole class presentation and discussion of the students' work that is challenged here, but rather the nature itself of the situation set up by the teacher, which strongly constrains the possibilities of really taking into account the students' reasoning.

The objective situation does not make it possible for the teacher to bring the students to:

- share with their peers the real reasons that have led each of them to construct implicit models of action and take some decisions in the framework of the corresponding models;
- grasp the reasons why the steps (modules) of the expected, standard solution are necessary;
- share the reasoning underlying each module of the standard solution.

If a situation provides the teacher with the possibility of devolving to the students an "autonomous" (or "self-contained") situation of action, then, according to the theory of didactical situations in mathematics, during the phase of analysis of students' solutions the teacher can refer to the objective situation. This is because the students can develop their personal strategies and forms of reasoning related to the situations with which they are confronted. The teacher does not have to have recourse to rhetorical didactical means to process students' forms of reasoning.

If, on the other hand, the teacher has no such possibility, the teacher cannot refer in his arguments just to the objective situation and must bring in information and provide feedback on the basis of a project that is not visible for the students; and this is why he is forced to use rhetorical didactical means.

ACKNOWLEDGMENTS

Translated by *Ginger Warfield*. The editor, Anna Sierpinska, asked for many clarifications and additions. She kindly translated into English the modifications she asked for. We sincerely thank them both for this work.

APPENDIX

Analysis of the complexity of Module 3

Module 3

Search for an organisation* (way of conceiving the purchased N passes, $N>N_p$, combining the different rates) that makes it possible to reduce expenditure to the minimum.

PROCEDURE 3.1

- Calculation, rounding the result down, of the number of packages required at the lowest rate, referred to as N_1; N_1 is the multiple of 216 immediately below (or equal to) N_p.
- Calculation of the number of children remaining after the first distribution.
- Calculation, rounding the result down, of the number of packages at the middle rate for the remaining children, referred to as N_2.
- Calculation of the number of children remaining after the second distribution.
- Calculation, rounding the result up, of the number of packages required at the highest rate for the remaining children, N_3.
- Additionally, calculation of the number of unused passes
- Calculation of the total cost J_1, $J_1=N1x1275+N_2x325+N_3x85$

PROCEDURE 3.2

- Calculation, rounding the result up, of the number of packages required at the lowest rate, referred to as N'_1, N'_1 is the multiple of 216 immediately above (or equal to) N_p.
- Calculation of the total cost of this purchase J_2,

PROCEDURE 3.3

- Calculation, rounding the result down, of the number of packages required at the lowest rate, N_1, N_1 is the multiple of 216 immediately below (or equal to) N_p.
- Calculation of the number of children remaining after the first distribution.
- Calculation, rounding the result up, of the number of packages bought at the middle rate for the remaining children, N'_2.
- Calculation of the cost of this purchase $J_3=N_1x1275+N'_2x\ 325$.

- Comparison of the respective costs for each of the procedures
- Choice of the procedure offering the lowest expenditure.

Series of procedures required to produce modules 3 (and 4) of the solution expected by the teacher.

NOTES

1. In spite of our efforts, this article could not be made more independent from the context of the larger project. It is based on a set of conceptions and results which have been presented in more detail in Patrick Gibel: "Fonctions et statuts des différentes formes de raisonnement dans la situation didactique en classe de mathématiques", Thèse de doctorat, Université Victor Segalen. Bordeaux 2, 2004.
2. "Didactical relation" refers to an interaction between two subjects, two institutions or two systems, organized by one of them with the aim of teaching a well-defined knowledge or behavior to the other, which is not able, by itself, to conceive it and perceive its necessity or reason for existence.
3. For simplicity, generic masculine pronouns will be used.
4. In a situation of formulation the student has to adapt his language to express his thinking, his implicit model of action. In a situation of communication he has to adapt it also to his interlocutor.
5. IMA in abbreviated form.
6. Gérard Vergnaud (1990) La théorie des champs conceptuels. *Recherches en Didactique des Mathématiques* 10(2/3), 133-170.
7. A situation whose solution can be invented by a subject who encounters it for the first time does not, in principle, require any didactical intervention: it is said to be a situation of a non-didactical nature. It is one in which an autonomous solution by the subject is possible.
8. Basic strategies are strategies, efficient or not, which allow them to start the solving process of the problem.
9. In the descriptions of situations of action in publications for teachers, there has necessarily always been a didactical analysis of the situation, containing a part of the *a priori* analysis of the behaviors, the difficulties and reasoning likely to appear in the lesson. On the other hand, these descriptions also contained a detailed analysis of the behaviors and students' work in the different phases.

REFERENCES

Balacheff, N.: 1982, 'Preuve et démonstrations en mathématique au collège', *Recherches en Didactique des Mathématiques* 3/3, 261–304.

Balacheff, N.: 'Processus de preuve en situation de validation', *Educational Studies in Mathematics* 18/2, 147–176.

Broin, D.: 2002, *Arithmétique et Algèbre élémentaires scolaires*, Thèse de doctorat, Université Bordeaux I.

Brousseau, G.: 1986, Fondements et méthodes de la Didactique des Mathématiques, *Recherches en Didactique des Mathématiques*, Volume 7(2), Edition La Pensée Sauvage, Grenoble, pp 33–115.

Brousseau, G.: 1990, 'Le contrat didactique: de milieu', *Recherches en didactique des Mathématiques* 9/3, 309–336. La Pensée Sauvage. Grenoble.

Brousseau, G. and Gibel, P.: 1999, 'Analyse didactique d'une séquence de classe destinée à développer certaines pratiques du raisonnement des élèves', *Actes de la X° Ecole d'été de Didactique des Mathématiques* (2), 54–71.

Brousseau, G.: 1997, *Theory of Didactical Situations in Mathematics*, Kluwer Academic Publishers.

Sinclair, J.M. and Coulthard, R.M.: 1975, Towards an analysis of discourse. The English used by teachers and pupils, Oxford University Press.
Grize, J.B.: 1974, Recherches sur le discours et l'argumentation, Droz.
Grize, J.B.: 1982, De la logique à l'argumentation, Droz.
Grize, J.B. and Pieraut-Le-Bonniec, G.: *La contradiction. Essai sur les opérations de la pensée*, Paris, Presses Universitaires de France.
Margolinas, C.: 1993, De l'importance du vrai et du faux dans la classe de mathématiques, La Pensée Sauvage, Grenoble.
Margolinas, C. and Steinbring, H.: 1994, Double analyse d'un épisode: Cercle épistémologique et structuration du milieu, *Vingt ans de Didactique des Mathématiques en France*, La Pensée Sauvage, Grenoble, 240–257.
Mopondi, B.: 1995, Les explications en classe de mathématiques, *Recherches en Didactique des Mathématiques*, 15/3, Edition La Pensée Sauvage, Grenoble, 7–52.
Moreira, M.: Le traitement de la vérité mathématique à l'école, Thèse Université Bordeaux I.
Oleron, P.: 1977, *Le raisonnement*, Presses Universitaires de France.
Perelman, C.: 1970, *Le champ de l'argumentation*, Presses Universitaires de France.
Perelman, C. and Olbrechts-Tyteca, L.: 1976, *Traité de l'argumentation*, Institut de Sociologie, 3° éd.
Robrieux, J.J.: 1993, Eléments de Rhétorique et d'Argumentation, Dunod.

ANNICK FLÜCKIGER

MACRO-SITUATION AND NUMERICAL KNOWLEDGE BUILDING: THE ROLE OF PUPILS' DIDACTIC MEMORY IN CLASSROOM INTERACTIONS

ABSTRACT. This paper is based on a long-term didactic engineering about division problems (only in a numerical setting) at primary school. Situations and students' work are analyzed by means of a double theoretical framework: the theory of situations and the theory of conceptual fields (Vergnaud 1991). The analysis focuses mainly on classroom interactions and on the didactic memory from both the teacher perspective and the learner perspective: in particular, it not only investigates how didactic memory is managed by the teacher, but also how students recall past events or reread those events in a-didactic situations.

KEY WORDS: concepts-in-action, didactical engineering, didactical memory, division problems, operational invariant, numerical knowledge, theory of conceptual fields, theory of didactical situations, schemes

1. Introduction

The concept of the *teacher's* didactic memory was first proposed in Brousseau and Centeno's work in the early 1990s, in relation to the theory of didactic situations. More recently, the concept was reconsidered in terms of the anthropological theory of didactics (Matheron, 2001). The concept of a *pupil's didactic memory* will be studied here in the dual framework of Brousseau's (1997) *theory of didactic situations* and Vergnaud's (1996) *theory of conceptual fields*. The idea will be to present the research that led to the definition and development of this concept. In line with Brun and Conne's (1991) work in Geneva, an initial study was conducted (Flückiger, 2000) to identify this pupil-initiated memory phenomenon. Then, as part of a project by the Franco-Genevese research team on comparative didactics, the data was reanalyzed to determine how teacher's actions can elicit this memory (Flückiger and Mercier, 2002). After a description of the main results of the initial study, and a demonstration of the emergence of didactic memory in the pupil, a functional structure of how numerical knowledge is built will be proposed.

Educational Studies in Mathematics (2005) 59: 59–84
DOI: 10.1007/s10649-005-5885-3

2. The Initial Study

The initial study dealt with the emergence of numerical knowledge in the classroom. At the theoretical level, the study was aimed at testing the relevance of articulating the two theories that supplied the framework for the present experimental research. The mathematical object under examination was written calculation algorithms, more specifically, how *problem divisions* are studied in a Genevese class of fifth graders (approximately age 10). The term "*problem divisions*" is used to refer to the fact that the divisions in question may be difficult for pupils who have not yet studied long division. This implies that finding the right answer will require approaching a genuine mathematical question. The term is also used to express the fact that these problems were not everyday *division problems* like the story problems commonly given in this grade. The pupils worked in a purely numerical context. Fifth graders already have knowledge of addition, subtraction, and multiplication algorithms, which they have been taught in school, and they also know about the equivalence between multiplication and division in simple cases like those found in multiplication tables (for example, 10 divided by 5 equals 2, because 2 times 5 equals 10).

Note that the goal of the study was not to lead the children to invent a division algorithm that would later be instituted in the classroom. Nor was it a question of testing a new teaching method for written division problems, currently learned in fifth grade. The goal was rather to devise a *research methodology* for studying the genesis of numerical knowledge over time, under didactically controlled conditions.

2.1. *Longitudinal study*

At the macro-engineering level, the idea was to create learning conditions, in which meaning could be controlled during the teaching of a division algorithm. The corpus of data that we analyzed was collected in a classroom in Geneva and included all classes over an entire school year where the concept of division was taught (about 50 sessions).

The methodology traditionally associated with the theory of didactic situations is called *didactic engineering*. In a didactic-engineering approach, unlike a "naturalistic" type of observation, empirical data are compared and related to theoretical models in an organized way. In the present case, the goal was to find out how numerical learning takes place while working towards the elaboration of an algorithm for long division. The aim here is to attach meaning to this learning in a setting organized for that purpose and based on a chosen theoretical framework, the theory of didactic situations. Artigue (1990, 1992) defined didactic engineering as follows:

Didactic engineering, seen as a research methodology, is, firstly, characterized by an experimental schema based on class[room] 'didactic sequences', by which we mean based on the design, the production, the observation, and the analysis of teaching sequences. Here classically two levels are distinguished, *micro-engineering* and *macro-engineering*, depending on the size of the didactic sequences involved in the research (Artigue, 1992, p. 44).

2.1.1. *Method of study*

The experiment was organized around weekly cycles in which the micro-engineering level corresponded to sessions held in the classroom, and the macro-engineering level – by virtue of its duration – corresponded to the general experimental device. The weekly cycles were composed of one or more teaching sessions, followed by a consultation session among the members of the team (researchers and teacher). This cycle was repeated throughout the school year. The regular link maintained between classroom experimentation and analysis sessions is shown in Figure 1.

The initial sessions were derived directly from Kamii's (1994) work, conducted in reference to Piaget's theory. These sessions, called "calculation" sessions, were based on pupils' inventiveness in the face of a new type of arithmetic problem (here, division), i.e., one for which no specific algorithm has been taught as yet. However, the pupils already knew, for example, that 12 divided by 6 equals 2 because 2 times 6 are 12. So new knowledge can be built on that already acquired, the equivalence between multiplication and division.

Inventiveness alone does not suffice to move forward in the learning process, and, in fact, only the teacher's actions (here, controlled by the study) and the manipulation of didactic variables enabled the pupils' procedures to evolve.

2.1.2. *Didactic variables*

Two types of didactic variables were applied in this study. These were the numeric variables for the problems of long division used in the experiment and the variables involved in the setting up of the classroom sessions. The list of classroom sessions is included in the annex.

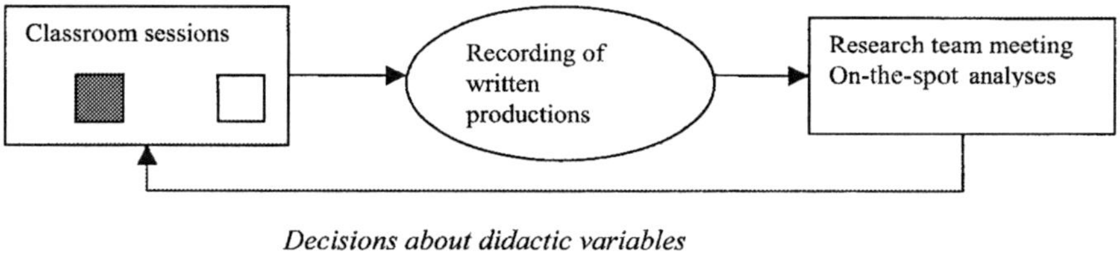

Figure 1. Weekly structure of the study.

Calculation sessions were organized temporally into "phases" (using the terminology proposed by Margolinas, 1993). During the *action* phases, pupils were given a division problem and had to work individually to find the answer. This was followed by a *communication* phase, during which the different answers (and procedures) found were presented to the class and compared. Based on the theory of didactic situations and a functional perspective on knowledge building, this is where the *formulation* and *validation* phases are articulated, with the conditions for moving from one to the other being among the questions raised in the study. For these sessions, the main didactic variable was, of course, the numerical variable. The researchers chose the numbers used in the sessions on the basis of previous research, which identified the difficulties connected with the numbers in long division and also according to the procedures elaborated by the pupils, which were analyzed at the end of each session.

For example, the first division problem proposed to pupils, $990 \div 9$, can be done "digit by digit", or by first representing 990 as a sum of 900 and 90 and then dividing each component by 9. The second division problem selected for use "$1818 \div 9$" obliged the pupils to find new ways of doing long division.

Besides the calculation sessions other types of sessions, described briefly below, were set up in accordance with the *on-the-spot* analyses conducted each week.

Journal-writing sessions marked off the progression of the pupils' work and the queries they raised. These sessions served as a support for the individual preparation of questions about the mathematical object "division". In a personal mathematical diary used for this purpose only (and referred to as "Journal" in this text), the pupils had to write their answers to questions raised by the teacher. They knew this book would never be marked or checked. The questions would be of a temporal or epistemological nature. For example, "How are you getting along with division problem?" "What do you find most interesting about division problem?"

Some of the responses were selected and given to the whole class for later journal-writing sessions or for use as topics of *debate sessions*. The first debate session was based on a statement taken from a pupil's journal: "My classmates' results are different from mine because they use a different method from the one I do." This idea was discussed first in small groups then by the whole class. It helped the pupils to consider the difference between method and result with respect to the uniqueness of the result of a calculation.

The debate-session setting was borrowed from Sensevy's (1998) work on the study of fractions in elementary school. In his study, which focused on the temporal dimension of knowledge production in the classroom,

Sensevy attempted to render the pupil's activity *chronogenic*, i.e., to have the pupils' productions move the learning process forward. This setting also creates conditions which make the pupil responsible for evoking past situations, thereby enabling the creation of connections between the private and classroom dimensions of memory. Also, after the fashion of what happens in a research community, *journal* writing can promote the cooperative dimension of scientific work in the classroom. It offers a medium for capturing the temporal dimension of knowledge production in a community, as stressed by Sensevy.

The methods tournament session was set up to compare the different procedures used in the classroom. It provided the opportunity for questioning the very notion of algorithm (efficiency, range of validity, etc.). In groups, the pupils demonstrated their methods of calculation to each other and debated them. Points were awarded for the speed, efficiency and variety of the methods each group put into play.

These different sessions supplied the variables that governed our study.

2.2. *A dual theoretical framework*

In the framework of a didactic system modeled by the teacher–pupil–knowledge triplet, the study focused on the elaboration of knowledge of division by the pupil subsystem, in a research-controlled didactic context. The question was, how do classroom interactions evolve in a situation where it is left up to the pupil to move the learning process forward and to discover new questions about the object under study.

In line with Brousseau's theory, the research methodology was engineered to create conditions that allow to trigger the dialectics necessary for a meaningful acquisition of the target knowledge. Brousseau modeled the different ways of functioning in terms of the *action*, *formulation*, and *validation* situations. When the pupil is interacting with the situation during the *action* phase, he/she is not necessarily capable of expressing the knowledge at play. This is achieved later in the course of the communication phases, where the pupil is led to *formulate* the knowledge and present it to others. At this point, the information must be understood and transformed by the interlocutor into a relevant *decision*.[1] In the theory, the *validation* situation represents the transition from empirical validation to an assertion recognized by all and integrated into known theorems. In Margolinas's (1993) terms, this involves creating conditions for moving from an *assertoric truth* to an *apodictic truth*,[2] which is brought about by scientific debate.

The idea in this study was to enrich the theory of didactic situations with the theory of conceptual fields. While the former served as a model for designing the experimental classroom setting, the latter was used to detect the operational invariants underlying the subjects' behavior. Vergnaud (1996) developed the theory of conceptual fields in an attempt to analyze the question of conceptualization, and the continuities and discontinuities that occur in the course of learning. Taking up Piaget's concept of *scheme* proposed in genetic epistemology, he centered his theory on the scheme–situation duality.

> The conceptual pair 'scheme–situation' is the keystone of cognitive psychology and of activity theory, for the simple reason that getting to know means adapting; it is the schemes that adapt, and they adapt to situations (Vergnaud, 2002; our translation).

Vergnaud defines a scheme as a fixed organization of activity for a particular class of situations.[3] A scheme is linked to the time course of the activity. The notion of *class of situations* is both innovative and essential in Vergnaud's theory, where a *conceptual field* is defined as a set of situations *and* concepts. A concept does not develop in isolation, but is part of an entire system of diverse concepts that develop jointly during the conceptualization of a notion. Hence, the notion of conceptual field.

> It is the situations that give meaning to the concepts, by way of the learner's activity; it is the concepts-in-action and the theorems-in-action contained in the schemes that enable these situations to be processed (Vergnaud, 2002; our translation).

Here, Vergnaud distinguishes two kinds of invariants that make the subject's action operational. *Concepts-in-action* permit the processing of information considered relevant to the situation at hand. By identifying these concepts, we can determine what the pupils have selected, in that situation, as appropriate information for processing the problem. *Theorems-in-action* are ones held to be true during action. While the concept of *scheme* is associated with that of *situation*, the notion of *operational invariant* is associated with *mathematical objects*, their properties, and their relations. In Vergnaud's theory, concepts-in-action and theorems-in-action are interpreted in reference to the concepts and theorems of mathematics.

The theory of conceptual fields, which is not a didactic theory *per se*, allows us to approach school learning in terms of its characteristic duration and nonlinearity. It allows us to see inside the pupil subsystem, the "black box" of the didactic system. It makes an essential contribution to understanding the cognitive facet of didactics. In the present analyses, the notion of scheme will be applied to the theory of didactic situations in order to analyze the pupils' productions.

Two levels of analysis are superimposed in our study. One reflects the level of didactic engineering in the context of the theory of didactic situations. There are two dimensions at this level. The macro-level deals with the long-term building of numerical knowledge about long division, and at the micro-level each session is divided into a phase of action followed by a phase of communication. The second level of analysis is that of the invariants in the context, with which the pupils had to deal in their calculations. The procedures invented by the pupils were then interpreted according to the theory of conceptual fields. This interpretation of the pupils' methods and their evolution allows the researchers to understand if the pupils' knowledge works well in registers anticipated by the theory of situations (action, formulation, and validation). The pertinency of the observations registered during this double evaluation allows the researchers to validate their theory.

Although we are unable to make a detailed list of every contribution to this research project, we shall indicate some of the most significant results concerning the object of division problem. This will be followed by further consideration of the pupils' didactic memory.

2.3. *Division as an object of learning*

Due to its complexity and the important place occupied in mathematical learning, division is particularly interesting for anyone hoping to understand numerical knowledge acquisition in elementary school children. Paradoxically, the fact that this subject matter is "didactically old," so to speak, is an interesting point in itself. The topic of division algorithms in school has been addressed from a number of angles, so it is possible to draw from results accumulated in the research over the years. The available studies have been conducted in the framework of the theory of didactic situations (didactic engineering, list of conceptions of division, etc.) or the theory of conceptual fields (error lists, integration of a scheme into an error explanation system, etc.). The body of findings obtained from different theoretical spheres offers a starting point for interrelating the two models of interest to us here by providing an original unit of analysis in didactics, the scheme. This approach is in line with the considerations brought to the fore by Brun, who emphasized the need to relate the original models of didactics to "lower" level models like those of developmental psychology.[4]

In the 1990s, the Genevese research team on mathematics teaching, headed by Brun, worked specifically on the issue of written calculation algorithms (Brun et al., 1991). The first series of studies dealt with the analysis of pupils' errors in written calculations. They showed that Brown and Van Lehn's (Brown and Van Lehn; 1980, Van Lehn, 1988) Repair Theory cannot account for all subtraction errors. According to Brun and Conne, systematic or recurring errors made by pupils are traces of the gradual construction of an algorithmic scheme. In this dynamic view of errors, pupils adapt their knowledge in order to progress in their calculations.

In a talk on written calculation algorithms presented in Geneva in 1996,[5] Brun showed how previously learned algorithms may or may not act as schemes. The criterion he used for schemes was adaptability. He raised the question of whether learned, automatized algorithms can be "decontextualized" and unraveled to uncover their numerical meaning. Another indicator used by Brun is the ability to communicate new procedures. To what extent can new procedures be imparted to, understood, or even actually put to use by others? The author's observations demonstrate the relevance of his scheme-based analysis of written calculation algorithms, and suggest that a previously learned algorithm can emerge in the form of a scheme that becomes available when a new situation is being processed.

This perspective was applied here by combining it with a more specifically didactic dimension, that of the conditions in which learning is taking place – i.e., the situations and how they are handled – and which trigger such adaptation.

3. Some results

3.1. *"Problematizing" division as an object of learning*

In this experiment, pupils were presented with a division problem on the blackboard, without ever having been taught any kind of algorithm to solve it. They had to think up ways to perform the calculation. How the pupils grasped the mathematical object of "division" and – far beyond the question of the answer obtained from the string of calculations they proposed – how they "problematized" their search for a solution, are interesting findings in themselves. In particular, the queries that emerged here about the existence or the uniqueness of a quotient are rare at this grade level! It is not just the numerical variable and the sequence of calculation sessions that enabled this to happen. As stated above – and this is a clear-cut result – the inventiveness of pupils does not alone suffice to move forward in the learning process. The *control variables* that dictated the experimental situation and determined the nature of the sessions proposed (journal, debates, etc.) allowed each pupil to share with the teacher the responsibility of advancing the learning process and undertaking the problematization of the question posed to the class as a whole. Along with the teacher's management of the dual didactic and research contract, the manipulation of the didactic variables is a condition for the emergence of true mathematical inquiry.

3.2. *Results of numerical knowledge building*

A scheme-based analysis of the data turned out not only to be compatible with the theory of didactic situations, the foundation of the engineering process, but also very fruitful from the standpoint of the results obtained. First, the results pertaining to numerical knowledge construction and to the theory of didactic situations itself will be presented. Then the issue of didactic memory will be addressed.

3.2.1. *The question of the remainder*

The analysis pointed to the invariants in the schemes that became operational during the pupils' activity. The detection of both the concepts considered relevant by the pupils and the theorems they used to treat the problem, served to pinpoint *topogenetic* shifts. This concept (extended in particular by Chevallard, 1985/1991) refers to the respective positions of the teacher and the pupil in their relationships to different objects. This dimension must be considered in conjunction with the *chronogenetic* dimension (Chevallard, 1995/1991), which accounts for temporal changes. The topics of classroom debates are indicative of how the pupil subsystem is evolving relative to the concept under study. For a given pupil, the operationalized invariants are a reflection of how his/her knowledge network is evolving in the conceptual field of, in this particular case, division.

For example, the question of the remainder gradually took over in the classroom debates. This question was first brought up by a newcomer in the class who had already studied the traditional division algorithm. He declared that the answer to "6 divided by 5" was "1 remainder 1". At the time, the class was unable to decide between the two answers proposed: "1 remainder 1" and "1 point 2". To conclude the session, a summary was made stating the disagreement and the two answers given, one supported by multiplication ($1.2 \times 5 = 6$) and the other by the Euclidean equation representing the division ($5 \times 1 + 1 = 6$). This became the topic of some highly interesting and heated debates in subsequent sessions.

The existence, relevance, and magnitude of the remainder with respect to the divisor are mathematical questions that supply the grounds for differentiating between the integer quotient and the decimal quotient. Associating a single quotient number to the dividend–divisor pair, or associating the quotient–remainder pair, is a choice which, in the absence of a concrete context, raises touchy questions like the uniqueness of the quotient–remainder pair, the nature of the numbers studied (integers or decimals), etc. These questions ended up leading the class to say to the teacher, "You have to tell us what set of numbers we're working in, N or R." Remember that this is grade 5 of elementary school!

Regarding the question of the division algorithm, approaching the data in terms of schemes pointed out the following: each partial quotient in the division algorithm is a Euclidean division, i.e., the entire quantity is not being divided. This was totally new compared to previously studied algorithms. Take the following production:

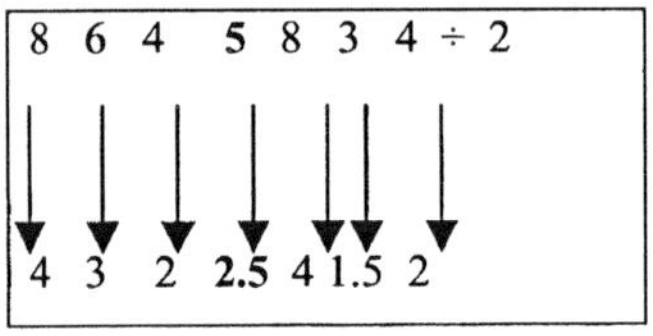

In the digit-by-digit processing done by this pupil, each successive number is divided in its entirety. By introducing several decimal points in the written quotient, the pupil failed to abide by the place notation system.

This example is reminiscent of Brun's observations regarding the transfer of previously learned schemes to a new algorithm.

In addition, subtraction, and multiplication problems, quantities are processed in their entirety (added, subtracted, or multiplied). In the case of the above division, the intermediate division "5 divided by 2" should produce the pair (2, 1), the remainder "1" being carried over to the next row to produce 18, which is then divided by 2. This is not what the pupil does; he produces the quotient 2.5. Providing a pair (quotient and remainder) as a result of division is indeed a completely new step in the calculation. The heated and recurrent debates between the pupils about the relevance of "leaving a remainder" illustrate the underlying mathematical difficulty of the division process, which in that case is perceived as uncompleted. This specificity of the division algorithm needs to be brought to the fore, particularly in teacher training programs. It is a difficulty that must be handled didactically, just like the virtual absence of subtraction in the procedures noted here. Although repeated subtractions form the basis for teaching the traditional division algorithm, the present experiment showed that the concept of subtraction may not be present in the conceptual field of division at the onset, and therefore has to be fully constructed, including for some teachers in initial training.

3.2.2. *Two major classes of procedures*

The systematic detection of invariants pointed to two major classes of procedures that contributed to the emergence of this numerical knowledge. Whether at the individual level or during interaction in the classroom, a key element lies in the link between the *calculation procedure* and the quotient *verification/invalidation* process. To understand the difficulty inherent in the nesting of the calculation and verification procedures, which are intertwined in division problem, the pupil's activity can be seen as referring to two different classes of problems (from the standpoint of the activated schemes).

The first, which concerns the search for the quotient, activates "calculation" schemes (in the strict sense of the term). The associated invariants enable the pupil to produce one or more numbers that belong to the field of possible answers for the numbers given (based on the way in which the division is introduced in this study, resolving $a \div b$ is equivalent for the pupils to finding the number q such as $b \times q = a$).

For instance, for the division of 2546 by 2, different approaches were observed. In one approach, the pupil transformed the numbers (dividend, divisor) or the operation. For example:

- 2546 became $2000 + 500 + 40 + 6$, then each term was divided,
- 2546 became a sequence of 2, 5, 4, 6, then each number was divided,
- 2546 divided by 2 became a multiplication with a "gap" e.g. $2 \times ???? = 2546$.

In all cases the aim was, *in fine*, to find a number which would be the result of the calculation.

In another approach, the pupils were concerned with the truth value. In this case, the result – partial or final – was assigned the value "right" or "wrong" with respect to the givens (initial numbers, operation to perform).

Here is an example of the second approach: the pupil is given the numbers to be divided, for example, 175 divided by 14. The list of the results found by the pupils is written on the board (here: 75; 63; 12.15; 12.5; 5.5) and each pupil can validate or invalidate the proposed quotients. For example, looking at the first quotient a pupil said "75 is too much, 5.5 is not enough." The teacher asked "Why?" The pupil replied "It's impossible, 10 times 75 already makes 750." By this theorem-in-action the pupil declares that the result of 75 is wrong.

The notion of scheme accounts for the general organization of the activity carried out in each of these situations, an activity which involves making the same pieces of mathematical knowledge work with various arrangements and variable levels of importance. It allows for constant reinterpretation of the situation, which is such that, whenever the givens change, the goals and subgoals are modified, along with the checking methods to use and the invariants that have become operational. The division scheme is based on the combination of two sub-schemes, one about the search for a result, the other about the verification/invalidation of this result. The selected invariants can be different, they may also be identical. It is the organization, the hierarchy of the procedures carried out in the activity which changes.

When observing the interactions in the classroom at a more macroscopic level, the procedure utilized for the division – and with it, the mathematical knowledge brought to bear – is assessed by peers verifying the correctness of a result made public. Inversely, a result thought to be the outcome of a valid procedure can be assessed either numerically (i.e., in terms of its nature, existence, or written notation) or as the solution to the division problem (Is this a quotient? Is it the correct quotient?). A genuine dialectical process that weighs the procedure against the result takes place here. Such dialectics seem to be a valid technique for building numerical rationality. Note also the necessity of having different procedures available, both for constructing the result and checking/invalidating it.

3.3. *Back to the theoretical models*

"Research modifies theory as much as theory determines experimentation" (Morf, 1972, p. 107; our translation). In the light of the results obtained here, let us now look at the model proposed by Brousseau in his theory of didactic situations. This theory, which served as a conceptual framework for setting up the teaching sequence tested here, offers a structure based on situations: action, formulation, validation, and finally, institutionalization.

3.3.1. *Verbalization phases*

In adidactical situations called action situations, a pupil's knowledge allows him/her to act upon the problem situation, which in return must provide the occasion either for modifying that action or for supporting a decision to act. The pupil's task is not to express, explain, justify, or even identify the knowledge at stake. By contrast, formulation situations are ones with a communicative goal. According to Brousseau, an original message that will be addressed to others (a single pupil or a group) is constructed from a known repertoire, but this dimension does not suffice for defining how knowledge functions. The key dimension is decision making. The knowledge formulated by the pupil must be converted into a relevant decision by the addressee who receives the information. To understand the adidactical calculation situations found in this study, it seemed useful to define what will be called the *verbalization phase*. The verbalization phase initiates the exchange phase, which is when communication takes place. Brun stressed the importance of the didactic organization of verbalization during the *didactic exchange* (Brun, 1994). As they verbalize, pupils talk about their own calculation procedure; they make the sequences of actions they performed known to others. This does not, however, make it a situation of formulation. From the standpoint of the theory, the verbalization phase is an action situation (see Figure 2).

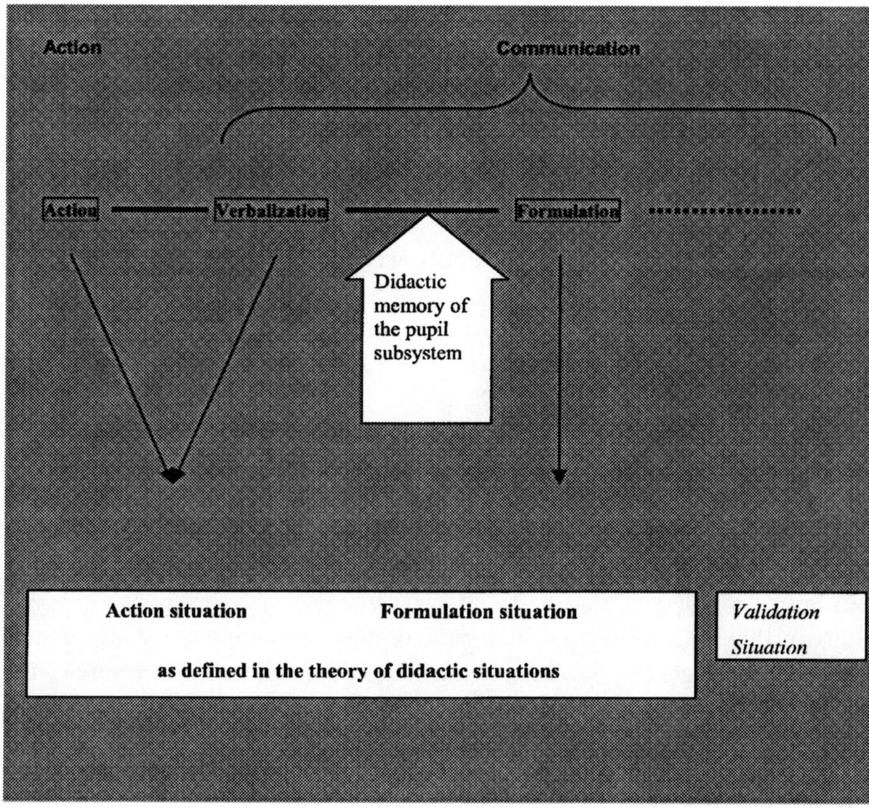

Figure 2.

Figure 2 depicts the locus of the verbalization phase, as it is understood in Brousseau's theory. The diagram also shows the point where the pupil's didactic memory is brought to bear in modifying the functioning of knowledge, a topic which will be addressed in the next section.

3.3.2. *From verbalization to formulation*

When pupils verbalize their procedure during a collective work session, they state the sequence of actions carried out to reach the goal of coming up with a numerical result. In this case, even if they can foresee some of their peers' reactions, they are not in a true position of expectation. They abide by the usual didactic contract in the classroom, where the pupil makes public some of his/her private actions. From the communicative standpoint, there is no real contract of collaboration, nor of opposition. It is the input provided in return by another pupil, based on the data given by the verbalizing pupil, that modifies the situation and grounds the exchange on the knowledge at play. Two critical positions can be found: either the pupil affirms the contradiction, or questions his/her own approach in the light of what was just said. In both cases, the prior situation must be reinterpreted to eliminate the discrepancy. During the reinterpretation process, which brings out the contradiction, a *decision* is made. Either the position evoked is maintained, or it is modified or even rejected altogether. In this case, the decision making occurs in a communication situation and no longer in an action situation. The action-verbalization situation is thus transformed into a genuine formulation situation. In our experimental system based on debates, contradictions play a central role in this transformation process.

Bringing a past didactic event to the foreground and comparing it to the present calls upon memory. Such episodes are pupil-initiated and are hence referred to as the *pupil's didactic memory*. One of its functionalities is to create the conditions for moving from an action situation to a formulation situation. This process involves comparing knowledge emerging at the current time with knowledge about a past situation (or point of view).

Now that we have identified this phenomenon relating to the adidactical nature of the macro-situation, a theoretical approach to the structure of the *didactic memory of the pupil* will be proposed.

4. THE PUPIL'S DIDACTIC MEMORY

Research in cognitive psychology offers us the basic functional characteristics of memory: recursive functioning, which makes it possible to postpone certain decisions, and the capability to anticipate. Because of its didactic nature, this study looks in particular at the conditions in the didactic system that permits the emergence of memory phenomena.

The concept of didactic memory was developed in the framework of the theory of didactic situations by Brousseau and Centeno. The question raised by these authors concerned how the didactic system handles the temporary knowledge of pupils (Brousseau and Centeno, 1991; Centeno, 1995), so they focused on the didactic memory of the teacher.

> Endowing an organism with a memory allows this organism to postpone certain decisions without losing information likely to influence it, and in doing so, to keep within its processing capacities, conditions that would tend to fall outside. Above all, it enables the reinterpretation of information, and consequently, through the nesting of transformation rules, all sorts of recursive functions. It is the inevitable instrument of anticipations. [...] Thus, locally, memory acts concurrently with adaptation since it permits its postponement or its avoidance.

In fact, in the middle term, memory promotes adaptation. Indeed, it appears itself to be the result of an adaptation to interactions in which the subject must survive, foresee, adapt, and learn (Brousseau and Centeno, 1991, p. 199; our translation).

These authors showed that the teacher's memory allows him/her to organize the change in status that knowledge undergoes. They also showed that when teaching takes place without the teacher's memory of prior situations, knowledge is connected both differently and to a lesser extent.

Perrin-Glorian (1993) defined the notion of *recall situation* as the "re-reading" of a situation treated in a former session. The re-reading process is carried out by a pupil when called for by the teacher. Our studies have shown that *pupils themselves can initiate re-reading of past events* in a context that authorizes a certain amount of adidactical functioning. Anticipations about relevant mathematical objects have also been noted. These phenomena are linked to the characteristics of the macro-situation set up, and to the teacher's management of it. The contract is unusual, since responsibility for memory processes is, in fact, devolved upon the pupils. Journal writing is particularly conducive to this recalling process. The teacher's actions can enable (or not) the individual queries evoked via recall to become shared by the class. If so, individual memories are integrated into the collective knowledge-building process. We therefore define the concept of a *pupil's didactic memory*. Indeed, this is a manifestation of true *didactic* memory because it is steered by the didactic situation; there is an intention to teach and a didactic contract specially set up for this study. Our research study highlighted different levels of recall. Using this as a basis, we now propose a structuring of a pupil's didactic memory before identifying the characteristics which make it an essential concept in the understanding of didactic phenomena.

4.1. *Structuring of didactic memory*

Three types of these memory manifestations (Recall 1, 2, 3) were identified here, each fulfilling a different function in the genesis of numerical knowledge.

- In R1, a result or an old calculation that has now become relevant is remembered.
- In R2, a past event that points to a contradiction is evoked.
- In R3, a new class of problems is created.

R1: Recalling a Result or an Old-But-Now-Relevant Calculation.

Memory recall serves to extend the repertoire of results applicable to problems. At the onset, the results repertoire available for division problems is the one found in usual schoolbooks, since division is introduced in this macro-situation as the inverse of multiplication. The repertoire is expanded by adding locally instituted results, which are then integrated as "subroutines" into new calculations. An example of this was noted here for "6 divided by 5", which came up several times with the numerical variables used.

> Division [205650 ÷ 5]
>
> Pupil LU is explaining his approach involving an additive breakdown of the dividend:
> 200000 ÷ 5 = 40000; 5000 ÷5 = 1000; 600 ÷5 = 120
> Two pupils spoke up when he came to 600 ÷ 5
> *ME: I just remembered that 6 divided by 5 is 1.2*
> *ER: So did I. I remembered that/since you also have to have about a hundred, that makes 120*

The connection made between "6 divided by 5" and the associated result 1.2 (highly problematic at this point in the elaboration process) was thus handled as a whole, as an operational invariant in new situations that could be combined with other mathematical knowledge, in particular, knowledge about decimal numbers being learned at the time.

This type of manifestation which exhibits a link with previous knowledge is usual in a classroom. It is even often expected or incited by the teacher.

R2: Evoking a Past Event Revealing a Contradiction.

The journal is where the pupil's didactic memory emerges. In the example below, FR saw and put back into question what he had already noted as an inconsistency.

> During session 21, the division [826 ÷ 14] was proposed to the pupils. Among the answers debated, LU proposed 86.5, which he justified by the following calculations (written on the blackboard in a column by the teacher):
> 10 × 80 = 800; 4 × 6.5 = 26, so 14 × 86.5 = 826
> This result was invalidated by performing the multiplication 14 × 86.5 = 1211

Several times, FR was surprised:

> (...)
> *LI: Already if you do/the answer he found, 86, you drop the decimal point and you multiply by ten, that already makes 860 so it has to be wrong*
> *FR:* **But that could convince us that it's ...**
> (...)
> *Teacher: What do we say for now about LU's?*
> *Pupils: It's wrong*
> *LI: Because from the beginning alone, you can tell it's wrong*
> *FR:* ***Not from the beginning alone, huh // because the beginning of the operation, I would have bet it was right***
> *Teacher: That's what you say*
> *FR: Yes*
> *Pupil: But that makes 1211 so it's wrong*

The result was checked by multiplication, and this concluded the debate. The answer 86.5 was declared wrong, but twice, FR manifested his reluctance. He repeated this during the journal-writing session. Below is FR's verbatim reply to a question about what had been a source of surprise during earlier lessons.

> *Why doesn't the division down here work? LU did this division*
> $826 \div 14$
> $80 \times 10 = 800$
> $6.5 \times 4 = 26.0$
> $800 + 26 = 826$
> $86.5 \times 14 = 1211.0$ (all operations were written in column format)

This remark can be understood by looking at session 20. At that time, the pupils were working on [$345 \div 23$]. Several procedures proposed by pupils had been validated, two of which were presented by their authors as follows (noted here in line format):

$5 \times 23 = 115$ and $10 \times 23 = 230$ so the correct quotient is 15,
$20 \times 15 = 300$ and $3 \times 15 = 45$ so the correct quotient is 15.

These two procedures are based on the distributivity of multiplication with respect to addition, and this is what FR thought he recognized in LU's procedure when he tried to find the quotient of 826 divided by 14. For him, there was an obvious contradiction, and the journal enabled him to express his doubts. He was reworking a mathematical invariant that was essential for performing the division algorithm. Breaking the dividend and divisor down into additions to isolate simplified quotients is a common erroneous procedure (e.g., 346 ÷ 123 becomes 300 ÷ 100 = 3 ; 40 ÷ 20 = 2; 6 ÷ 3 = 2; result 322). Of course, the teacher then made the didactic decision about whether or not to pass on FR's query to the class as a whole.

Finding contradictions, which generates a twofold process of reinterpreting the past and anticipating the future, is a frequent phenomenon in the present teaching context. Such contradictions may be verbalized by the pupil him/herself, who formulates certain invariants and then puts them back to work. Or a peer may point out a contradiction between two decisions separated in time.

This type of memory manifestation is part of the chronogenetic dimension of knowledge construction. The specific didactic conditions, and particularly the journal-writing sessions allowed the public emergence of this type of questioning. This work of invariants, which is a central aspect of the conceptualization, remains usually private to the pupils and is not publicly visible in the classroom.

R3: Creating a New Class of Problems.

The creation by a pupil of a new class of problems is indicative of a significant topogenetic change in the organization of the conceptual field, in this case, division. This reorganization which structures in a new way the total number of classes of situation connected with the concept of division has an impact on the memory function. This new structuring of situations will lead to a different practical approach to division problem because types of situations and schemes are dialectically connected.

Let us look at the following example:

> Division [147097 ÷ 7]
> During the debate, a pupil named VIA proposed:
> VIA: *You have to put a decimal point and then nines to infinity, as LI did for a hundred divided by three*

This pupil saw that the current situation was analogous to one previously encountered in "100 divided by 3," a division proposed by LI in the journal. To this calculation situation, VIA associated the production of a quotient of the type "infinite decimal," and based his answer on this scheme. In Vergnaud's theory, the notion of scheme is closely tied to that of class of situations. In a given situation, if a pupil makes the connection between a certain problem and an identified class of situations, the associated schemes can be activated, even combined, for the processing at hand. The creation of such classes contributes to overall cognitive efficiency and is therefore a fundamental process. The lengthy duration of our engineered teaching plan is one of the conditions that promotes the detection of cognitive reorganizations like these. On the basis of a remembered event, a new problem can be related to a prior problem, sometimes publicly in the present teaching setting.

When this happens, we can see how a pupil's knowledge is reorganized and how this allows him/her to anticipate the results of an entire series of calculations from the same class. This type of knowledge reorganization is a topogenetic event that temporarily isolates a class of situations associated to a quotient-finding scheme. In the present case, the scheme did not become a routine because[7] it was shown to be incorrect in later calculations. This approach is similar to the one found in neuroscience, where memory research focuses on operational processes and memory is seen as a generator of new categories (Rosenfield, 1988).

This third category of pupil's didactic memory manifestation shows a reorganization of the conceptual field of the pupil. A previously constructed scheme is associated with a category of new questions. This phenomenon is rarely so clearly formulated as in the example above, and is essential in the construction and the structuring of the knowledge.

4.2. *Characteristics of the pupil's didactic memory*

Two characteristics of the pupil's didactic memory should be underlined:

- firstly, that it is developed in a classroom community,
- secondly, that it is guided by the teacher's teaching activity.

Clearly, the phenomenon of the so-called collective memory exists in other contexts, in the same way as the intention to teach pupils exists outside of the classroom. However, it is the interplay between the two above-defined points which is at the core of didactic memory.

Of course, didactic memory is built on the basis of personal memories but this aspect does not completely cover the area studied. The didactic memory of the pupils' sub-system is built collectively. During the sessions the pupils did not only recall personal memories of how their learning had evolved. They based their comments on remarks made by a peer. The pupils' recall revealed the similarities and differences and even their disagreements regarding their knowledge. These discussion sessions more often brought to light not the disagreements of position between two pupils but the change in position of a peer. A frequently heard comment was: "before he/she agreed and now he/she disagrees." The pupils insisted strongly that any change in position regarding the learning process should be justified by the authors. The way the teacher dealt with such contradictions in the research project was essential.

For the researcher to understand how didactic memory played its role in speeding up the process, she had to analyze the teacher's classroom management in two ways.

- The intention to teach pupils a part of a body of knowledge during their scholastic period. The concept of a didactic contract is there to explain what cements the teacher-pupil relationship around the acquisition of a body of knowledge. Everyone in the class knows the learning at stake goes beyond the single classroom session.
- The classroom management through which the teacher first organizes the succession of mathematical concepts to be acquired by the pupils over a given period (chronogenesis) and then defines the teacher's and pupils' respective positions in relation to this knowledge (topogenesis)

In the study under examination a deliberate choice to use an a-didactic form of classroom management was made, to allow pupils' contributions to occur more freely including manifestations of didactic memory. The term "a-didactic" indicates that the pupils were allowed to assume certain responsibilities usually assumed by the teacher. For example, the responsibility of saying what is true, or of indicating, which objects of knowledge are the most pertinent. In a macro a-didactic situation such as the one under consideration, the progress in didactic time is partially devolved. This leaves them free to back-track or go "fast-forward" in public. Both these processes are encouraged in their journal writing by the teacher's questions.

Example: Journal no. 3

Question 1: What have you found out about how to do divisions?

Question 3: What would you like to know now about division?

In ordinary classroom organization, the teacher is the one who reminds the pupils of a certain lesson or exercise, or who says what they are going to learn next. In our study, the pupils are allowed to decide because of the special macro-research situation which was established. This places the pupils progressively in an unusual position regarding the acquisition of knowledge. More specifically by turning the class into a research group, the pupils are involved in a collaborative working environment in the sense that they have to built collectively what they have to learn.

Here the question at stake is to establish a written algorithm for performing division. However, the general path the pupils need to follow is marked out by the teacher due to his/her choice of didactic variables (numerical or instructional, for example, the instruction which turns a calculation session into journal writing). This "marking" became clear during classroom debates. The teacher's strategy involves the following steps:

- Marking out a field of investigation for the pupils. The teacher needs to check what the class has learnt previously (for example, how to do an addition). He/she then allows the class to discuss, which new mathematical objects they decide to work on.
- Highlighting any contradictions. The teacher encourages conflicting debate relative to the acquisition of the particular knowledge at stake (here, numerical knowledge involved in division).

Didactic memory is relative to this common object at stake and is not only the individual development of each pupil's memory. Regarding the time required by each pupil to complete a learning process, there is a process of acceleration connected to didactic memory. The collective examination of individual discoveries accelerates the learning process. Each pupil can be helped by the observations made by his/her peers.

5. Conclusion

In a didactic engineering approach based on the theory of didactic situations, the concept of scheme and the detection of invariants in the pupils' interactions proved useful in determining how knowledge evolves. Schemes and the accompanying analyses thus provide a relevant supplement to the theory of didactic situations. Concerning the conceptualization problems found in the pupil subsystem of the didactic system, using the scheme as a unit of analysis avoids the pitfall of taking a linear approach to knowledge building. The scheme turns out to be a good *didactic*-analysis instrument for taking into account the recursive and anticipatory functioning of didactic memory. By combining the theory of didactic situations and the theory of conceptual fields to analyze how knowledge evolves over time, it is possible to rethink the concept of didactic memory from the standpoint of the pupil subsystem. This concept cannot be dissociated from the knowledge-imparting process. Its role in transforming verbalization problems into true formulation situations is essential to the construction of numerical knowledge.

By relying on the notions of scheme and class of situations, the present study provided insight into how the conceptual field of division is elaborated under specific didactic conditions. Understanding this elaboration process requires considering the duration of school learning and didactic memory phenomena specific to the pupil subsystem. This study demonstrated the existence of the pupil's didactic memory, and showed how it is necessary in this type of didactic contract.

ACKNOWLEDGEMENT

I would like to thank Vivian Waltz and Mary Stuttart for the English version of this work.

NOTES

1. The theory of didactic situations was couched in terms of game theory, where *decision* is a key concept.
2. Mathematical propositions are *apodictic* and not *assertoric*. Here, the terms *assertoric* refers to Kant's modalities.
3. In the theory of conceptual fields, "class of situations" is understood as "class of problems."
4. On this subject, see Brun (1994).
5. Piaget-Vygotsky Congress, Geneva, 1996.
6. Though this is unusual in Genevese elementary schools, the teacher introduced the symbols N and R to refer to the natural and real numbers.
7. Term borrowed from Saada-Robert and Brun (1996).

ANNEX: LIST OF CLASSROOM SESSIONS

	Divisions		
Session (*N*)	*n*	Dividend	Divisor
1	1	990	9
	2	1818	9
2	3	2546	2
	4	2592	6
3	5	345	23
4	6	720	20
	7	426	2
5	8	425	5
	9	175	14
6	10	180	15
	11	427	7
7	12	4500	150
	13	633	3
8	14	8645834	2
9	15	35 787	3
10	J1	Journal no. 1	

(*Continued on next page*)

(*Continued*)

	Divisions		
Session (*N*)	*n*	Dividend	Divisor
11	16	8324782	2
	17	645357	3
12	18	405	45
	19	315	45
	20	3780	45
13	21	2000	16
14	J2	Journal no. 2	
15	22	3618	18
16	Debate session		
17	23	4824266	2
	24	4962	2
	25	645354	6
18	Posting results		
	26	525575	5
	27	124892	4
	28	35791	2
19	29	8324782	2
	30	55155	5
	31	12143	1
20	32	816328	8
	33	345	23
	34	71	71
21	35	826	14
	36	6	5
22	37	205650	5
	38	465	155
23	39	17	5
	40	4854816	4
24	41	14 497	7
	42	301 725	5
25	J3	Journal no. 3	
26	43	304515	5
	44	369	246
27	45	1268148	4
	46	1545005	5
	47	180	12

(*Continued on next page*)

(*Continued*)

	Divisions		
Session (*N*)	*n*	Dividend	Divisor
28	48	100	3
29	List of divisions		
to	List of methods		
32			
33	49	65724	2
	50	3780	90
	51	3612	12
34à36	Tournament of methods		
	52	147097	7
37	J4	Journal no. 4	
	53	25	2
38	54	81822	3
	55	995	9
39	Debate regarding the remainder		
40			
41	56	6412	8
42	57	3615	12
	58	224	7
43	59	167	8
44	60	75035	25
45	61	266	4
46	62	2186	5
	63	801	20
	64	375	12
	65	368	7

REFERENCES

Artigue, M.: 1990, 'Ingénierie didactique', *Recherches en Didactique des Mathématiques* 9(3), 281–308.

Artigue, M.: 1992, 'Didactic engineering', in R. Douady and A. Mercier (eds.), *Research in "Didactique of Mathematics". Selected papers*, La Pensée Sauvage, Grenoble.

Brousseau, G.: 1997, 'Theory of didactical situations in mathematics', in N. Balacheff, M. Cooper, R. Sutherland and V. Warfield (eds.), *Didactique des mathématiques 1970–1990*, Kluwer Academic Publishers, Dortrecht.

Brousseau, G. and Centeno, J.: 1991, 'Rôle de la mémoire didactique de l'enseignant', *Recherches en Didactique des Mathématiques* **11**(2–3), 167–210.

Brown, J.S. and Van Lehn, K.: 1980, 'Repair theory: A generative theory of bugs in procedural skills', *Cognitive Science* 4, 379–426.

Brun, J.: 1994, 'Évolution des rapports entre la psychologie du développement cognitif et la didactique des mathématiques', in M. Artigue, R. Gras, C. Laborde and P. Tavignot (eds.), *Vingt ans de didactique des mathématiques en France*, La Pensée Sauvage, Colloque ARDM, Grenoble, pp. 67–83.

Brun, J.: 1996, 'Algorithmes et schèmes dans les calculs écrits', in *Paper presented at the Symposium "Concepts pragmatiques et scientifiques dans le fonctionnement et le développement des schèmes"*, Conférence Vygotsky-Piaget, Geneva.

Brun, J. and Conne, F.: 1991, 'L'étude des algorithmes de calcul dans la transmission et la constitution des connaissances numériques', PME XV, Fond National de la Recherche Scientifique, Assisi (research conducted by J. Brun, F. Conne, J. Retschitzki and R. Schubauer).

Brun, J., Conne, F., Lemoyne, G. and Portugais, J.L.: 1994, 'La notion de schème dans l'interprétation des erreurs des élèves à des algorithmes de calcul écrit', *Cahiers de la recherche en éducation* 1, 117–132.

Centeno, J.: 1995, *La mémoire didactique de l'enseignant (thèse posthume inachevée: Textes établis par C. Margolinas)*, LADIST, Bordeaux.

Chevallard, Y.: 1985/1991, *La transposition didactique. Du savoir savant au savoir enseigné*, La Pensée Sauvage, Grenoble.

Flückiger, A.: 2000, 'Genèse expérimentale d'une notion mathématique: La notion de division comme modèle des connaissances numériques', Doctoral Thesis, Université de Genève.

Flückiger, A. and Mercier, A.: 2002, 'Le rôle d'une mémoire didactique des élèves, sa gestion par le professeur', *Revue Française de Pédagogie* 141, 27–35.

Kamii, C.: 1994, *Young Children Continue to Reinvent Arithmetic (in 3rd Grade). Implications of Piaget's Theory.* Teachers College Press, New York.

Margolinas, C.: 1993, *De l'importance du vrai et du faux dans la classe de mathématiques.* La Pensée Sauvage, Grenoble.

Matheron, Y.: 2001, 'Une modélisation pour l'étude didactique de la mémoire', *Recherches en didactique des mathématiques* 21(3), 207–246.

Morf, A.: 1972, 'La formation des connaissances et la théorie didactique', *Dialectica,* 26(1), 103–114.

Perrin-Glorian, M.-J.: 1993, 'Questions mathématiques soulevées à partir de l'enseignement des mathématiques dans les classes faibles', *Recherches en didactique des mathématiques* 13(1.2), 95–118.

Rosenfield, I.: 1988, *The Invention of Memory, A New View of the Brain*, Basic Books, Inc., Publishers, New York.

Saada-Robert, M. and Brun, J.: 1996, 'Les transformations des savoirs scolaires: Apports et prolongements de la psychologie génétique', *Perspectives* XXVI(1), 96, 26–37.

Sensevy, G.: 1998, *Institutions didactiques, étude et autonomie à l'école élémentaire*, PUF, Paris.

Van Lehn, K.: 1988, 'Toward a theory of impasse-driven learning', in H. Mandl and A. Lesgold (eds.), *Learning Issues for Intelligent Tutoring Systems*, Springer-Verlag, New York, pp. 19–41.

Vergnaud, G.: 1996, 'La théorie des champs conceptuels', in J. Brun (ed.), *Didactique des mathématiques*, Delachaux & Niestlé, Lausanne, pp. 197–242 (Reprinted from *Recherche en didactique des mathématiques,* 10 (2/3) 133–170, 1990).

Vergnaud, G.: 2002, 'Piaget visité par la didactique', *Intellectica* 2001/2, 23, 106–123.

ANNICK FLÜCKIGER
Psychology and Education Sciences,
University of Geneva, 40 Boulevard du Pont d'Arve,
Geneva 1211, Switzerland
Telephone: +4122 705 9831
E-mail: annick.fluckiger@pse.unige.ch

PATRICIA SADOVSKY and CARMEN SESSA

THE ADIDACTIC INTERACTION WITH THE PROCEDURES OF PEERS IN THE TRANSITION FROM ARITHMETIC TO ALGEBRA: A *MILIEU* FOR THE EMERGENCE OF NEW QUESTIONS

ABSTRACT. The purpose of the present article is to give an account of the emergence of knowledge pertaining to the transition from arithmetic to algebra in the course of a debate in a grade 7 classroom. This debate follows two other instances of work: (1) the adidactic interaction between each student and a given problem, (2) the adidactic interaction of each student with the procedures generated by other students during stage 1. The two kinds of processes – adaptation to a *milieu* and social interactions – play a critical role in the change of "rationality" required for the move from arithmetic to algebra. Both the design of the initial mathematical problem given to the students and the organization of the interactions leading to the debate under study in this article are based on this hypothesis. The research presented in this article is set in a broader work of didactic engineering that aims at studying didactic conditions for making a connection between arithmetic practices and algebraic practices.

KEY WORDS: interaction with procedures of peers, milieu as generator of new questions, theory of didactic situations, transition from arithmetic to algebra

1. THE THEORY OF DIDACTIC SITUATIONS AS A FRAME OF REFERENCE FOR OUR WORK

Our work is set within the perspective of Guy Brousseau's '*Theory of Didactic Situations*'. This theory proposes a model, which considers teaching as a process focused on the production – implying transformation and validation – of mathematical knowledge in a school environment. Although the Theory of Didactic Situations – as pointed out by Perrin-Glorian and Laborde in the Introduction – has emerged with the purpose of modeling teaching from a certain perspective, an important theoretical work has been carried out in recent years to extend Brousseau's concepts and adapt them to the study of ordinary classroom situations. Since our work is set within the perspective of didactic engineering – as part of the investigation, we designed the teaching units whose realization we study – we will outline below those aspects of the theory that refer to conditions concerning the didactic situations conceived based on this theoretical framework.

Brousseau makes two fundamental assumptions: (a) the student elaborates knowledge through interaction with a problem that offers resistance

Educational Studies in Mathematics (2005) 59: 85–112
DOI: 10.1007/s10649-005-5886-2

and feedback, which, in turn, affects the mathematical knowledge at stake, and (b) the didactic intentionality of the teacher is an aspect, which is inherent to both the process of classroom production of knowledge and to making a connection between such knowledge with cultural knowledge. Based on these assumptions, Brousseau formulates the need for a *milieu*,[1] which is conceived and sustained by a didactic intentionality.[2] The interactions between the student and the *milieu* are described in terms of the theoretical concept of *adidactic situation*, which models a student's activity of knowledge production independently of the teacher's mediation. In this situation, the student acts more as a cognitive subject than as a "student", or subject of the school institution. The subject engages in an interaction with a problem, not only putting his own knowledge at stake, but also modifying or rejecting it or producing new knowledge depending on the interpretations he makes about the results of his actions (the feedback he gets from the *milieu*). Thus, the concept of *milieu* includes the mathematical problem the subject faces, as well as a set of relations – also essentially mathematical – that are modified as the subject produces knowledge in the course of the situation, thereby modifying the *reality*[3] with which he interacts.

The interactions between the teacher and the student concerning the interactions of the student with the *milieu* are described and explained using the notion of *didactic contract*, which refers to the more or less explicit assumptions and expectations underlying the way the teacher and the student interpret each other's actions. The personal processes of learning are embedded in the weave created by the two types of interaction – subject/*milieu* and teacher/student – which are separate only in the theoretical analysis.

Brousseau points out that the theoretical necessity of a *'milieu'* is determined by the fact that the didactic relationship will eventually come to an end and the student will have to face situations with no didactic intention (1986). We would like to mention as well, that a learning process based primarily on the interactions with the teacher without confronting the student with some portion of the *"reality"* that can be known – and therefore, modified – through the tools that mathematics offers, does not allow enough space for the student to test his anticipations against the feedback of the "reality" he interacts with, and to learn, through this confrontation, to control this reality and to recognize the limitations of the relations he noticed and used. From our point of view, without interactions with a *milieu*, both the role of mathematical concepts as means of solving problems and the possibility of bringing into play the validation tools pertaining to the discipline become blurred.

Through the *a priori* analysis of potential interactions between the subject and the *milieu*, the theory of situations aims at accounting for the possibilities of action for the subject when confronted with a problematic task, the feedback from the *milieu* and the means of validation that the subject may generate within those interactions. From the methodological point of view, the *a priori* analysis of the didactic situation that the researcher wishes to study enables the construction of a set of observable issues that will constitute a framework for an interpretation of data in the experimental work.

Teaching conceptions that would be centered exclusively on the processes of knowledge production in an autonomous interaction with a *milieu*, would ignore the social and cultural nature of the construction of knowledge at school. But, from Brousseau's perspective, the classroom is conceived as a space of production, in which social interactions are a necessary condition for the emergence and elaboration of mathematical issues. The cultural framework of the class imposes restrictions, which determine, to a certain extent, the knowledge that is elaborated. The teacher's inevitable reference to the erudite mathematical community plays the role of regulator in the constitution of this cultural framework. The regulations of the teacher that have as a reference both the classroom and the mathematics as a discipline – conceived as a body of organized knowledge – are explained through the theoretical notion of didactic contract.

2. Some points of contact between the theory of didactic situations and other theoretical perspectives

The perspective, succinctly described in the previous section, conceives of mathematical learning at school as a process, in which the cognitive and the social adaptations are interwoven. From this point of view, this perspective is close to those put forward by Cobb and Bauersfeld on the one hand, and by Steinbring on the other. Cobb (1996) and Bauersfeld and Cobb (1995) point out that learning can neither be described as a conciliation of the individual mind, which tries to become adapted to an environment, nor reduced to a process of enculturation to a pre-established culture. Cobb states:

> Bauersfeld, however, takes the local classroom microculture rather than the mathematical practices institutionalized by the wider society as his primary point of reference when he speaks of negotiation. This focus reflects his concern with the process by which the teacher and students constitute social norms and mathematical practices in the course of their classroom interactions. Further, whereas sociocultural theorists give priority to social and cultural process, analyses compatible with Bauersfeld's perspective propose that individual students' mathematical activity and the classroom microculture are reflexively related (Cobb, 1989;

Voigt, 1992). In this view, individual students are seen as actively contributing to the development of classroom mathematical practices, and these both enable and constrain their individual mathematical activities. Consequently, it is argued that neither an individual student's mathematical activity nor the classroom microculture can be adequately accounted for without considering the other.

(Cobb, 1996)

On the other hand, Steinbring (1998) also refers to a double play, which combines social and epistemological aspects:

I would like to characterize the core problem of understanding school mathematics as follows: *students have to decipher signs and symbols!* There are two different types of signs and symbols: *mathematical signs and symbols* (ciphers, letters, variables, graphics, diagrams, visualizations, etc.), and *social signs and symbols* (in the frame of the classroom culture: hidden hints, remarks, reinforcements, confirmations, reflections, etc. made by the teacher, and comments and remarks of similar types made by other students). (Steinbring, 1998)

The ideas of the quoted authors have points of contact with the Theory of Situations in as much as they emphasize the epistemological and social components of school mathematical learning. From our point of view, however, the works to which we have just made reference are focused on processes of knowledge development as a product of social interactions and do not account for the role played by the work of each individual student with a specific mathematical problem in these social interactions. Neither do they mention which characteristics of the proposed mathematical task would promote a certain type of subject/problem interaction.

3. Differences between the various productions in class: Some issues that help clarify the purpose of this article

To represent the social nature of mathematical knowledge, Brousseau describes didactic situations that are organized in such a way that the interactions between the students and the *milieu* require exchanges between peers. These are situations of formulation and validation, in which the students exchange information and argumentation, respectively.

In these models, the interaction between peers is a necessary condition to tackle the mathematical problem and this necessity is given by the way the situation is organized: an asymmetry is deliberately generated in the resources made available to different students to force an interchange. It could be said that the mathematical problem, with which the students are dealing, is embedded in a social organization with the purpose of provoking the production of information or propositions referring to the object of teaching at play (Laborde, 1991).

Now, in dealing with a given mathematical problem, the interactions amongst the students in a class may go beyond the social organization that has been assumed about the situation. Inevitably then, the ways in which some students tackle the problem may modify what other students decide to do. Moreover, as we will see in the present work, the differences between the various productions are a potential source of imbalance in the class, which necessarily leads to the emergence of new questions. We surmise that the formulation of new questions, which broaden the scope of possibilities in relation to the mathematical work, is an inherent aspect of knowledge production in a class, when students have to face a drastic change in mathematical practices, as is the case with the arithmetic–algebra transition.

What are the conditions that have to be satisfied to prevent the different contributions made by individual students from "escaping" the classroom before they have served their productive function for the class as a whole? Is it possible to grant the productions of each student a certain objectification so that they are accepted or rejected based on mathematical grounds and not only for social reasons?

On the background of these questions, our aim in this paper is:

1. To specify the conditions that enable the creation, in the classroom, of a *milieu* for generating new questions that are constitutive of the mathematical knowledge required for getting involved in algebraic practices.
2. To analyze the knowledge related with the arithmetic–algebra transition, which emerges when grade 7 students (12-year-old, last year of primary education) discuss the productions of others, after being invited to take a position concerning such productions during an adidactic phase of work.

The debate analyzed in the present article is a part of the implementation of a didactic design, conceived in the framework of a broader project of didactic engineering (Sadovsky, 2003) aimed at exploring the possibilities of making a connection between arithmetical practices and algebraic practices.

The assumption of the existence of a rupture between these practices played a central role in the didactic choices that were made in the project. We elaborate on these choices below, by explaining the general intentions of the project and our assumptions about the arithmetic–algebra transition.

4. SOME ASSUMPTIONS ABOUT THE ARITHMETIC–ALGEBRA TRANSITION

Numerous scholars worldwide have studied the epistemological and didactic rupture involved in the transition from arithmetic to algebra – from the perspectives of both teaching and learning – and many results of these investigations have become part of the common culture of researchers who continue addressing the didactic issues in this field. We will refer here only to those results that are relevant to our work.

Both in the school culture and the society at large, when the term arithmetic is associated with primary schooling, what is meant is a type of activity that involves the solution of ("word") problems that require one or more arithmetical operations with specific data and, generally, with a unique solution. As Vergnaud et al. (1987) point out, solving these problems consists basically in choosing data to start operating with those, possibly identifying intermediate unknowns, operating with them and reaching a solution.

The intervening relation or relations are never dealt with all at once but they are used in a step-by-step fashion. Contrary to that, an algebraic practice involves making explicit the relationship between the unknown and the given data to then move on to a relatively automatic manipulation of these to reach a solution. Roughly speaking, the "object" of arithmetic in primary school is numbers, whereas elementary algebra focuses on relationships between quantities.

In the postscript to a book containing a compilation of recent works on this problem, Balacheff (2000) mentions that the control one has over the activity of solving a problem is part and parcel of knowledge. Summarizing the perspective of different authors, Balacheff emphasizes the difference between the modes of control necessary in arithmetic and algebraic work. He points out that, in the case of solving more complex problems, reference to extra-mathematical contexts does not provide elements for proving the existence or the completeness of a solution. It is at this point that algebra appears as necessary.

In the above quoted book, and establishing from our point of view a point of contact with Balacheff's position concerning the constitutive role of justification in (the construction of) knowledge, Campos Lins (2000) argues that, for a subject, knowledge is a pair made of a statement that the subject considers true and a justification that the subject develops to bear out that truth. Thus, if, with respect to a certain statement, different subjects put forward different justifications, they possess different knowledge. In relation with transformations of a certain algebraic expression, the author states that the transition from justifications based on extra-mathematical contexts to the production of meanings based on only algebraic properties entails a rupture, which is facilitated by the pressure exerted on the student by the didactic intentionality of the teacher (or the researcher). In this perspective, an idea of cognitive adaptation overlapping with social adaptation appears and intervenes in the production of algebraic meanings.

Summing up the above, we could say that while Balacheff postulates that the reference to the extra-mathematical context suggested by a problem does not offer elements to control central aspects of algebraic practice such as, for example, the question of the number of solutions, Lins points out that leaving behind this context is possible only if the student becomes involved in a game that he ignores but dares play due to his trust in the teacher as a representative of knowledge.

This poses a problem that we have considered when designing the engineering put to work: how to conceive of a didactic scenario in which students could elaborate criteria to validate their productions around problems which, as they include the notion of variable, entail a certain degree of rupture with arithmetical practice. We will outline below only some aspects of the didactic engineering that we consider necessary to justify the didactic choices made for the classroom episodes under study in this article.

5. Discussion of options we faced in a larger didactic engineering project

The larger engineering project that we have developed aimed at exploring the type of knowledge produced by grade 7 students confronted with problems requiring the construction of notions of variable and dependence. How do the students use their previous arithmetic knowledge to deal with the new problems? What ruptures are involved in the resolution of these problems? What aspects of the former relationship with operations are revealed now in the face of the ruptures that the new problems entail? What new issues are raised? What is the role of the introduction of the notion of variable in these new issues? What demands, with respect to the production of new semiotic tools, are posed by the need to represent several or an infinite number of solutions? What is the function of the writings produced by the students? How is this written production validated? What is the position of the students in the face of the need to attribute values to one of the variables to generate solutions? What criteria do they elaborate to count the number of solutions to a problem? What types of justification do they propose? These central questions oriented the development of two teaching units, on which the didactic engineering project was based. The second unit was structured around word problems involving two variables with one degree of freedom.[4] Without loss of the general sense of the situation – we believe – we will analyze the potential of this type of problems, based on one of them, which is the one we used in the lesson plan we study in this paper.

6. Organization of Interactions between Peers, Taking as a Starting Point the Uncertainties and Ruptures Generated by Two-Variable Problems with One Degree of Freedom

One of the problems given to the students, as a starting point, was the following: *Marisa has 20 pesos in coins of 10 and 50 cents. How many coins of each kind can she have*? At the beginning the students had to obtain some solutions; then, they were asked to state the number of possible solutions and propose a procedure that would enable the generation of them all.

Several factors set this problem widely apart from the old, traditional arithmetic problems. By introducing one degree of freedom, the students must set in motion – with a greater or smaller degree of awareness – the notions of variable and dependency; each solution is now a pair of numbers and the solution set is made of several pairs. As a general procedure to obtain all the solutions is required, what is promoted is that the students systematize all the solution pairs around one relationship and this contributes to a process of generalization, the deepening of which leads to the objects of algebra.[5]

When we analyzed the possible strategies of the students to tackle the production of solutions, we took into consideration the level of generality involved in them. We considered, on the one hand, procedures through which the students could engage in a systematic search for solutions and which would ensure they have explored all (correct or incorrect) solutions. On the other hand, we considered strategies, through which the students would obtain "loose" solutions, without establishing any relationships between them and without proposing criteria to analyze if all possible solutions have been considered or obtained. When contrasting these two types of strategies – whose nuances we will not discuss here – we were able to understand better the transformations entailed in moving from one type to another.

Tasks such as considering the issue of the number of solutions, analyzing the domain of the variables, producing a procedure in response to a problem, are new for the students.[6] Many of them may find these tasks odd. There are no shared meanings for these notions in the class yet. Working on this problem aims precisely at establishing them publicly in the classroom. The emphasis is therefore laid on putting in place a change of practices concerning mathematical work, rather than on introducing a new concept.

As the resolution of the problem demands finding out the number of solutions and ensuring that all have been found, it is now necessary to elaborate criteria for the validation of these aspects. This validation cannot emerge only from the interaction with the problems. Why not? The students are in a position to make their own procedures work, that is, to obtain solutions through this interaction and to check, by going over the formulation of the problem again, whether they are right or not. However, even if they would like to verify the fact that the procedure can generate every solution, the relationships they would bring into play would probably be the same they have used to conceive the procedure. The contributions of peers, considering the framework their own productions offer, may act as feedback to decisions made by individual students. We can see that the emergence of this knowledge has an unavoidable social dimension.

Establishing a procedure in order to generate every possible solution will require putting at stake the mathematical relations inherent in the problem and that will be represented through new semiotic tools. Under these circumstances, without a thorough knowledge of the conventional use of these tools, decisions regarding the formulation of the procedure each student suggests are made with a great degree of uncertainty. What didactic response can then be offered to the problem posed by the validation of the students' productions?

Our option was that each student's production is put to the test and assessed in terms of the effectiveness that the "others" – those who did not produce it – confer to that production in the generation of every solution. Accepting each proposal will then be the product of an extensive discussion between the teacher and the students within the class' collective space.

The previous considerations show the need to generate the conditions that will offer each student the possibility to interact with the productions of other students. Therefore, we have designed the following organization for the problem we have just described. The work with the problem is organized in four stages: (1) the students work individually following the specific instruction to obtain solutions, decide how many there are and describe a procedure to obtain every solution; (2) students work in small groups and their task is to choose one procedure among those suggested by the individual members of the group or, if none is accepted, to produce a new one; (3) the teacher transcribes on the board the procedures proposed by each group and the small groups analyze the different procedures that have been exposed; and (4) collective debate about the procedures.

The first stage constitutes a period of autonomous work with the problem, in which the student produces relations that will serve as a frame of reference for the subsequent stages.

During the second stage, each student – as the producer of a particular procedure – must face the challenge of explaining his/her work to the small group and try to convince the others of the effectiveness of his/her products. The need to choose a single procedure to represent their group forces the students to consider different procedures as an object of work. By being forced to accept or reject a procedure, each student will – implicitly, in part – turn to arguments concerning both the relevance of each procedure and to more personal criteria such as those related to its clarity. The former kind of arguments are more likely to convince another student to give up his/her production than the more subjective criteria, even if they are publicly expressed. These negotiations, which take place among students without the teacher's intervention, depend to some extent on such features of the social functioning of the group as leadership, underestimated students, etc. These considerations allow us to anticipate that each student in the group will have a different rapport to the procedure that is finally proposed to represent their group.

In the course of the third stage, each group establishes an adidactic interaction with the written production of the other groups, in a position that combines both evaluation and validation. An opportunity is thus offered to put each student's own production to the test, both as a framework for analysis and as a control strategy for the work of others. When the moment comes to analyze the results presented by the other groups, relationships established with their initial individual work and the different arguments presented by the other members of the group during the second stage, will operate differently for each student. At the same time, as a result of the analysis, each student will produce a series of judgments relative to the procedures analyzed. However, many of these statements will not be validated by the authors and therefore will have for them a status close to that of a question. The others – other students and the teacher – will be required to deal with these questions.

In the fourth stage, the collective debate on the procedures brings about the public negotiation of a number of issues such as, for example: how to establish the number of possible solutions, what criteria could be used to validate this matter, what criteria must be implemented to accept or reject a certain position. The collective debate is established on the basis of judgments, about which there is no agreement. The adidactic interaction of the previous stage confers strength to the positions of the different students.

When we analyze the four stages as a whole, we can see how at each consecutive stage, the *milieu* is based on the choices and constructions made in the previous stage. The third stage is crucial for the students to find other limitations for their deeply rooted arithmetical practices and the fourth one is indispensable to start producing answers for the questions generated in the previous stage and those that are elaborated in the course of the discussion. In this sense, stages three and four constitute a *milieu*, which is, for each student, a generator of questions about his own work and not just a mere feedback for his solution.

7. THE PRODUCTION PROCESS IN THE CLASSROOM

We shall consider some episodes of the events that took place in a class,[7] from the very moment the different procedures chosen by each group were written on the blackboard, which corresponds to the two last stages of the organization of our plan.

The selected episodes seem to illustrate well the emergence of specific knowledge supported by the adidactic interaction with the procedures of others. These episodes are structured around the discussion of the procedures elaborated by three of the groups.[8] To make our analysis clearer, we present the three procedures as they were written on the board. The names of the authors of each procedure were not originally written on the board but we have included them here to facilitate the reading of the episodes transcribed below.

Procedure 2 (Guido, Juan Alejo, Manuel and Juan Manuel)

We decided to take any number between 0 and 200; 34, for example.

34 X 10 → amount of cents in 10C coins

↘ 2000 – 340 = 1660

1660 ÷ 50 = 33.2

33 50C coins and 35 10C coins

Procedure 3 (Gastón, Alejo, Martín and Ilan)

(50C *x n* lower than 40 – 2000 C) ÷ 10 = number of 10C coins = A

(20 – A . 0.10) ÷ 0.5 = number of 50C coins = B

Procedure 5 (Silivna, Pilar and Sebastián)

Number of 10C coins	Number of 50C coins
0	40
5 = 50	39 = 19.5
10 = 50	38
15	37
20	36
200	0

Explanation: in order to work out every solution starting from zero 10C coins and forty 50C coins, you have to progressively add five 10C coins and subtract one 50C coin, until you get to two hundred 10C coins and no 50C coins.

We shall first analyze, through an example, the role the adidactic interaction with the procedures of others can play for a student with respect to his/her own process of understanding in the third stage.

The other examples analyzed refer to the fourth stage, in which all the procedures are discussed collectively.

In Section 7.2 we will analyze two issues that arise from the debate about procedure 2. On the one hand, all along the discussion an element of normative knowledge[9] emerges: a procedure must establish a series of action rules for every possible value, both in the initial variables and in the results that arise from intermediate steps. On the other hand, when there is a need to justify an action rule that is formulated in order to complete the procedure, the class gets involved in a new mathematical problem.

The two points developed next (7.3 and 7.4) refer to interactions that bring, onto the classroom scene, some other knowledge which constitutes the notion of procedure and which also raises normative issues.

7.1. *Procedures of the others: a new opportunity to understand*

The authors of procedure 3 call the teacher, at first to discuss the fact that they have discovered a mistake in their own writing (we can see that the terms in the subtraction of the first formula are inverted). Then the interaction with the teacher brings about a discussion concerning the other procedures. Below is a transcript of the discussion starting from the moment Gastón refers to procedure 5:

Gastón:	The fifth one, I can surely say that that one is well explained, I understood it quite well. . . . . . you begin from one end till you get to the other, that is from zero 10C coins and forty 50C coins till you get to two hundred 10C coins and no 50C coins.
Teacher:	And are there any reasons why you would prefer the fifth [method] rather than yours?
Gastón:	No, I don't prefer the fifth, ours is better. (Very self-confidently)
Teacher:	Why?
Gastón:	Because it is ours. We thought it out.
Martín:	But it seems to me that the fifth one is better explained than ours. . . . I think.
Teacher:	Why do you think it is better explained, Martín?
Martín:	Well . . . for example, there, you already know it is wrong, because, well, it is 2000 minus the result, but if you read our method and you also read the fifth, I think you will understand the fifth one better.

Both Gastón and Martín can see the clarity of procedure 5, but only Martín considers it is better explained than their own. Martín is a "weak" student – we detected some signs of it during the whole research – and in the first stage of individual work he only suggested "loose" and unstructured solutions. Even if we did not witness the process of choosing a particular procedure in the group he belonged to, we have enough evidence to assume that his participation at that stage was not very active. The episode transcribed above allows us to infer that Martín is not really attached to his group's production, which puts him in a position to evaluate all productions fairly, without bias, putting them on an equal standing. We understand that it is not until now – in this phase – as he places himself as *an evaluator, after he has tried, without great success, to be a producer*, that Martín *understands*.

It should be mentioned that the circumstantial presence of the teacher during this group's work, as she inquires about the individual student's preferences, enables Martín to legitimate his weak adherence to the group's procedure.

Putting this episode more in perspective, we can observe that making a statement about the validity of another procedure, after having produced one on one's own, is a task which is essentially different and which enables the student to access relationships that may not have been elaborated at the time of his or her own solution of the problem.

7.2. *The obscurity of the second procedure: an opportunity for the emergence of new problems and new norms*

In the writing corresponding to procedure 2 (transcribed above), the authors transform value 34 – which they initially take to be the number of 10 cents coins – and the result 33.2 for the 50 cents coins, into the answer "33 fifty cents coins and 35 ten cents coins". There is not one word in the text that explains this transformation. The interpretation is then for the reader to work out. This widens the distance between the readings different students make. The version of the procedure that is finally written on the board has not been "touched up" by the teacher's contributions, which helps to generate a degree of uncertainty that invites debate.

We will look into two issues already mentioned by making reference to the following transcript.

Teacher:	Let's move on to procedure 2. Estefanía, please.
Estefanía:	I did... er... we can take any number from 0 to 200. I took for example 52, I multiplied 52 by 10 and the result was 520 and then, 2000 minus 520 was 1,480; 1,480 divided by 50 was 29.6, but I cannot have 29.6 fifty cents coins.
Sabrina:	Yes, you can.
Teacher:	Now they had said ...
Pilar:	It is not possible to do that with every number.
Sabrina:	But it says [on the board] below that they add a 10C coin.
Teacher:	They add a 10C coin.

Sabrina:	No, every two coins, you add one 10C coin.
Teacher:	Where does it say so?
Sabrina:	It says so there: 0.2 and I add one more 10C coins; 0.4 that is two 10C coins.
Teacher:	Where does it say so?
Alejo:	It doesn't say so, but it is so.
Teacher:	Then, Estefanía – with the procedure just as it is written here – gets to the point where she has a problem that the procedure cannot solve. Has anyone found a way to improve the procedure so that Estefanía wouldn't have the problem she has now?
Sabrina:	You would have to specify that every 0.2 50C coins, you must add one 10C coin.
Teacher:	(writes that down on the blackboard) Do you understand what I wrote?
María Sol:	What is the specification?
Teacher:	Every 0.2 50C coins, [it means here you get 'point 2'], you add one 10C coin.
Sebastián:	And what if you get 0.3 as a result?

i)The demand for precision in a procedure on the part of the class as a whole

The transcript shows an interaction between those who propose and those who oppose. The "others" are in charge of controlling those aspects that are difficult to take into account when one is in the position of a producer.

Within this tension, an action rule gets to be specified to transform the intermediate decimal point results. This rule is implicit in the text. At the beginning of the episode, Estefanía makes the procedure work but says that "*it is not possible to do that with every number*". This contribution brings out the need to make explicit the transformations the authors probably thought about but never wrote down. It is Sabrina, another reader of the procedure,[10] who sets out – at first in an imprecise way – to defend the validity of the procedure. The teacher's intervention aims at getting her to formulate explicitly a more general rule: *"every 0.2 50C coins, you must add one 10C coin"*.

Concerned about "completing" the writing down of the procedure with the rule Sabrina formulated, the teacher does not ask for the justification of the rule and then, implicitly, she is endorsing the procedure. Probably many students did not understand why this rule works, though it is a question they could have tackled with their numerical knowledge.

Immediately after the writing down of the rule, Sebastián's contribution *"And what if you get 0.3 as a result?"* expresses a concern the whole class agrees with: the need to give an answer to the possibility of obtaining intermediate results not contemplated in the rule as formerly expressed.

When Sebastián asks his question, the rest of the group sees this as a "natural" event. In some sense, it is evident for the students that what Sebastián is asking "deserves" an answer. It does not contradict any "previous" idea. It belongs in the area of "what was not thought of", not in an area of "what was thought of differently". At the same time, it presents a demand that refers to those issues that need to be considered in order to construct a procedure. The notion of procedure is put at stake here.

Even though the teacher is the moderator of the activity, it is the students, through their interactions, who demand more and more adjustments: proponents and opponents cooperate to obtain a more precise formulation of the procedure. These precisions were not produced in the previous phase, when they analyzed individually the procedures of others, but the relations construed by each student in that phase, when they had to assume responsibility for the study of the procedure, are present in the collective production.

ii), The other as a source of new mathematical problems

Sebastián's question gives rise to a numerical issue whose formulation is completed by Juan Alejo. The teacher takes up the problem and asks the whole class to try and sort it out. This activity is sustained during a significant period of time. In the following transcript we present the beginning of the discussion and how the teacher closes it.

Sebastián:	And what if you get 0.3 as a result?
Sabrina:	You cannot get 0.3.
Teacher:	Why not?
Juan Alejo:	Because... you can never get it... you tell me a number and you can't get 0.3, 0.5, 0.1, 0.7 or 0.9
Teacher:	Why not? Have you tried with all numbers?
Juan Alejo:	No, but there is a way to prove it.
Teacher:	How can you be absolutely certain?
Juan Alejo:	(to Guido) Come on; you tell her.
Guido:	Look, whenever I divide by 5, I always get "0 point. . . an even number".
Sebastian:	I got 0 point 4.
Teacher:	You got 0 point 4, the problem is to know if one will always get "0 point 2, 4, 6, 8", or if one can get "0 point 1, 3, 5, 9".

Sabrina: (Because) when you take an odd number, if you put "0 point odd", you multiply it by 10 and you'll get an even number, then, when you divide it by 5, once more you'll get an even number.

The students go on with the discussion without getting to a conclusion, until the teacher decides to postpone the debate.

Teacher: Well, let's leave this matter aside for a while. We will see later if one gets an even or an odd number. The idea, then, is the following: if it is true that when I divide I always get "0 point even number", then, for each 2 I have after the point I must add 1 to the number of 10 cents coins, that is to say, if I get for example 33.6, I would say it is thirty-three 50 cents coins and I would add 3 to the 10 cents coins. Is that right? This is what it says on the board now. The procedure has changed now that I have written this here.

We can notice that the numerical problem faced (how to tell for sure that the result of a calculation is always "point even number"), appears in an absolutely contextual manner within the original problem about the coins. It is difficult to imagine that such a problem could be posed in isolation, at least at this stage of schooling. Now, this is not a problem, which the teacher could have anticipated or whose emergence could have been foreseen through an *a priori* analysis. There is, in this episode, a class working as a production community: they have formulated a rule and, with the teacher's support and demand, they are trying to justify it. A moment of what we would call "*living mathematics*" is manifested, with all the unforeseen aspects a creative process implies. These considerations lead us to affirm that, inevitably, the *a priori* analysis cannot include all the mathematical issues that may arise in relation to a problem.

The numerical problem that emerges takes the teacher by surprise but, unlike what she did during the previous episode, when she did not ask for the justification of a rule, she now takes up the problem and allows a discussion to develop, even though she cannot provide elements that will allow her students to find a justification. We think that one of the reasons why a conclusion regarding this problem is not reached through the debate, is that it was difficult for the teacher to conceive, on the spur of the moment – the problem was not foreseen, it is a student who poses it – a mathematical explanation that would be adequate for her students' knowledge. This prevents her from orienting the discussion adequately and that is why she decides to postpone it. The episode refers us to Bloch (1999), who believes that the mathematical activity of the teacher in the classroom is an indicator of her students' mathematical activity. The teacher's mathematical activity refers primarily to the reconstruction of validation processes adapted to the students' level of knowledge, and this implies for the teacher a reorganization of his/her own knowledge. In other words, teachers learn mathematics when they think about how to justify something, bearing in mind what the students know. We believe that the "0 point odd number" episode is an example of Bloch's tenet because the teacher faces the mathematical problem of having to think out an argument adapted to her pupils so that, afterwards, a conclusion can be drawn in the mathematics lesson. She takes her time (the time between this class and the next one) to re-launch the debate.

The questions and queries that emerge during the debate make way for the production of original mathematical problems. In these cases, the teacher faces the demand to decide within the context of the lesson if she can make room for them or not. To attribute to each new problem that arises a certain viability as a problem for this class will be conditioned by the teacher's own possibility to find an answer and fundamentally to conceive an explanation adapted to the knowledge of the students.

7.3. *Can values be assigned? An issue that is settled in the collective space*

We will analyze here a part of the class discussion concerning procedure 3. Gastón, the principal author of the procedure, starts by correcting the order of the operations in the first "formula" and immediately points out the need to have the second formula. Another student objects to this and the debate starts. Below is the transcript of the classroom observation record at this point:

Teacher: That is to say, that it is actually 2000 – 50 multiplied by a number lower than 40 divided by 10. That is the idea you had.

Gastón: But, then you have to continue that. You get a result and then, you have to calculate 20–A $\times$ 0.10 divided by 0.50. Everything else is right; you have to change the order.

Matías: The calculation that is written below is unnecessary.

Teacher: Why?

Matías: Because when you multiply 50 by a number lower than 40, you already get the number of 50C coins.

Gastón: Sure, but...

Matías: And there (pointing to what is written on blackboard) you already find the amount of 50C coins (pointing again) [it says so] down there.

Teacher: Let's see, Gastón, is there anything you would like to say?

Gastón: That in this case you are providing a formula, not an example. If it is an example, you simply calculate 2000 minus 50, multiplied by the number.

Teacher: Right, but what he says is: what should we include the second calculation for if I already know that through that calculation I will find the number of coins I included at the beginning?

Gastón: Yes, I know, at first you include $2000 - 50 \times 37$, because you have that [number] already...

Teacher: That is what they [the members of the group] say... you choose a number smaller than 40, for example 15

Gastón: But in that case, it is necessary yes... because you are giving an example of how to calculate the number of 50C coins.

Teacher: Let's see, Gastón, can you explain more clearly why is it necessary to include both formulas?

Gastón: Because if you use the formula that is written below you can work out how many 50C coins you have to...

Matías: But you already know how many 50C coins there are.

Teacher: Can everyone else understand what Matías and Gastón are discussing?

The rest of the students, even though they do not participate in an explicit way, have been carefully following the debate from the very beginning. The teacher has been assigning turns for Matías and Gastón to speak, and they contribute on very similar terms, restating their points of view. Finally, the teacher asks:

Teacher: Gastón, why do you say it is necessary to include the second formula if we want to agree on a general procedure?

Gastón: Because otherwise, you have to say that you chose a number at random.

Teacher: Wouldn't it be correct to say that you took a number at random?

Gastón: I do that almost all the time, but when you are asked about reasons and specific details, the way you did, that would be the formula.

Teacher: And what would be wrong about taking a number at random?

Gastón: I don't know, but the instructions for the problems always say "justify, what you did" blah blah....

Teacher: And saying that you chose a number at random does not constitute a justification?

Gastón: You work it out, in one way or another... (He pulls an angry face)

Several students discuss if one can or cannot solve the problem by choosing a number at random. They all speak at the same time. With a series of questions, the teacher "corners" Gastón, who must finally admit that, in order to bring his procedure into play, he must begin by choosing a number at random.

Teacher: Well then, if I start by randomly choosing this number, then I make this little calculation and I get to know the amount 50C coins... then, this procedure here... what do I use it for?

Gastón: To occupy more space.

Another student (at the same time as Gastón): To justify.

At this point, the teacher closes the debate.

Teacher: It is valid to choose at random. Otherwise, we will move in circles. If I cannot take this number at random and I have to work it out from this calculation, but this calculation is impossible to be made .... I end up by not being able to make any calculation. Somewhere along the procedure, there is a number that I have to "invent". I choose it randomly but that does not mean that anything goes. You have just clarified something, you said the number must be lower than 40.

Gastón's work is affected by several tensions operating simultaneously. During the first stage, he produced examples by taking arbitrary values for one of the variables and calculating the other. However, this proves insufficient when it comes to proposing a general procedure – which will be presented before the rest of the class for discussion. He chooses to control the uncertainty generated in him by the nature of the solutions to this problem (pairs of numbers linked by a relationship within a certain domain) by including in his procedure two equations: one for each variable. He insists that there is a need for the two formulas, based on two algorithms of calculation and the *belief* that the values, which are part of the result, *must* be obtained through the operations using the data given. We believe that Gastón finds support for his position in clauses of the didactic contract constructed through the work on arithmetic (one calculation for each result, operating with the data, step by step and eventually with intermediate variables) (Vergnaud, op. cit.).

Matias' intervention breaks the balance that Gastón tried to achieve through his two formulas. As Matías' interpretation of the solutions is close to the one of pairs of numbers dependent upon each other, he cannot accept Gastón's proposal.

Although the discussion is centered on these two students, the attention with which the whole class follows it accounts for the point of rupture that the students confront when they tackle a problem, in which there is a free variable.

The debate does not really advance. On the one hand, Matías does not know enough mathematics to justify his position; this issue is indeed about the construction of knowledge through this sequence. On the other hand, Gastón feels uncomfortable in the face of the evidence provided by the other student and the teacher and so resorts to what is normative (*you cannot pick numbers at random)* as a place of safety to make up for his inability to couch in mathematical terms the need for the two formulas. Gastón's position responds to an issue determined by the didactic contract – he is self-conscious and very much aware of what he believes is expected of a student in relation to an arithmetic problem – and his convictions are strong enough for him to continue defending his position despite the fact that it is clear from the start that the teacher favors Matias' position. This example illustrates how the introduction of new norms that call for a revision of those construed through a very sustained practice requires a certain time that largely exceeds that of a particular sequence of lessons. The didactic contract is constitutive of thought and the constructions, which it supports, are frequently used by students in a variety of circumstances (Schubauer-Leoni quoted by Sensevy, 1998).

The teacher restricts herself to Gastón's arguments and points out the need to formulate a new norm – *it is necessary to choose numbers randomly, otherwise a vicious circle is created.* Her option was to deal with Gastón's procedure without fully tackling at this stage the nature of the solutions.

Independently from Gastón's failed explanations to sustain that both formulas are needed in order to avoid "*that the numbers are chosen at random*", it is true that each one of the formulas he wrote represents a different calculation procedure: the "input data" are different in each case. The equivalence of the two expressions can only begin to be considered when the students accept the formula as a two variable equation, the solutions to which are pairs of numbers (which are precisely the solution pairs for the problem).

In that sense, the norm "*it is necessary to choose numbers randomly* " is only a part of what the students must elaborate on.

The discussion between Matías and Gastón is – though not exclusively – the manifestation of a distance in the way in which each student conceives the solutions and in that sense, the confrontation of the different writings is a way of raising the problem of the nature of the solutions, though in this example the teacher did not make room to deal with the question specifically.

Within the class, diverse ways of conceiving the formula will surely co-exist, and the collective debate on the subject should promote the students' growing flexibility about their points of view, thus connecting the aspect "calculation procedure" – probably widely accepted – with the aspect "two variable equation, the solutions to which are pairs of numbers". Presenting tasks that will get the students to adopt a reflective position with respect to their productions will contribute to this process of "acquisition of flexibility". This implies transformations in the functions that are attributed to algebraic writing. (Panizza et al., 1999).

7.4. *Different procedures, the same solution set: a new idea for the students*

During the last stage of the debate, it becomes clear that the class as a whole affirms that two procedures can be correct and that both can offer a different number of solutions:[11] the students say that through procedure 2, they can get 201 solutions, and that by means of procedures 1, 3 and 4 they can obtain 41 solutions.

At this point, the teacher becomes aware that new knowledge needs to be constructed: any procedure, if it is correct, must provide all the solutions, which implies that two correct procedures must offer the same solutions. On the teacher's side, there is, in this episode, a breach of the didactic contract, because she had not anticipated that the students could accept two procedures as being correct and, at the same time, associate a different number of solutions to each of them. Is it that students believe that the solution to a problem can vary according to the solving procedure chosen? Is it that they accept a procedure as correct because it provides some of the solutions, even though it will not provide all possible solutions? There are not enough elements to answer these questions but, nevertheless, it becomes evident that what is required now is to work on a new norm: the procedure *must* be exhaustive. Furthermore, it is necessary that the students should elaborate strategies to establish the number of solutions for a problem. These are general matters that, at this level of schooling, can only be tackled with respect to particular problems.

Determined to carry on, the teacher decides to pose a new question, which aims at questioning the possibility of there being 201 solutions. The students must first work individually and then move on to a collective phase.

"*Last time in class, we got to the conclusion that by means of procedure 2 we can find 201 solutions, and 41 by means of procedures 1, 3 and 4. If you agree with this statement, suggest a solution that can be obtained by means of procedure 2 and cannot be found through procedure 3. If you do not agree with this statement, explain why you disagree*"

It should be clarified here that this problem was not planned in the original design, and that it arose from the social interaction generated by the activity of analyzing the different procedures. The question the teacher poses is highly unlikely to appear in the framework of an individual activity.

In order to establish the number of possible solutions, almost every student had counted the values the "independent" variable could take in the procedure they themselves had considered. Faced with the problem the teacher poses, they all put procedure 2 to work and begin to realize that the solutions repeat themselves, which brings about a reconsideration of the former criterion: it is valid only if two different values of the selected variable provide different solutions. This understanding leads them to either restrict the domain of the selected variable in order to get a one-to-one application, or to select the other variable as an independent one.

Estefanía's notes clearly illustrate the production of the class.

$152 \times 10 = 1520$ $2000 - 1520 = 480 \div 50 = 9.6$
10C coins: 155
50C coins: 9
There are 41 solutions because
Example: 11 50C coins – 145 10C coins
10 50C coins – 150 10C coins

If I take one of these numbers, e.g. 147, it will lead me to 150, because every 0.2 50C I add one 10C coin, because I will get "point something".

$147 \times 10 = 1470$ $2000 - 1470 = 530 \div 50 = 10.6$
150 10C coins and 10 50C coins

This episode enables us to see how the scenario that is being formed makes it possible for the teacher to formulate new questions, which are necessary to go through the change of practices involved in the move from arithmetic to algebra. The teacher's questions find their "place" in the class from the moment all students have appropriated a set of diverse procedures starting from the promoted spaces for interaction.

In the frame of the discussions the problem generates, strategies are elaborated that will allow students to count the number of solutions. With respect to this, it is interesting for us to focus on the interaction between the teacher and a very weak student, Luana, concerning this activity.

Luana: What I do not understand is that, here (on the board), there is a procedure and, down there, it says there are 201 solutions.
Teacher: That's what the boys said yesterday.
Luana: But, how do they know there are 201 solutions?

Teacher: I'm telling you that that is what they say, I'm not saying it is right. Because, since x can have any value between 0 and 200, if one assigns to it the value 0, you get one solution; if you assign value 1, you get another solution, and so on. Then, since there can be 201 possible values for x, they say there are 201 solutions.

Luana: Ohhhhh (in a "Now I get it" tone)

With considerable effort, it is only at this point that Luana seems to discover that it is possible to know the number of solutions. We should notice that Luana asks for a criterion to help her think and not for a rule that will help her "do" something. We believe that the different contributions of the students have helped Luana see that there is a way of realizing how many solutions there are, a way she does not yet have access to but is eager to understand. "How do the students know" is a key question for Luana, which allows us to identify, in the individual construction of knowledge, the role played by the productions and the statements of others.

8. CONCLUSIONS

In the context of the Theory of Situations, we have assumed that the processes of adaptation to a *milieu*, as well as the social interaction, play a critical role in the transition from arithmetic to algebra. The study that we conducted enabled us to determine, in several directions, how these two types of processes are interwoven.

- *The emergence of novel questions which promote the transition from arithmetic to algebra*

  The drastic changes involved in being introduced to algebraic practices need the input of the teacher, who should supervise and monitor the new ways of approaching the job.

  Now, we claim that it is possible to think of a space of problems, which are generators of insufficiencies, requirements and uncertainties, on which to base these changes. Our work shows that it is possible to obtain an adidactic *milieu* which generates questions. In this way, the move from arithmetic to algebra is nurtured by questions framed in the social space of the classroom as a consequence of the work proposed by the teacher.

- *The fruitfulness of the adidactic interaction with the procedures of others*

  Our analysis enables us to state that the students have seriously considered the work of their peers as worthy of study. We could say that the fact of inserting a space of adidactic interaction with such

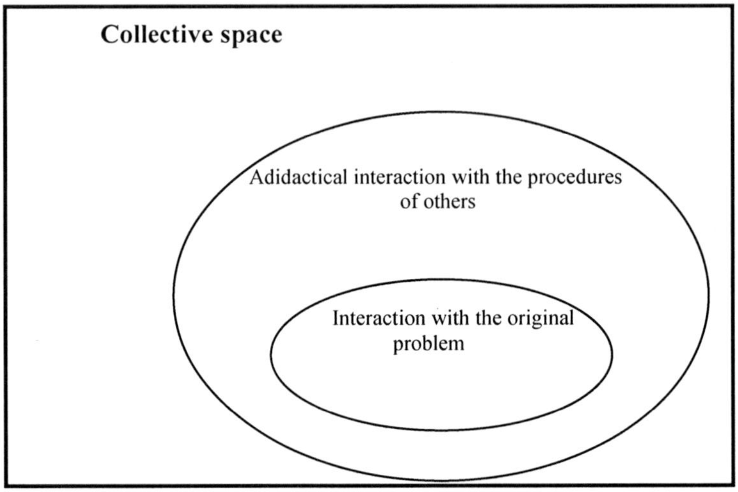

Figure 1. The adidactic interaction with the procedures of others enriches the mathematical relations, which are put into play in the collective debate.

works, as they are turned into problems to be considered, in a way, objectifies them. Interaction with the work of the others influences each student's own line of questioning. The limits that each student finds to validate both acceptances and rejections of the procedures of others, generate good conditions to raise new questions. The layer of mathematical relations and norms of work, which are put into play in the collective debate, becomes thicker. In many cases the formulation of new questions is the teacher's job and can only take place in the frame of the debates generated as a result of confrontation (see Figure 1).

– *Elements of knowledge which emerges in the wake of the introduction of problems with one degree of freedom between their variables*

In the description of the process of knowledge production in class, we identified the emergence of "micro-knowledge" implied in the dealings with two variables with one degree of freedom. This micro-knowledge can be considered as a necessary "fuel" to put into motion a notion essential for algebraic work, that is the notion of variable. Some elements of this knowledge can be stated as follows:

- Each solution for the problem is a pair of numbers linked by a condition. All the solutions can be generated starting from a single formula (this is progress with respect to arithmetical knowledge where each number in the result must be obtained through a specific calculation).
- It is legitimate to attribute values to one of the variables to start with the process of producing solutions.

  - A procedure must be thorough in its formulation. This means that it must contain enough information to narrow down the value of independent variables and establish rules of action for all the possible values both of the initial variables and of the results that emerge in the intermediate steps.
  - A procedure is correct only if it allows obtaining all solutions for the problem.
  - To establish the number of solutions by counting the number of the possible values that the independent variable can take on, it is necessary to make sure that the function, which assigns a solution to each value of that variable, is injective.

- *About the position of the student when studying the procedures of others*

  The task of evaluating the procedures of others after having produced one of one's own, places the student in a position of control, in which his critical analysis is enhanced. Validating a production acquires then a recognizable aim for the student. This would be presenting a possible option regarding the complex problem – pointed out by several authors – of making the students assume validation as part of the mathematical work, which remains under their responsibility.

  From this position of controller/agent of validation, the student can contribute to the future mathematical activity of the class and, at the same time, reconsider his own past. In fact, on the one hand, some issues that the producers could have missed out may become visible for them and that, as long as they put them forward for class discussion, could become new mathematical problems that the class as a whole may assume. On the other hand, the analysis of the work of others can offer some answers to the uncertainty caused by individual production, leading the students to modify their decisions.

- *The teacher allows work on unanticipated issues*

  All along the present work we have emphasized the role of peer interactions in the emergence of new mathematical problems. It is time now to stress that those issues would not have taken arisen in the classroom, had the teacher not kept them open for a while.

  In some cases, the discussions of the students were the expression of the distance between the norms that governed each individual's work. As of those discussions, the teacher became aware of the need to work in an explicit manner on normative aspects.

  To characterize the complexity of the teacher's work – when her teaching project includes students' interchange regarding their own production as a response to a problem – is a didactic question that needs to be addressed in greater depth. This implies studying:

- The decisions the teacher must take regarding the duration of the work when it comes to coordinating the time allotted to the teaching of an element of the curriculum by the institution (the *time capital*, see Assude, this volume) and the time requirements imposed by the cognitive functioning of the students.
- The decisions that lead the teacher to adjourn certain issues till the next classroom session, deal with others on the spot, and ignore some at the present moment considering the future of her teaching project.
- The different degrees of the teacher's tolerance of those students' productions, which are far from her project.

We consider this study of the complexity of the teacher's work to be fundamental if one aims to contribute, from the field of research, to making mathematics classes true communities of knowledge production.

NOTES

1. "The student learns by adapting himself to a *milieu* which is a source of contradictions, difficulties, imbalance, in a way which is similar to how the human society has had to adapt so far. This knowledge, arising from the student's adaptation, reveals itself through new answers that are proof of learning" (Brousseau, 1986) (own translation)
2. "A *milieu* without didactical intentions is manifestly insufficient to induce in the student all the cultural knowledge that we wish him to acquire" (Brousseau, 1997).
3. We are taking into consideration here a "reality" – which can be intra- or extra-mathematical – in which a problem to be solved has been identified, and this already presupposes a knowledge system interacting with it.
4. The first one aimed at the conceptualization of division as an object in itself, with a collection of problems, in which two terms of the relationship $a = b \cdot x + r$ were given and the other two were unknown.
5. Rather than thinking that there is an "event" that marks the beginning of school algebra, we believe there is a process of algebraization of school mathematics. This process involves work of generalization which becomes gradually more and more explicit (Bolea et al., 2001).
6. The problem we analyze here is the third one of the sequence presented to the students.
7. The class was observed by two people who took notes, audio-recorded the small groups' work as well as the debates, and finally produced a written record. These records constitute the basis of our analysis.
8. Altogether, five distinct procedures appeared in the class.
9. "Normative knowledge" refers to knowledge, which regulates mathematical work, such as what is allowed in mathematics and what is not, what is considered sufficient for validating a statement or a procedure, what are the criteria to be met to establish that strategy is "mathematically relevant". Yackel and Cobb (1996) give an account of the processes of emergence of norms in the mathematics class.
10. Sabrina is a "good" student, who was absent the day stages 1 and 2 were implemented; she is joining in the work only at this stage by analyzing the procedures her classmates produced.

11. Stating an opinion about the number of solutions is a novelty for the students, since the problems that have only one solution – which, so far, have been the essential part of their arithmetical activity – have made this issue "transparent". The term "transparent" is often used in two opposite senses. In this case, we mean that the question "cannot be seen" from the point of arithmetical practices.

REFERENCES

Balacheff, N.: 2000, 'Symbolic arithmetic vs. algebra', in R. Sutherland, T. Rojano, A. Bell and R. Campos Lins (eds.), *Perspectives on School Algebra*, Kluwer Academic Publishers, Dordrecht, pp. 249–260.

Bauersfeld, H. and Cobb, P.: 1995, *The Emergence of Mathematical Meaning: Interaction in Classroom Cultures*, Lawrence Erlbaum Associates Publishers; Hillsdale, New Jersey, U.S.A.

Bloch, I.: 1999, 'L'articulation du travail mathématique du professeur et de l'élève dans l'enseignement de l'analyse en première scientifique', *Recherches en Didactique des mathématiques* 19(2), 135–194.

Bolea, P., Bosch, M. and Gascón, J.: 2001, 'La transposición didáctica de organizaciones matemáticas en proceso de algebrización: El caso de la proporcionalidad', *Recherches en Didactique des mathématiques* 21(3), 247–304.

Brousseau, G.: 1986, 'Fondements et méthodes de la didactique des mathématiques', *Recherches en Didactique des mathématiques* 7(2), 33–115.

Brousseau, G.: 1997, *Theory of Didactic Situations in Mathematics: Didactique des mathématiques 1970–1990*, N. Balacheff, M. Cooper, R. Sutherland and V. Warfield, (trans. and eds.), Kluwer Academic Publishers, Dordrecht.

Campos Lins, R.: 2000, 'The production of meaning for algebra: A perspective based on a theoretical model of semantic fields', in R. Sutherland, T. Rojano, A. Bell and R. Campos Lins (eds.), *Perspectives on School Algebra*, Kluwer Academic Publishers, Dordrecht, pp. 37–60.

Cobb, P.: 1996, 'Where is the mind? A coordination of sociocultural and cognitive constructivist perspectives', in *Constructivism: Theory, Perspectives, and Practice*, Teachers College, Columbia University, pp. 34–52.

Herbst, P. and Kilpatrick, J.: 1999, 'Pour Lire Brousseau', *For the Learning of Mathematics* 19 (1), 3–10.

Laborde, C.: 1991, 'Deux usages complémentaires de la dimension sociale dans les situations d'apprentissage en mathématiques', in C. Garnier, N. Bednarz and I. Ulanovskaya (eds.), *Après Vygotski et Piaget. Perspectives sociale et constructiviste. Ecoles russe et occidentale,* De Boeck Université, Bruxelles.

Mercier, A.: 1998, 'La participation des élèves à l'enseignement', *Recherches en didactique des mathématiques* 18(3), 279–310.

Panizza, M., Sadovsky, P. and Sessa, C.: 1999, 'La ecuación lineal con dos variables: Entre la unicidad y el infinito', *Enseñanza de las Ciencias* 17(3), 453–461.

Sadovsky, P.: 2003, 'Condiciones didácticas para un espacio de articulación entre prácticas aritméticas y prácticas algebraicas', Unpublished Doctoral thesis, Universidad de Buenos Aires.

Sensevy, G.: 1998, *Institutions didactiques. Étude et autonomie à l'école élémentaire*, Presses Universitaires de France, París.

Steinbring, H.: 1998, 'Mathematical understanding in classroom interaction: The interrelation of social and epistemological constraints', in F. Seeger, J. Voigt and U. Waschescio (eds.), *The Culture of the Mathematics Classroom*, Cambridge University Press, Cambridge, 344–374.

Yackel, E. and Cobb, P.: 1996, 'Sociomathematical norms, argumentation, and autonomy in mathematics', *Journal for Research in Mathematics Education* 27(4), 458–477.

Vergnaud, G., Cortés, A. and Favre-Artigue, P.: 1987, 'Introduction de l'algèbre auprès de débutants faibles. Problèmes épistémologiques et didactiques', *Actes du colloque de Sèvres. Didactique et acquisition des connaissances scientifiques*, pp. 259–288.

PATRICIA SADOVSKY and CARMEN SESSA
Facultad de Ciencias Exactas y Naturales – Universidad de Buenos Aires
Secretaría de Educación – Gobierno de la Ciudad de Buenos Aires
Argentina
E-mail: patsadov@mail.retina.ar; pirata@dm.uba.ar

MAGALI HERSANT and MARIE-JEANNE PERRIN-GLORIAN

CHARACTERIZATION OF AN ORDINARY TEACHING PRACTICE WITH THE HELP OF THE THEORY OF DIDACTIC SITUATIONS

ABSTRACT. In this paper, we use the theory of didactic situations to characterize a mathematics teaching practice, currently used in secondary schools in France, which we have called *interactive synthesis discussion*. We have studied this practice in ordinary classes, i.e. classes where the researcher intervenes neither in the preparation nor in the management of the lessons. We have looked at the didactic situations the teacher chooses, and how he manages his teaching project, the students' work in the classroom and at home, and classroom interactions. We present two case studies of experienced teachers, one in grade 8, and the other in grade 10.

KEY WORDS: adidactic milieu, classroom interactions, didactic contract, distribution of responsibility between the teacher and the students, evolution of the status of knowledge, graphical solution of equations, proportionality, mathematics teaching practices, ordinary teaching, theory of didactic situations

1. Problematique, Theoretical Framework and Methodology

1.1. *Problematique*

We assume that the goal of teaching is for the students to acquire a certain established and culturally recognized knowledge, which they will be able subsequently to use without the teacher's help. Constructivist theory, which has spread among teachers and textbooks authors, suggests that students give meaning to knowledge and can use it by themselves only when they have developed this knowledge as an answer to some problem considered their own. Consequently, teachers, students and institutions now consider the traditional practice of exposition of knowledge (by the teacher or a textbook) followed by its applications (in the form of exercises to be done by the students) as not appropriate for the secondary school.

However, constructivist ideas are not easy to bring into practice and their influence on teachers' practices appears rather superficial. For about 20 years now, French curriculum guidelines recommend introducing mathematical notions by means of preparatory activities and grounding teaching on students' activity. However, teaching situations, which would make it possible for the students to produce new knowledge without the help of the teacher, are difficult to develop and very constraining. Moreover, leaving

Educational Studies in Mathematics (2005) 59: 113–151
DOI: 10.1007/s10649-005-2183-z

the initiative to the students increases the teacher's uncertainty level; even if the problem was carefully chosen, the teacher cannot anticipate and therefore prepare beforehand for all the procedures the students can come up with. He[1] will have to make decisions on the spot, taking into account the different students' solutions while making sure his teaching project stays on course, because the knowledge produced as an answer to a problem must be recognized as the knowledge, which is aimed at by the curriculum. The teacher is thus caught between two, possibly contradictory, constraints: the constraints of the curriculum and the constraint of grounding his teaching on the knowledge used by students in solving the proposed problems. It is probably in response to these different constraints that secondary school teachers have started replacing the traditional "exposition-exercises" practice by a kind of "dialogue courses", based, largely, on short interactions between the teacher and the students. The nature of these interactions and their functions in the teaching and learning process may vary in relation with other features of the teaching situation, its management by the teacher and the knowledge at stake.

In this paper, we will describe and analyze a particular case of such "dialogue courses," which we have called the "interactive synthesis discussion" practice (ISD, for short). In brief, the practice consists of problem solving sessions in small groups of students followed by whole class discussions of the solutions to the problems. An important feature of the practice is that the problems are chosen so as to partly require what the students already know, but they include questions or can be extended to problems whose solution requires some new knowledge. During the whole class discussion of students' solutions or attempts at them, the teacher extends the problems in the direction of the new knowledge. He helps the students to *synthesize* the solutions obtained by different students using old knowledge and he extends the problems by highlighting or asking those additional questions, which are then solved collectively in class. This way, new knowledge is introduced as a solution to these new questions in the same context and linked to the old knowledge. Thus, problem solving and problem discussion replace the exposition-exercises teaching practice.

Our research on classroom interaction is a qualitative study, done from the perspective of "*Recherches en didactique des mathématiques (RDM)*,[2] " as defined by Bartolini Bussi (1994, p. 122). The aim of the research is to gain knowledge and understanding of teaching phenomena; it is not to produce immediate action or to improve teaching in a direct way. Moreover, our project is not one of didactic engineering (Artigue, 1992; Artigue & Perrin-Glorian, 1991). Indeed, the researcher intervenes neither in the design of teaching nor in its realization. We aim at understanding the teacher's practice, including his choices of exercises as well as his class management decisions aimed at developing both his teaching project and the students' knowledge.

Much research worldwide aims at improving the teaching of mathematics, but the evolution of ordinary practices seems slow and their effectiveness is not obviously growing (Ball et al., 2001). We hypothesize that teaching practices are very complex and that researchers generally do not take sufficiently into account the "economy" of ordinary practices, that is, the teachers' attempts to balance the various professional constraints under which they work and the degree of freedom they have. Clarification and understanding of ordinary practices is, for us, an essential issue and a first step towards research on teacher training.

Hence, our aim in this paper is to characterize the ISD practice from this standpoint and not to study its effects on students' learning. However, interviews with some students and one teacher give an idea of the way students understand this practice and lead us to raise new research questions.

1.2. *Using the theory of didactic situations to analyze an ordinary practice*

To account for the regulations carried out by the teacher in managing his class, we use mainly the framework of the theory of didactic situations (in short TDS) and, especially, its latest developments concerning the didactic contract (Brousseau, 1996). This choice is based on our belief that class interactions are affected by the knowledge at stake and its status (old, new) so that we cannot study them separately from the teacher's project. For an introduction to this framework, we refer the readers to (Brousseau, 1997) and (Herbst & Kilpatrick, 1999), and also to the Introduction (Laborde and Perrin-Glorian, this volume). Here, we only mention a few basic ideas and concepts needed for the purposes of this paper.

1.2.1. *Use of TDS to study ordinary teaching practice*

We think that there are various ways of learning. In some cases it is enough that knowledge be presented and explained; this is then the more economical way to teach. However, many students have difficulties in learning some important mathematical concepts. In this case, we think it is necessary to elaborate teaching situations, which give the students a chance to make sense of this concept. A way for the teacher to do that is to design (or adopt, or adapt) a situation including both a problem whose optimal solution involves the concept in question, and an *objective*[3] milieu (in the sense of TDS). This milieu should include some material or symbolic objects that are able to provide feedback to the students' actions on them. To solve the problem, the student has to engage in actions on the milieu, to formulate hypotheses, to validate them or not, to elaborate strategies (as if trying to win a game) and to take into account the feedback from the milieu. This kind of situation, named, in TDS, the *adidactic situation*, works, ideally, with almost no input from the teacher. Still, the teacher is responsible for obtaining that the students assume responsibility for solving the problem: this is called the *devolution* process. Upon the completion of the action, the teacher must change the rules of the game, i.e., the milieu, thus defining a new situation aimed at the formulation by the students of the knowledge they have developed by acting on the milieu. The teacher must also help students to make a link between their new experience and an existing and established knowledge, useful to solve other problems: this is the *institutionalization* process.

In TDS, the model of a didactic situation includes an adidactic situation (with an objective milieu) and a *didactic contract*. The didactic contract is a way of regulating the mutual expectations of the teacher and the students with respect to the mathematical notions at stake. Devolution and institutionalization are two important ways of regulation of the didactic contract.

This model has proved its relevance for the design and study of situations, which give the students a chance to achieve a better understanding and learning of difficult mathematical concepts. But, as emphasized by Herbst and Kilpatrick (1999), TDS does not provide the teacher with a model of "good practice", nor keys to improve his practice. It is mainly a tool for analyzing teaching:

> [...] didactique does not turn the constructivist hypothesis (...) into a pair of handcuffs to shackle some forms of pedagogy [...] The constructivist hypothesis is instead used as a tool to find the possible meanings that the learner may be attaching to a declared piece of knowledge being taught, given the characteristics of the situation in which the transmission takes place. Therefore, although the notion of didactic contract may help the teacher understand his or her practice, it's not a technical tool for acting on that practice. Instead, it is a technical tool enabling the researcher to study practice. (Herbst and Kilpatrick, 1999)

In this paper, we show that this theoretical model is relevant for the study and understanding of ordinary teaching in a way, which takes into account the progress in students' learning of the knowledge at stake. Indeed, it does not suppose that the student only learns when acting. It even implies that some knowledge can only be transmitted directly during institutionalization. Therefore, mainly with the help of its recent developments (Brousseau, 1996; Margolinas 1995; Comiti and Grenier, 1997; Perrin and Hersant, 2003), the theory is able to account for different kinds of situations and different ways of teaching.

The ISD practice we are studying here is an intermediate practice leaving some space for student's own research and their production of new knowledge; but this space of freedom left to the students is rather limited for various reasons we will try to specify. The theory of didactic situations will help us in identifying these constraints. Indeed, with the help of the notions of milieu and didactic contract, we can distinguish what the students are in charge of, given their previous knowledge, from what the teacher is in charge of, concerning both solving the given problem and the progress of the students' knowledge.

1.2.2. *Milieu and the potential of adidactic work for the students*

The existence of a problem,[4] which models some knowledge (in the sense that this knowledge is a better way to solve the problem), does not ensure the possibility of building a didactic situation allowing the learning of this knowledge under a constructivist contract. In ordinary teaching, actual adidactic situations are rare, but one can observe situations that have some *adidactic potential*.This means that there is a milieu, which provides some feedback to the actions of the students, but the feedback alone may be insufficient for the students to produce new knowledge on their own. In this case, the teacher may have to intervene to modify the milieu, for example, so that the student becomes aware of an error. We say "potential" because the teacher may ignore this potential and manage the situation without using it, evaluating by himself the students' answers, instead of waiting for the students to react to a feedback of the milieu. But if the situation has no such potential the teacher can do nothing but react by himself to students' actions.

We speak of the ISD practice only when the students are actually allowed the time to work on a problem on their own, so that some elements may be interpreted with their previous knowledge and may bring some feedback to their actions independently from the teachers' interventions. These elements are modeled by an objective milieu (external to the students and the teacher). Of course, students need personal knowledge[5] to interpret the feedback but, in the model, personal knowledge is on the students' side, not in the milieu. Nevertheless, the milieu may include some institutional knowledge, recalled in the text of the problem. For example, the text may provide some formulas useful for solving the problem; students don't have to know them, but they have to understand and be able to use them. Thus, the concept of milieu makes it possible to account for the potential of adidactic work in the situation.

1.2.3. *Didactic contract and management of the situation by the teacher*

The concept of didactic contract was introduced in the theory at the very beginning of the 1980s, and has been widely used ever since by many authors. But it is only in 1995, at a summer school on didactics of mathematics (Brousseau, 1996) that, within the more general framework of the transmission of knowledge between two systems (the transmitter and the receiver systems; in our case, the teaching system and the taught system, respectively), Brousseau distinguished and characterized different types of contracts, taking as the principal criterion the distribution of responsibilities between the two systems. He thus considered different didactic contracts, from very weak contracts, where the teaching system is hardly responsible to the taught system, to very strong such contracts where the former is totally accountable for the learning outcomes of the taught system.

In the case of strong didactic contracts, we propose to refine their characterization in order to better explain the interactions between the local level of the actions of the teacher and the students (e.g. within the framework of the resolution of a concrete problem) and the more global level of the management of the teacher's project and students' acquisition of knowledge. For that, we consider a number of additional dimensions of the didactic contract, on which the teacher and the students can act. Identifying these dimensions and their evolution in the course of a teaching project is a way to achieve a better understanding of the teacher's practice.

1.2.4. *Structure of the didactic contract*

We distinguish four *dimensions*[6] of a didactic contract. The first two are related to the knowledge to be taught and learned: the mathematical field or *domain* and the *didactic status* of the knowledge. The third dimension is related to *the nature and characteristics of the ongoing didactic situation*; the last dimension concerns the *distribution of responsibility*, with respect to the knowledge at stake, between the teacher and the students. These dimensions are not independent. Indeed, in general, the distribution of responsibility regarding the knowledge at stake is related to the didactic status of this knowledge and the characteristics of the didactic situation. Let us briefly specify what is covered by each of these dimensions.

We consider the *domain* as one of the dimensions of the contract because the fact of situating a problem within a certain mathematical field guarantees that certain techniques will appear natural and will be favored whereas others will be improbable. Moreover, it is one of the elements on which the teacher can play: he can voluntarily call on a mathematical field the students hadn't thought of. Douady (1987) pointed to the possibility of enhancing learning a mathematical concept by solving a problem involving it through translating this problem into a different mathematical context (*settings interplay*). Taking into account the mathematical domain in describing the didactic contract enables us to identify the initiator of these changes. This dimension may appear on a relatively global level (arithmetic, algebraic or geometrical contract, for example) but can intervene also locally in the changes of the mathematical context.

The didactic status of knowledge corresponds to distinctions close to those made by other authors (Brousseau and Centeno, 1991; Douady, 1987; Chevallard, 1999). We distinguish, for our part, three main kinds of knowledge. At one extreme, there is the entirely new knowledge, and, at the other extreme – the old knowledge, which, in principle, is no longer a teaching objective. Between the two there is the knowledge in development, for which we may consider three states: recently introduced knowledge, knowledge in the course of institutionalization and institutionalized knowledge, which must be consolidated. This dimension is related to the *distribution of responsibility between the professor and the students*, because, generally, the teacher leaves more responsibility to students in the case of old knowledge. Nevertheless, it is a distinct dimension because, especially in the case of new knowledge, while the teacher may delegate a great part of responsibility to the students, he can also keep for himself all the responsibility with regard to the validity of students' solutions.

The teacher can delegate responsibility to students with respect to new knowledge only in a didactic situation whose milieu is endowed with a *feedback potential*: we say that the situation has an *adidactic potential.* We think that it is interesting to identify this feature of the didactic contract as a dimension in its own right, because the meaning of the other dimensions is not the same depending on whether there is such a milieu or not. Another reason is that students can themselves recognize that the teacher's expectations concerning their own activity vary according to the type of situation proposed to them.

On the other hand, we distinguish *three levels* in the structure of the didactic contract: the macro-, the meso- and the micro-contract. These levels correspond to various time scales and didactic aims. The *macro-contract* is mainly concerned with the teaching objective, the *meso-contract* – with the realization of an activity, e.g. the resolution of an exercise. The *micro-contract* corresponds to an episode focused on a unit of mathematical content, e.g. a concrete question in an exercise. On each level, some dimensions remain relatively stable, but the dimensions as a whole are stable only at the level of the micro-contract, and few dimensions, except possibly the mathematical domain, stay stable at the level of the macro-contract. The nature of the macro-contract will therefore be deduced from the analysis of dimensions on a more local level. In this sense, our analyses will go up from a local level to a more global level, so that the macro-contract is mainly characterized by the meso-contracts and micro-contracts whose existence it makes possible, and for whose organization it allows.

Epistemological hypotheses concerning mathematics learning, underlying the types of contract, are in the background of our analysis, at the level of the macro-contract. We did not identify them as a dimension of the contract but they do appear in two of them: the choice of a milieu with a feedback potential and the distribution of responsibility. Thus, epistemological hypotheses are not defined in an independent way but are deduced from contracts.

Micro-contracts are defined mainly based on the distribution of responsibility between the teacher and the students. Meso-contracts are deduced from two dimensions: existence of a milieu with a feedback potential, and the status of the knowledge at stake. We cannot, within the constraints of this paper, define and characterize all the different meso- and micro-contracts that we have observed. Information about these can be found in a previously published paper (Perrin-Glorian and Hersant, 2003). We will specify only the meso-contracts or micro-contracts met through the case studies that are presented here.

Our analysis of the structure of the didactic contract has been summarized in Figure 1.

Moreover, in our analysis based on TDS, the teacher's role appears essentially through the two basic processes of devolution and institutionalization. The ISD practice will be specified based mainly on the institutionalization process, but this practice also presupposes the devolution of some problem, in which students invest personal knowledge. The achievement of the devolution of the problem is conditioned by the existence of an objective milieu, and therefore by the adidactic potential of the situation. Institutionalization is a condition for the advancement of the teaching project. Institutionalization of new knowledge changes the didactic contract.

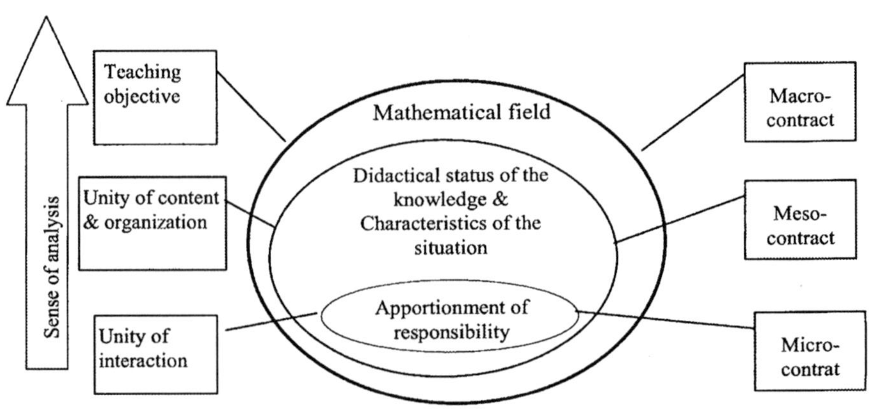

Figure 1. Structure of the didactic contract.

1.3. *Research methodology*

Our research relies on long term (at least a month long) observations of ordinary classes. This allowed us to obtain a great deal of information about the whole teaching project related to a given curricular objective.[7] The classes were audiotaped and the records transcribed and we regularly interviewed the teacher between classes. The teacher's project was then derived from his declarations and also from an analysis of the situations reconstructed on the basis of the observation. To keep the context of the teacher's actions, we carry out our analysis with a succession of zoom-ins. First, in the transcript of the recording, we identify *sequences* of lessons devoted to a common teaching objective. Each such *sequence* of lessons is then divided into *phases* relative to the unity of content and classroom organization. Each phase is further divided into *episodes*, relative to the unity of interaction. Finally, inside an episode, we zoom in on some particular interactions, which we study in detail.

For each sequence, if possible, we identify the new knowledge by modeling one or several situations using TDS. In particular, we define the aims of teaching, an objective milieu, and we identify the knowledge necessary to interpret correctly the possible feedback of this milieu to actions or decisions of students. Eventually, we identify some gaps in this milieu and the need for teacher interventions. This constitutes an *a priori* analysis. Then, in the *a posteriori* analysis, the observed development of the lesson is interpreted by reference to this *a priori* analysis and we describe the different dimensions of the didactic contract.

Moreover, in the grade 10 class, at the end of the school year, we conducted interviews with the students (in groups of two or three) and with the teacher to get some more information concerning the way they experienced the ISD practice along the school year.

2. Presentation of two case studies

We now present an overview of the observed classes in grades 8 and 10, trying at the same time to emphasize some features of the ISD practice appearing in the examples. In each case, we analyze more precisely some significant sequences of lessons. In the case of grade 10 classes, we focus on the way the teacher managed knowledge with different kinds of status at the meso-contract level. In the grade 8 classes, we will see more precisely how the teacher managed the distribution of responsibilities at the micro-contract level.

2.1. *Quadratic functions and solving equations in grade 10*

The class is a rather weak class of tenth graders. The teacher is experienced but he has been teaching in this grade for only about four years (before this, he taught in a middle school from grade 6 to grade 9). The mathematics classes are allotted about four hours per week. There is one problem solving session of one hour and fifty minutes with half the class (about fifteen students working usually in groups of three). These are followed by two whole class sessions of fifty-five minutes each, generally devoted to an interactive synthesis (ISD) of the results obtained in the small groups.

2.1.1. *Some features of the ISD practice observed in this case*

We will see how, in the whole class sessions, the teacher:

- *reinforces previous institutionalization* by revising knowledge already introduced or in the process of being learned;
- *prepares a milieu*[8] which would be stable enough to allow him to ask a question introducing a new aim of learning; during a discussion of the solution to an exercise concerning old knowledge, he enriches it with a new question to be solved collectively;
- *manages ruptures in the didactic contract* and simultaneously reduces the uncertainty of students with questions which guide the students' reflection without, however, completely eliminating it (*regulations of the didactic contract),*
- carries out an *institutionalization by small alterations,* asking the students to formulate a general property from the example done, which they will have to use again in other cases.

These elements, which refer to our theoretical framework, will be marked by italics in the description of the episodes.

The unit on quadratic functions and equations we discuss here (Appendix I), took five classroom sessions (about 7 h) divided in two sequences as follows: for the first one (Appendix I, part 1), problem solving in small groups and two whole class discussion sessions; for the second sequence (Appendix I, part 2) problem solving in small groups again, and another whole class session. The objectives of both of them included solving quadratic equations using factorization[9] in the algebraic setting, and using graphs in the setting of functions; to see the respective advantages of the two settings, and to convert from one to the other.

2.1.2. *The development of the first sequence of lessons*

In the first session, the teacher handed out a list of exercises (Appendix I, part 1). The students were introduced to functions from the beginning of the school year but we could observe that recognizing a function given in a graphic form and knowing how to connect points on the graph (in the order of abscissas and not in the order of calculations, with free hand and not with a ruler) were still among the teaching objectives. During the session, in spite of their calculators, the students got through, at most, the first exercise. In the weakest groups, they did not reach the questions dealing with the intersections of the curves with the axes. Therefore they did not get the opportunity to renew their acquaintance with the characteristics of the graph of a function. They were mainly concerned with calculations (substitution) and connecting the points. The students were assigned to finish the exercises at home but, at the beginning of the next (whole class) session, they had not all developed the same experience with the problems.

The work of synthesis on these problems took place during the two consecutive whole class sessions. The teacher used a computer and a data-show to project the needed graphs on the whiteboard. In this synthesis, we distinguished three large phases corresponding to the "enriched" discussion of solutions to each exercise. The learning phase is the third one so we elaborate on it a little more. Accounts of Phases 1 and 2 are useful to describe the background milieu of the third phase.

Phase 1: While discussing the solution to the first question, the teacher asks the students to state an explicit process to find some points in E and some points not in E, where E was defined as the set of points of the plane whose coordinates (x,y) are related by the equation $y = (x + 3)(8 - 2x)$. By doing this, he makes them recall and verbalize an element of their old knowledge: for each value of x, there is only one value of y, so it is a function. He stresses that the points in E are the only ones whose coordinates are related by this formula. He thus *reinforces the institutionalization* of *knowledge in development.*

Concerning the finding of points on the horizontal axis, the students were only asked to highlight, i.e. to recognize them. The teacher *adds a new question*: he asks the students to write down: "could we predict this result?" This question leads the students to propose an algebraic method of solving, which is *carried out collectively*. Next, the teacher *tries to institutionalize* the respective advantages of graphing and calculating, but it is difficult to make a point with the example at hand (since the solutions are integers). Therefore the students do not react and the teacher defers this institutionalization till later. The same work develops with regard to the questions on inequality (search for points with positive or negative ordinates) and, in the last question; it is connected with the search for antecedents.

Phase 2: The end of the session is devoted to the discussion of exercise 2 up to question 3a, again leading to an algebraic solution. For the occasion, the teacher recalls that it is better to come back to the formula $x^2 - 9 = 0$

and then to factor: this way, it is easier to avoid the risk of forgetting some solutions. But students read the solutions off the graph before, so they did not forget the solution -3. For the next session, the teacher asks the students to review the exercises not yet discussed and to answer the following new questions: "Could we predict the results? In which cases are we going to use the graph?"

The next session begins with finishing off the discussion of exercise 2, following the same plan, but adding the question of the symmetry of the graph. Here, the teacher is *reinforcing the institutionalization of the use of factoring in easy cases (old knowledge used for a new purpose), and he is preparing a milieu for the learning, in phase 3, of the use of factoring in a non-trivial case.*

A priori analysis and development of Phase 3: This phase was focused on the discussion of exercise 3, where the curves $y = (x + 3)(8 - 2x)$ and $y = x^2 - 9$ were brought together and compared. In this exercise the students were asked, first, to calculate the values of y for integer values of x from -7 to 6. Next, they had to represent the two curves in the same coordinate system, and answer the following two questions:

1. Do (E) and (P) have common points? If yes, give, if possible, the coordinates of these points and highlight them in blue.
2. Are there some points of (P) that have the same abscissa as a point of (E) but whose ordinate is higher? If yes, highlight them in red.

The objectives were to solve an equation or an inequality that is not given in a factored form, to link the algebraic and the graphical solution, and to make explicit the advantages of each method.

To analyze this question we use the TDS model.

The objective milieu that is likely to provide some feedback to students' actions consists in:

- two graphs drawn separately as well as in the same Cartesian coordinate system;
- their equations;
- previous calculations carried out in order to solve the equations - as an answer to the new questions

The students' knowledge, which is, in principle, available and, moreover, stimulated during the discussion of the solutions, includes:

- reading from the graph;
- ability to perform algebraic calculations, especially factoring;
- knowing the principle that a product is null if and only if at least one factor is null;

- knowing that an equation is an equality which is true for some values of the variable x; solving it consists in searching for the values for which it is true; to do that, first you have to transform it into an equality whose right hand side is zero (this students' knowledge is derived from the didactic contract underlying the solution of linear equations).

The new question is: "Are there some points which are common to the two curves?"

One point, $(-3,0)$, is accepted unanimously. The teacher does not waste time on this and approaches the real question: "What is the x-coordinate of the other point?"

Reading from the graph appears insufficient, on the computer screen as well as on paper. A student suggests calculating. Directed by the questions of the teacher, the students collectively produce the equation $x^2 - 9 = (x+3)(8-2x)$. The teacher allows the students to develop their first idea, which is to replace x by numbers: 3.6, 3.7, 3.75, 3.65; then he makes the students restart the search *(devolution of the new question)*. Here is an excerpt from the interchange:

T: Perhaps we could try another way. You proposed this way of solving it... If it doesn't work, you can perhaps propose another way... Well you found numbers... or you found another way? What's your idea? But, in any case, you must do something, everybody looks for something, everybody proposes. Well, who has an idea, what could we do?

And then a student replies, "We could solve the equation". The teacher seizes the opportunity to make the students recall what it means to solve an equation and, guiding them by some questions, he makes them say that, here, a solution is already found and they are looking for another one because they saw on the graph that there were two. Finally, he makes the students recall the techniques available for solving and he insists on the approach he wants to favor. It is a *first institutionalization of an algebraic method in a new case*:

T: Well, what technique do you need to solve this equation? We need?

S: To factor

T: To factor, that's to say? Wait... before, first?

S: To set all equals to zero

P. Well, you have to set all equals to zero, that is to say? How do you do to set all equals to zero? Why, yes, you subtract this number from the two sides of the equation...

The collective algebraic solving goes on with the formulation by the students, upon the teacher's request, of all the steps and all the identities used. The exact solution (11/3) is formulated as well as its decimal approximations. Finally, the teacher asks: "did you see what the graph was useful for?" *This way he institutionalizes the graph as a milieu to anticipate and to check.*

The session goes on with solving the inequality and the two practice exercises: solve the equations $x^2 - 9 = 7$, and then $x^2 - 9 = x + 3$. For the first, it is an opportunity for the teacher to insist on the algebraic method: to change the equation to get one of the sides equal to zero, and then to factor, which avoids omitting solutions; and to show that the graph allows one to check the solutions.

The second equation is solved by the graphical method after being interpreted as the intersection of the same parabola with a straight line (that needs to be drawn). The solutions are read from the graph and checked by calculation.

A posteriori analysis of phase 3: Three kinds of knowledge are distinguished here: old knowledge for solving linear equations, factoring and reading coordinates; knowledge in development about functions and reading properties from their graphs; new knowledge for factoring to solve quadratic equations and for using graphs to solve such equations. The teacher reinforces the institutionalization of the old knowledge delegating to the students the responsibility to use and to justify it; he guides them to find how this old knowledge may be used to solve the new problem (*knowledge in development to be used in a new context*). Thus the teacher informs the students of the existence of another method by asking them to find it, and he, moreover, suggests it by acting on the domain dimension of the contract. Factoring is quite easy for the first two equations; the third one brings about a break in the didactic contract (x appears on the two sides of the equals sign). Nevertheless, the uncertainty of students is reduced by calling on the domain dimension of the didactic contract concerning the context of linear equations, where students were used to move all terms to one side of the equation. At the same time, the teacher encourages the students to use information provided by the milieu (two solutions, approximate values of the second one) to conjecture and to check. In this phase we see both the reinforcement of the institutionalization of old knowledge (or knowledge in development) used by the students themselves in the problem solving sessions (such as reading from the graph) or collectively during the first phases (factoring) and the support of the milieu for the institutionalization of the new knowledge. The fourth phase will help us to explain those issues.

2.1.3. *The development of the second sequence of lessons: Expanding or factoring?*

The fourth session is a problem solving session focused on exercises presented in Appendix 1, Part 2. The fifth session is a whole class discussion session with the aim of synthesizing the solving of these exercises and might be considered as the fourth phase of our example of ISD practice in grade 10.

The aim here is to formulate and to institutionalize algebraic methods, using the usual standard algebraic identities (i.e., $a^2+2ab+b^2 = (a+b)^2$, etc.) to factor, or expand to eliminate the terms of second degree, or the constants, so that factoring becomes possible.

Here, there is no *objective milieu* capable of feedback except for the instructions themselves: you have to factor or expand. Then the teacher relies on the previously developed practice of using the standard identities in expanding and factoring. By doing this, he uses and reinforces a knowledge derived from the didactic contract: in order to expand or to factor, we have to use the standard algebraic identities. But the last equation ((j): $(3x + 1)(x - 2) = (5x - 3)(x + 2)$) marks the limits of this kind of knowledge and calls for new knowledge.

We consider two episodes, related to solving parts b) ($(3x - 1)(x + 2) = (x + 2)(2x + 5)$ and j) above. The first one is an opportunity to go over the expected method, using an error observed during the problem solving session. The second one is an opportunity to see a case, where the didactic algebraic contract fails. Only a graphic solution and search for values approaching the solutions are feasible.

Episode focused on Exercise b: Most groups used a possibly correct but hazardous method: to divide the two sides of the equation by $x+2$, but no student talks about it spontaneously during the whole class discussion period. The teacher wants to address this question with two main objectives in mind: to reinforce the safer method that he wants to encourage the students to use, and to recall that they may not divide by a number about which they are not sure it is different from zero. Therefore, he asks a student to explain how he previously solved this equation and why he abandoned this idea. The student says that he took off $x+2$ but found only one solution. The teacher asks then what happens when $x = -2$, thus the student finds the second solution. Then, by his questions, the teacher makes him formulate that taking off $x + 2$ is dividing by $x + 2$ and so there is a risk of dividing by zero.

When the equation is finally solved, the teacher returns to this issue: "I would like you to take note of this: when you divide... Because you will always divide and forget to write that the number is not zero". This way the teacher *renews the institutionalization of the standard method as a way to avoid errors, in the context of the problem.*

Episode focused on Exercise j: The students say that they found neither a common factor, nor a standard identity. They expanded and ended up with $x^2 + 6x - 2 = 0$, which they could not solve. The use of the graph is not spontaneous because the students have just solved a long list of equations using algebraic calculations (*the mathematical domain dimension of the didactic contract*). Thus, the didactic contract is broken down. Again, it is by pressing his questions that the teacher deals with this rupture and leads the students to approach the problem using graphing. Students say they cannot solve the problem because the equation is of second degree. The teacher specifies that they know how to solve such equations only if there is a common factor or if it is a standard identity and that they will learn how to solve this kind of equations algebraically later. He adds that now they can guess if there are solutions or not and get an idea of these solutions in a quick way. It is sufficient for the students to think of the graph.

Here, the teacher points out the students' lack of knowledge; it is a first devolution of a new problem: The students who have a graphic calculator use it; others use the curve drawn on the computer (and projected on the board); others search approximate values with their numerical calculator. The last ones, in fact, are looking for a value between 0 and 1. Finally, a student, looking at the graph of the computer, notices that there are two solutions. The teacher asks the students to explain why it is so. They respond by referring to the parabola, named during some previous exercises. The parabola is indeed one of the functions to study in this grade: students have to know how to identify the vertex and the axis of symmetry. The number of roots according to the parabola has not been studied yet but students might have recognized the form and expect two roots.

In those episodes, we see again how the teacher makes an institutionalization with small alterations of the knowledge in development, remaining within the context of the problem. Discussing the solutions to each exercise, the teacher comes back to the method he wants to favor, comparing it, if necessary, to others. He also seizes all opportunities to reinforce the institutionalization of other results (for instance, not to divide by zero), which he knows can help avoid recurrent errors. At the same time, he prepares the introduction of new knowledge by introducing a problem not solvable with the methods presently available to students. The teaching gives much importance to problem solving, but then there is a risk that the new knowledge stays very dependent on the context of problems in which it was introduced.

We now turn to the study of our second example: the grade 8 classes.

2.2. *Proportionality in grade 8*

The following two situations were observed in the same grade 8 class; the teacher is experienced at this level. In this class, lessons about proportionality occupied eleven 50-min classroom sessions. In seven of them (1st to 5th, 7th, and 8th), students worked individually, either on a computer program containing a bank of problems[10] on proportionality, or on exercises designed by the teacher. These latter exercises were compulsory for all students, but exercises to solve on the computer could be selected by the students. The sessions 6, 9, 10 and 11 were devoted to whole class interactive synthesis discussions of the solutions to exercises. We analyze the development of the 9th and the 10th session. In these sessions, two topics were discussed: percentage of increase/reduction (based on an exercise given by the teacher), and then the slope of a straight line (based on an exercise from the computer program).

We shall see how, in discussing the solution of the exercise on percentages, the teacher both reinforces the knowledge in development and manages a milieu with some feedback potential, shared by the whole class, to introduce a new question. Like the grade 10 teacher, he carries out institutionalization by small alterations, seizing all opportunities. We can also notice that he always allows for student's interventions and often uses them to move his teaching/learning project forward. In the case of the slope exercise, the teacher does not know exactly what the students did on the computer, and he does not succeed in proceeding in a similar way.

2.2.1. *The "5% interest" situation in grade 8*

The declared objective of the teacher is that students see the application of a percentage of increase as a single multiplication by a decimal number, obtained by factoring out a number (final price = initial price + $k \times$ initial price = $(1+k) \times$ initial price). For this, the teacher uses an exercise relative to an interest situation the students have begun to solve previously in class and finished at home. In this exercise only the relation between the initial amount (S) and the interest (I) intervenes, in arithmetic and algebraic settings. Thus, the students have to recall old knowledge (application of a percentage to calculate the interest) and more recent knowledge (expression of the interest as a function of the deposit).

A priori analysis: what is the adidactic potential of the lesson? At the very beginning of the 9th session, the teacher declares that they will discuss this exercise but also adds a "total sum" row in the table and requests students to express this sum as a function of the initial amount. We cite the formulation of the exercise below, marking in bold the additional questions of the teacher.

A bank pays 5% annual interest for every deposit. The annual interest is proportional to the deposit.

1. Complete the following table

Deposit (Francs)	100	200	350	550	660	780	896	*S*
Amount of the interest (F)								*I*
Total sum (F)								***T***

2. Let "S" represent the deposit, and let "I" represent the interest. Express the interest as a function of the deposit.
3. **Express the total sum as a function of the deposit.**

The third question, asked orally, corresponds to the adidactic situation aimed at obtaining the factor of proportionality between the initial and the total amounts. The filled in rows of the table ("interest", and the "total sum" which is obtained by addition of the two first rows) may help the students to check their proposals of factors for passing from the first row to the second and from the first to the third. Moreover, the algebraic expression (previously obtained, question 2) of I as a function of S may help them to express T as a function of S. This is what allows us to say that those elements constitute the objective milieu of the situation and that this situation has some adidactic potential.

The development of the situation: The discussion of the "enriched" exercise proceeds in four phases, the first three of them taking place in the 9th session. During the 1st phase the teacher adds the "total amount" row and the students fill it in using addition. We will not detail this phase, focusing instead on the second one, which plays an essential role in the consolidation of knowledge requisite for the situation and for the setting of the objective milieu to solve the new question in the 3rd phase.

Phase 2: this collective phase concerns the calculation of the interest values and the algebraic expression of I as a function of S.

In the first three episodes, involving the students' old knowledge, the teacher does not intervene in the production of the answers; he only intervenes in their validation. However, relying as much as possible on the students' answers and using a set of questions, he helps to introduce the use of a decimal factor, which is a very important element for the continuation of the lesson. Thus, *he sets up the milieu of the "5% interest" situation.* Indeed, when Romain proposes to divide by 20 to calculate the interest for 200F, which is a linear procedure very relevant to the problem, the teacher immediately asks other students to recognize the procedure ("So, what kind of method is he trying to use?") and to express the factor in a decimal form ("Could anyone express [. . .] the factor by something else than divided by 20?"). For the last question, although the expression of a factor of proportionality between the deposit and the interest is an old knowledge for the students, the teacher relies essentially on two students' answers because this knowledge doesn't seem to be available for the whole class. Then he asks other students to decide about the validity of their statements. This way, he gives students, who did not actually contribute to producing the solution, a part in its validation. We name this distribution of responsibilities a *micro-contract of agreement*.[11]

Then, for the remaining part of the discussion of the second row, the teacher deals again with the whole class, keeping a part of the responsibility in the evaluation of the answers. Digressions are perceptible. Some of them are related to students' mistakes, on which the teacher capitalizes to *institutionalize the use of a decimal factor for the calculation of the interest*, stressing this procedure. The following interaction is a typical example of *institutionalization by small alterations*: the teacher evaluates the proposed answer, without seeking to rectify the error, but shows that he values the use of a decimal factor, by referring to the safety of the method.

P: and thus, here (896), how many is it?

Floriane: 46

P: forty. . . ah, no.

Ee: 44. . . 46

P: 44.80. So, you have to explain to Floriane how you did it. We'll see it. And I think that perhaps there are not many methods. Which is the one that could . . . help us avoid making mistakes? Jérémie.

Jérémie: 896 divided by 20.

P: yes, it's so or multiplied by. . .

Ee: 0.05

P: well, I think that, here, the factor is useful. Because eight hundred four . . . except if one has some number sense, I don't know. Did you add here, some cases to find 896? Did you multiply one of the cases by a number to find 896? No. So, here we are stuck, we must use the factor.

Some time later, the teacher also seizes the opportunity to institutionalize the factor in decimal form:

P: So, for you, before going on with the problem, because I am not going to be hypocritical, I prefer this expression (×0.05). Why? (. . .) Thus Romain's proposition was correct. Why are we going to prefer [0.05]?

Emilie: because it's easier to calculate.

P: ah, bah, no because here you divided by 20. There is perhaps another... Why?

Floriane: because it's a multiplication.

P: maybe because here it's a multiplication, actually. And why else? Laëtitia

Laëetitia: because 5% is 5 for 100.

P: Good. It's because the idea of percentage is hidden behind it.

Moreover, he reinforces this institutionalization by proposing to calculate the factor for other values (when the interest is 8%–15%).

Other incidental moments consist in interactions between the teacher and one student, allowing to increase the repertoire of available procedures. For example, when a student proposes to calculate the interest using the deposit of one franc, the teacher lets him explain his calculation.

After the discussion of the 3rd row using an additive procedure, the teacher gives the students the whole responsibility in the production and the validation of the expression of the interest as a function of the deposit. He first plays the role of a secretary, writing down the students' statements and then he moderates a debate between students, which leads to making correct statements. Here, we say we have a *micro-contract of collective production*.

Phase 3 and 4: During the first two episodes of the 3rd phase, the teacher asks the students to express T as a function of S, then, once the factor of 1.05 is produced (it is easy because of the 100 in the 1st row), he asks for a justification and then he asks the students to write 1.05 in the table. During these two episodes, the teacher relies on the few students whose understanding permits a sufficiently fast progress in the course and asks other students to accept and validate the statements. Then, delegating to the whole class the responsibility for producing other factors in the same standard way, *he gives other students an opportunity to appropriate the new knowledge and, by doing so, he institutionalizes it.*

In the 4th phase, during the 10th session, the teacher poses again the same problem and asks the whole class for the justification of the factor 1.05.

Thus, episode after episode, we observe that the distribution of responsibility varies according to the didactic status of knowledge and the teacher's objectives (see the table in Appendix III). These fluctuations are not very surprising insofar as all the students are not able to produce and to formulate immediately the target knowledge. However, they are typical of the ISD practice: the teacher tries to obtain, as much as possible, the knowledge from most students but often obtains it from only some students in the class and then makes the whole class use it. Phase 2 is characteristic in this respect.

2.2.2. *The "slope" situation; necessity of an objective milieu*
The objective of the lesson was a relationship between the visual slope of a straight line and the coefficient of the associated linear function: the bigger the coefficient, the steeper the slope. For this, the teacher wanted to rely on one of the computer exercises, *The trains* (Appendix II), and on the graphical representations and their explanations given by the software for this problem (one can find the faster train by comparing the steepness of the slopes of the straight lines). In the initial situation planned by the teacher, the objective milieu is made of the data of the problem and the graphical representations of the lines. The question is: "How can we see on the graph which train is faster?". A vertical cut of the graph and knowing that "the order of speeds is the same as the order of distances covered in a given time" would allow the validation of the answers. So this situation affords the students with some possibilities to act on an objective milieu and to interpret its feedback if they are able to read from the graphs the distance covered by each train in one hour.

In fact, most students did not attempt the problem, and the ones who did, did not read the explanations. So, the references constituting the milieu of the situation were not available and the teacher could not rely on the students to bring out the knowledge he wanted to institutionalize. So, to manage the class in the way he is used to, the teacher immediately establishes a new situation, giving him a way to question students and to draw the target knowledge from their answers. For this, he uses linear functions previously given by the students ($y = 1/2x$ and $y = 2x$) and their graphical representations drawn on the blackboard. He asks the students to explain the relative position of the two lines and, later, to locate two others lines ($y = 3x$ and $y = 5x$) in connection with them. Next, he goes on to proposing similar exercises with negative coefficients and he comes back to the comparison of speeds.

This lesson highlights yet another feature of the ISD practice: the necessity for the teacher to organize his lesson around a problem, which had previously been solved by the students. He needs a minimal objective milieu, here the graphic representations drawn on the blackboard and other lines added on the same graph to leave the students the responsibility to produce answers and validate them in this milieu. Indeed, although the adidactic part of the last situation is very light, the teacher chooses to urgently build this new situation instead of taking on the responsibility to articulate the target knowledge himself.

3. CHARACTERIZATION OF THE INTERACTIVE SYNTHESIS DISCUSSION PRACTICE

One of the aims of our research was to define and characterize the specific practice of ISD as a macro-contract by the dimensions of micro-contracts and meso-contracts, the relation between these dimensions in the succession of micro-contracts or meso-contracts and also by some features concerning devolution and institutionalization. The main issues concern, on the one hand, the relation between the status of knowledge at stake and the distribution of responsibilities between the teacher and the students and, on the other, the relation between oral interactions and written work. We will illustrate this characterization with the above-presented case studies. Others examples can be found in (Hersant, 2004).

3.1. *Choice of situations and the didactic status of knowledge*

In the ISD practice, the teacher chooses situations with some *adidactic potential*, but this choice does not imply that the teacher actually manages them in an adidactic way. However, the teacher needs a minimal milieu shared by all the students, as it could be seen above. The choice of such situation concerns the meso-contract, while the management affects the micro-contract.

The *a priori* analysis allows us to foresee the kind of feedback that the milieu may provide according to the students' knowledge, as well as the necessity of some teacher interventions if the feedback is lacking. For instance, in grade 10, Phase 4, episode b, there is no graph; the students realize that there are two solutions because some of them found two using another method. The objective milieu is insufficient but the comparison with the results of other students enriches the milieu and helps the students to find the two solutions. But it is not yet sufficient for them to be able to explain why there are two solutions; the teacher needs to intervene and guide the students on this point.

The *a posteriori* analysis permits us to identify a possible gap between the knowledge expected from the students and the knowledge actually invested by them in the solution of the problem. We could also identify the way the teacher manages this gap. In this practice, he generally chooses to reinforce old knowledge and to defer the introduction of new knowledge. For instance, in the "5% interest" situation in grade 8, the teacher expected that the students knew how to use a percentage, by multiplying by a decimal number. But this knowledge had to be reinforced so that the actual objective of the lesson is to reinforce this knowledge rather than to introduce the new. These shifts and the insufficiency of the milieu explain why the distribution of responsibilities between the teacher and the students can only be stable at the level of the episode and not at the level of the phase.

3.2. *Distribution of responsibility between the teacher and the students*

This distribution is complex and strongly linked to the status of knowledge involved and to the progress of the lesson.

3.2.1. *Distribution of responsibility and status of knowledge*

- The students are awarded some time to investigate the problem; therefore there is some devolution of the problem to the students but this devolution is concerned mainly with the use of old knowledge or knowledge in development.
- After giving the students some time to solve the problem, individually or in groups, in class or at home, the teacher begins a whole class discussion of the solutions of the problem based on the actual productions of the students. On this occasion, he introduces new questions permitting them to go further and aimed at the actual objectives for learning. During this phase, at the same time, he institutionalizes and reinforces the knowledge in the course of being learned, and he poses and "devolves" to the students a new problem allowing the introduction of a new knowledge or a new point of view on an old knowledge. It is for this new problem that we use the model of TDS.
- The students generally solve the new problem collectively, without having more time for investigation.
- The teacher relies, as much as possible, on the knowledge and the productions of students, so the actual distribution of responsibility concerning knowledge is not stable and can only be identified at the level of one episode, therefore at the level of the micro-contract.

3.2.2. *Management of taking turns talking and succession of micro-contracts*

The progress of the lesson arises from a particular management of taking turns talking in class tied to the distribution of responsibilities concerning knowledge and thus to the succession of micro-contracts. The teacher provides some information through the formulation of his questions but he rarely shows a definite position about a question. Particularly in the oral solving of the new question he favors the students' speaking as much as possible, yet controlling it: he allows one student or another to speak according to what he knows about their presumed or observed knowledge.

First, in discussing the previous work, the teacher encourages all students to speak, especially the weakest ones. It is a way to extend and strengthen the knowledge, which is being learned, to the whole class; it is also necessary to establish the new knowledge. According to the status of knowledge, the teacher delegates to the class the responsibility concerning knowledge in two possible ways. If the knowledge is available for most students, we have a micro-contract of *collective production*: many students raise their hands to speak and the teacher allows one student, then another one and so on. Other students agree; the teacher repeats or formulates in a clearer way and also structures the students' answers by his questions. He may also ask a student something to check his knowledge; others wait and may intervene in case of error (micro-contract of *individual production* for the student who speaks, inside the micro-contract of *collective production*). If the old knowledge is still currently being learned and available to a few students only, the teacher lets those students speak with a micro-contract of *agreement* for other students.

Next, the teacher asks a question introducing the new target knowledge, and he settles a micro-contract of *collective production* with a distribution of responsibilities between the whole class and himself. But, as his objective is now to solve the problem and make explicit the new method, he often relies on a few students and the contract turns into a micro-contract of *agreement*. When the new problem is solved, again, all students are encouraged to speak about a similar problem in a *collective/individual production* micro-contract.

For instance, in grade 10, in Phases 1 and 2, the teacher makes sure that the students can recognize a function in a graphic form, read and interpret coordinates on a graph and solve quadratic equations by factorization (question added during the discussion). Reading from a graph is known for most of the students but the way to solve equations or inequalities is not available to all of them, so the teacher relies on a few ones and guides them to give all conditions.

Another illustration may be found in the succession of episodes of lesson 1 in grade 8 summarized in the table in Appendix III. When the knowledge is rather old and shared by most of the students (second phase, episodes a, b, c, e, g, i) the teacher lets all students speak and encourages the weakest ones to speak in a micro-contract of *collective production*. But in others (d, f) he favors the factor, then the decimal factor with the help of a few students (micro-contract of *agreement*). In episode f, again in a micro-contract of *collective production*, he trains the students to use this method with other examples, and he lets the students speak if it may reinforce an interesting method (for instance episode 2h, unit value). In episode j, students have to express the interest as a function of the deposit, which is knowledge in development. After accepting all propositions, by way of questions, the teacher helps the students to justify the correct expressions and to reject the others. The response to the new problem is found by very few students (episodes 3a, 3b) in a micro-contract of *agreement* but the teacher immediately asks other students to do the same about other examples (episode 3c) setting again a micro-contract of *collective production*. Later (Phase 4), the teacher poses again the question to the whole class.

3.3. *The relative status of writing and speaking*

In Section 2, we presented phases of work corresponding to oral phases mainly, because we are especially interested in the way the teacher manages institutionalization. However an interactive synthesis discussion is always based on the students' research, carried out individually or in small groups, at home or in class. It generally involves old knowledge or knowledge in development. In grade 10, the teacher asks the students to write a report from the problem solving session and he also provides the students with some written elements at some point during the discussion.

The oral collective synthesis relies on this previous work and on these writings. During the discussion of the solutions, the teacher relies on the actual answers of students and eventually on the written elements he provided (see for instance grade 10, Phase 3). He can check the old knowledge of students and they can appropriate this knowledge, which is useful for the progress of the lesson. But the teacher also seizes the opportunity of the discussion to introduce new questions permitting to go forward and achieve the actual learning objectives (for instance grade 10, Phase 3 and grade 8, Phases 2 and 3). So, the problem involving new knowledge is solved orally and collectively without more time for personal research of students.

During this oral phase, the teacher indicates some issues to remember, but neither during this phase nor after it, does he dictate to the students what to write down in their notebooks. The students keep few notes of this part. The new knowledge will be reinforced in other problems where students will have to use it, but to get a decontextualized knowledge, they have to use the textbook.

3.4. *Progressive institutionalization*

Institutionalization is carried out by small alterations, and remains in the context of the solved problems, even if the teacher says that this new knowledge will be useful for other problems. Beyond the correction of exercises, time is not provided for students to take careful notes and the teacher does not explain precisely what to write down in their notebooks. Occasionally, however, the sheets of exercises have explanatory notes.

We saw in the examples earlier, how the teacher institutionalizes in a diffuse way, using several modes, and in several contexts. For this, he:

- stresses the importance of some method (for instance, grade 10, it is safer to avoid dividing by zero);
- asks the students to make explicit the expected procedure on several examples (for instance in grade 10 to change the equation and arrive at something equals zero and then factor; in grade 8, to multiply by a decimal number);
- uses an error of a student to depreciate some procedure and to value another one (in grade 10, the division by $x+2$; in grade 8, the error for 896);
- proposes similar exercises to encourage all students to use the target method.

There are no course notes written in a notebook; it is the use of the new knowledge in different problems that leads to a decontextualization of knowledge; therefore it is important, in the case of the ISD practice, to see institutionalization as a process including some reinvestment of new knowledge.

3.5. *Is interactive synthesis practice different from other forms of teacher-students dialogue?*

The teacher–students dialogues, currently observed in the classrooms, are often based on "disguised ostension" (Salin, 1999): the teacher gives the students a problem to introduce the topic, but he does not actually allow the time for them to find a solution by themselves. There is a collective process of solving the problem, and then the teacher makes an exposition of the theory more or less illustrated by the problem they have just solved; then he gives practice exercises to apply the knowledge exposed in his lecture.

The practice we study resembles "disguised ostension", but there is one important difference: the existence of an actual milieu capable of some feedback to the actions of students and the performance of actual actions on this milieu by the students. So we chose the name of the interactive synthesis discussion practice and not "interactive lecture" to stress the importance, in this practice, of a real personal work of students, work actually taken into account by the teacher.

In the last section of the paper, we will compare our study with other research on classroom interactions.

4. Interactive Synthesis Discussions as Seen by the Actors

We will now discuss this practice starting from some questions concerning the teacher's work or the students' learning. These questions stem from the two case studies above and also from interviews carried out at the end of the school year with the grade 10 teacher and with some of his students.

4.1. *The students' point of view*

In this type of practice, there is room for various procedures of students. Moreover, during the interactive synthesis discussion, there are moments inside the progress where the teacher seizes the opportunity to relate the different kinds of knowledge, old and new, and different exercises. This seems to enhance learning. However, at the end of the year, the grade 10 students who were interviewed on their work raised two main difficulties: starting by research problems for which they had no technique and the lack of structured written lecture notes. We give some examples of comments from students identified by the initials of their first names.

4.1.1. *Appreciation of the richness of the research problems and experience of difficulty in starting to solve the problem, sometimes for the same student*

L: If you do not understand a theorem, you cannot apply it; in middle school, first we were exposed to the theory[12] and then we solved exercises [. . .] It's interesting to try our own ideas, after that we understand better our mistakes and we don't make them anymore.

No: It was not easy because it's very different: we were used to having a lecture and then the exercises; here we immediately attack the exercises. [. . .] We had nothing to rely on to do the problems; we knew nothing, we had to do everything by ourselves.

Se: We enter directly into the heart of the matter.

Na: We have to enter the chapter directly without having studied it before so we are confused.

Y: It upsets me because if I don't understand the problem, I don't think I will understand the course.

F: It's difficult. . . He was asking us actually to reinvent the theory; if we were given the theory we would have it easier. [. . .] Sometimes it's good . . . because I could find a solution. Indeed, it's good for those who are able to find a solution, because then we have control over it and we remember what we have found [. . .] That's why it's mainly useful for good students.

R: [a middle ability student] I prefer to study the theory before and do exercises after to apply the theory. I don't see how we can do exercises if we have no theory.

Cl: [a good student] It's interesting but we haven't got enough time, so, for some chapters we did only that and after we had to look things up in the book and it was not so clear... The problem is that, as it is something we don't know well, there are things that we have a feel for, but we don't know well, we can't find the solution and then there are things that are not clear...

4.1.2. *The difficulty, in discussing the solution to an exercise, to go ahead*

F: For instance, I do an exercise; I come to class; the teacher discusses the solution of the exercise, but he adds more questions and those, which he adds, I can't understand...

Na: There are some extra little things the teacher puts in and that was just the trouble for me [...] If you don't understand the theory, you don't understand the little "extras" anymore.

4.1.3. *Lack of course notes*

Many students said they miss having course notes. Some find it difficult to take notes; they use the textbook but find it difficult to do because the work in class does not follow the book and they do not know how to synthesize the book and the work in class.

L: I found the course not sufficiently structured. We did only exercises that were corrected in class... [...] When I missed two days of school, I could not catch up... If you borrow a notebook from another student, there are only the corrected exercises, so if you did not hear the oral discussion, it's impossible to catch up.

C: When you arrive in class it's better to have worked on the exercises before at home. What we do in class completes what we do at home, if we do nothing at home, we cannot follow in class.

Sy: We did exercises; we had no theory. There is nothing we can refer to; we are lost... Which was good, we had a lot of different methods, how to search...

O: [good student] For vectors for instance we had solved exercises, I looked at how we did it and then I synthesized, I modeled [...] It's necessary to have understood the exercises.

Ju: For me the theory is above all a reference. The problem is that, there, I have trouble understanding what he wants; he wants us to learn methods but I don't see where he is going. We begin an exercise, if you let go, it's finished...

Je: We have no theory in class... In the book, the theory is not sufficiently developed; we need to work on exercises; I am stuck... I waste a lot of time to reconstruct the theory by myself. After the teacher discusses the solutions it's a little better; but then he goes on to other exercises, and it's too fast, I'm lost.

4.1.4. *Taking notes: Some found it difficult, others easy enough*

C: You either listen or you write; at home you can more or less manage with the book.

No: The teacher tells us what to write down. We feel by the way he says it what is essential.,

Na. He is very clear on that. I take notes, I write half words and in the margin, I write "theory" [in French: "*cours*"] in big letters.

We can see that students express both positive and negative feelings (e.g., L, No, Na). Many of them see mathematics as something that is first learned, and then applied. They find it difficult to do problems if they do not know what to apply, especially because there is not enough time. Their main concern is the lack of course notes, which implies the need to be very attentive in class and to work a lot at home. Moreover, it makes it difficult not to miss anything (for instance due to an absence).

4.2. *The teacher's point of view*

The ISD discussion practice has become quite common in the French middle and high school although it is not the only one. The grade 10 teacher explained to us why he adopted this practice. His explanations suggest that the practice exists because it corresponds to some guidelines and recommendations of the official curriculum, which emphasize students' activity in learning mathematics. Moreover, it is supported by the existence of materials in textbooks and in the IREM[13] publications for constructing mathematics activities for students. Another reason is that, since the creation of a single middle school for all students and the larger openness of high school, the lecture format of teaching is no longer suitable for all students. Below, we cite some of the things the teacher told us:

For me, the "exposition of theory followed by application exercises" format was much easier to teach. If I changed, it is for several reasons. The first influence was a book published by the Grenoble IREM. Its style was more suitable for the new clientele, because I had all types of students. [...] There was no more selection or streaming at the end of grade 7. There was no more scientific stream in grade 10 so in middle school students were less motivated. [...] And also, I have taught in CPPN[14] and there, the teacher has to adapt himself. [...] And after, in middle school, the official texts asked the teachers to listen to their students. When I came to teach in a high school, I thought I'd do a traditional course but I had very weak classes; the only thing that worked was problem solving in small groups. [...] I cannot make an exposition of the theory because students disconnect...

From this point of view this practice gives some comfort to the teacher insofar as he feels it is appropriate to include more pedagogic ideas in his course. Moreover, the managment of institutionalization through the whole class discussion of solution to exercises may prevent the students from "disconnecting" and provides the teacher with a better control of their understanding.

However, the success of this practice depends upon the teacher's ability to manage the students' answers, to analyze their productions from the point of view of what they know but also of how it is possible to move on with the course, and to construct new situations on the spot. He is often in a state of uncertainty, much more than he would be in the lecture-exercises format. This uncertainty may be a source of difficulties in time management, as we saw in the second lesson in grade 8.

4.3. *An additional question about learning*

This practice seems a way to make new pedagogical methods compatible with the constraints of teachers' work and their habits. Our previous research (Perrin-Glorian, 2001) already showed the influence of institutional constraints on the organization by the teacher of the mathematical contents of teaching and of the students' work.

Students' remarks suggest that most of them need to be strongly supported in their learning and they are not ready to assume a large responsibility in the production of new knowledge, even in a very orchestrated way. They need something to refer to, such as course notes, and they are not able to achieve, by themselves, a clear synthesis of what they have to retain from the work done in class. In a traditional course, the teacher first gives an oral exposition of knowledge and indicates to the students what exactly they have to write down in their notebooks. They listen and take notes and then try to understand it by doing exercises. The real effect of the ISD practice on the learning, beyond students' discourse, remains an important open question intersecting with those discussed by Sfard, Nesher, Streefland, Cobb and Mason (Sfard et al., 1998) about the possibility of learning mathematics through conversation. For instance, in this class, there were many "reflexive shifts in discourse" in the sense of Cobb (Sfard et al., 1998) apparently productive for learning, but students, even if they did appreciate them, still asked for course notes and being clear about what to learn. Even if action, reflection, formulation and discussion help the students to learn, it seems they are not sufficient for all of them to know. Moreover, as Sfard says (ibid., p. 47) "Orchestrating a productive mathematical discussion or initiating a genuine exchange between children working in groups turns out to be an extremely demanding and intricate task". Therefore the question is: for a better learning of all students, if it is desirable to engage them in investigating problems, to what extent is it desirable indeed, and how can the teacher help them structure and learn the knowledge encountered in this investigation?

5. Discussion

5.1. *The theory of didactic situations to analyze ordinary lessons*

TDS helped us analyze ordinary lessons, the concept of milieu accounting for the responsibility concerning knowledge possibly left to the students in the observed situation and the notion of contract accounting for the management of this situation by the teacher in connection with the productions of students. Indeed, in this kind of practice, we found a milieu, sometimes very small, provided to redefine the situation based on the objectives of teaching and not only on the problem to solve. In other classes (Hersant, 2001) we could not find anything to be a milieu susceptible of feedback. Anyway, in all cases, the definition of the dimensions of the didactic contract, mainly the status of knowledge and the distribution of responsibilities helped us analyze the interactions in terms of progress of teaching and progress of learning.

An important theoretical and methodological question is how to develop indices helping to characterize the teacher's actions in relation with his teaching objectives. In particular, we need means to select lessons to analyze at a local level but allowing us to obtain information concerning the development of class knowledge and the actual mathematical activity of students at a more global level.

5.2. *Relations with other research*

Teachers' and students' beliefs, their representations of mathematics, their work, and students' position in the class stay in the background of our research; they are not considered as such. However, it appears that in this practice, the teacher gives a great importance to the generic example (Balacheff, 1987) perhaps with the idea that the general method emerges from generic examples and that it is a better way to learn to use it, to develop operational knowledge. The question is then to know if all students are able to transfer and use this knowledge.

5.2.1. *Relations with other research on interaction*

In our research, we focus on mathematical interactions: the use we make of the adidactic milieu and of the didactic contract is a way to take into account the mathematical dimension of interactions in relation to the progress of knowledge in class. Much research concerning mathematical interactions in the classroom (Steinbring, Bartolini Bussi and Sierpinska, 1998) focuses on students' learning and the role of language and interactions for learning. In our approach, we focus mainly on the teacher and the progress of the course; the learning of students is in the background. However, the notion of 'milieu' is a way to emphasize an objective reference for interactions and, in this sense, may be compared to a 'learning environment' in the sense of Steinbring (2001).

This characteristic of the situation (the milieu) that we take into account helps us interpret teacher's actions referring to the situation and knowledge at stake. Other authors studied ordinary teaching and defined formats or patterns of interactions from different theoretical points of view (for instance Voigt, 1985; Bauersfeld, 1994; Sierpinska, 1997; Krummhauer, 2000). Patterns of interaction have a similarity with the didactic contract in the sense that both are the result of an implicit negotiation (Voigt, 1985, pp. 88–95, Sierpinska, 1997, p. 3). More precisely, Voigt (1985, p. 93) considers the concept of working consensus (mutual assumptions and obligations) extended to the contents communicated and linked to patterns of experience as similar to the concept of didactic contract. Moreover, Voigt (1985, p. 82) marks as an essential point for a pattern of interaction to focus on a topic and in a certain sense, he takes into account the status of knowledge at stake when he characterizes the elicitation pattern as "one certain organization of classroom discourse within introducing new mathematical subject matters" (Voigt, 1985, p. 95). To some extent, interactions in the ISD practice, when the teacher introduces a new question, look like an elicitation pattern, but they differ significantly in that, in ISD, the task is well defined from the beginning. Moreover, the new question is included in the discussion of a problem already solved by the students; the solution of previous questions brings on some elements of an objective milieu, which gives to this interaction a sense that it will not be able to get without any objective milieu. Another convergence point of our work with Voigt's is that he considers that patterns of interaction are produced under institutional conditions and through routines. But he presents patterns of interaction "from the perspective of symbolic interactionism and ethnomethodology which considers regularities in interactions as processes interactively constituted by the participants" (Voigt, 1985, p. 82). This point of view is developed in Bauersfeld (1994). From this perspective, the study of class discourse is an essential issue. From our perspective, class discourse is important, but other considerations are also very important to define the didactic contract, for instance tacit choices of the teacher, such as the choice of some particular exercise instead of another one, especially for assessment, strongly affects the didactic contract. Moreover, our perspective intersects with a constructivist perspective and a social perspective (entering into an institutional culture). Indeed, a main issue is that knowledge to learn and knowledge to teach are not taken for granted but problematized; only relations to knowledge can be considered (Chevallard, 1992; 1999). This relation to knowledge is defined inside institutions. For students, knowledge to learn is defined through teaching choices and through interactions; the teacher guarantees the institutional conformity of taught knowledge, so the teacher and the students are not in the same position regarding knowledge. Moreover, this knowledge is always evolving along with teaching. With the notion of didactic contract, we take into account this evolution through the status of the knowledge at stake. Our interest is not only in the characterization of a pattern of interaction in itself but in its relations with the progress of knowledge of the class and of individual students, knowing that interactions are only a part of this progress. Therefore, for us, it is very important to relate the analysis of local interactions to analyses on more global levels. We hope that a deeper discussion of this issue will be possible in further research.

NOTES

1. Generic masculine pronouns will be used throughout the text to alleviate the style.
2. Or "didactique" (Herbst and Kilpatrick, 1999).
3. Independent from the teacher and from the students.
4. Or a game in the sense of TDS.
5. We use here "personal knowledge" to translate the French word "connaissances" and "institutional knowledge" to translate "savoirs".
6. In Hersant (2001) and Perrin-Glorian and Hersant (2003), we used the term "component" but it could suggest a vision of the contract as an union of subsets; we hope that "dimension" is clearer.
7. The teachers with whom we are concerned in this paper were observed during long periods in several classes, over two or three consecutive years.
8. In this case, the objective milieu is mainly made of knowledge which is not questioned, such as the given data, previously accepted results, graphs obtained on calculators . . .
9. Using usual formulas is not an aim of the curriculum of this grade, but only in 11th grade. In 10th grade, students learn to solve those equations putting them in the canonical form $(x + d)^2 - k^2$, but the students have not seen it yet.
10. 'La proportionnalitéàtravers des problèmes', software produced by IREM (Research Institute on Mathematical Teaching) of Rennes and CNED. For a presentation of this software, see Hersant, 2003.
11. In French "contrat d'adhésion". It means that a new idea is produced by a few students and that the others seem to agree with it; the teacher is attentive to this agreement of other students.
12. In this section "exposition of theory", or "theory" is a translation of the French "cours" (Editor's note).
13. Research Institute on Mathematics Teaching.

14. Classe Pré-Professionnelle de Niveau: It was a special class for fourteen years old students in high difficulty at school, a class preparing the students for entering the job market soon.

APPENDIX I: GRADE 10

Part 1. First problem solving session in small groups

Exercise 1

Draw a Cartesian plane with x-axis and y-axis. Take as units OI on the x-axis such that $OI = 1$ cm and OJ on y-axis such that $OJ = 0.5$ cm.

We are interested in the points of the plane whose coordinates (x, y) are related by the relation $y = (x+3)(8-2x)$. We name by set of these points "(E)".

1. Give 5 pairs of coordinates that belong to (E) and 5 that do not belong to (E).
2. Represent as many points of (E) as possible on the graph.
   (a) Are there some points of (E) on the x-axis? If so, highlight these points in blue; if not, explain why.
   (b) Are there some points of (E) on the y-axis? If so, highlight them in yellow; if not, explain why.
   (c) Are there some points of (E) whose ordinates are positive? If yes, highlight them in green; if not, explain why.
   (d) Are there some points of (E) whose ordinates are negative? If yes, highlight them in red; if not, explain why.
   (e) Are there some points of (E) with the same abscissa? If yes, give examples; if not, explain why.
   (f) Are there some points of (E) with the same ordinate? If yes, give examples; if not, explain why.

Exercise 2

Same questions as in the exercise 1 about the relation $y = x^2 - 9$. The set of points is named (P).

Exercise 3

Fill in the following table.

x	−7	−6	−5	−4	−3	−2	−1	0	1	2	3	4	5	6
$y = (x+3)(8-2x)$														
$y = x^2 - 9$														

Represent sets (E) and (P) in the same coordinate system.

1. Do (E) and (P) have common points? If yes, give, if possible, the coordinates of these points and highlight them in blue.
2. Are there points of (P) that have the same abscissa as a point of (E) but whose ordinate is higher? If yes, highlight them in red.

Part 2. Second problem solving session in small groups: Factoring? Expanding?

1. Here is a list of equations. For each of them, say whether factoring or expanding seems to be the tool for solving the equation. Explain why.

   (a) $(3x + 1)(x - 2) - (x - 2)(2x + 3) = 0$
   (b) $(3x - 1)(x + 2) = (x + 2)(2x + 5)$
   (c) $9x^2 - 1 = 3x + 1$
   (d) $(3x + 1)^2 - (5x - 2)^2 = 0$
   (e) $4x^2 - 12x + 9 = 0$
   (f) $(2x + 3)^2 = 4x^2 + 4x + 1$
   (g) $4 - (7x - 1)^2 = 0$
   (h) $(2x - 3)^2 = x^2 + 3x + 9$
   (i) $3x^2 + 5x - 1 = 3(x^2 + 4)$
   (j) $(3x + 1)(x - 2) = (5x - 3)(x + 2)$

2. Make use of the method you selected in 1) to solve the equations, if possible.
3. Explain how you can check the results by reading from the graph.

APPENDIX II: GRADE 8

The trains

The "Goeland" train covers k_g km in a_g h and b_g min. The "Mistral" train covers k_m km in a_m h and b_m min. The "Evasion" train covers k_e km in a_e h and b_e min. The "Libert$_e$" train covers k_l km in a_l h and b_l min. Order these trains from the fastest to the slowest.

For the students, there are numbers instead of parameters in the problem; there are several versions of this problem in the computer program, each with different values of these numbers.

The explanations for these problems, given in a graphical form, consist of six consecutive screens. Each screen is made up of a text and a graph.

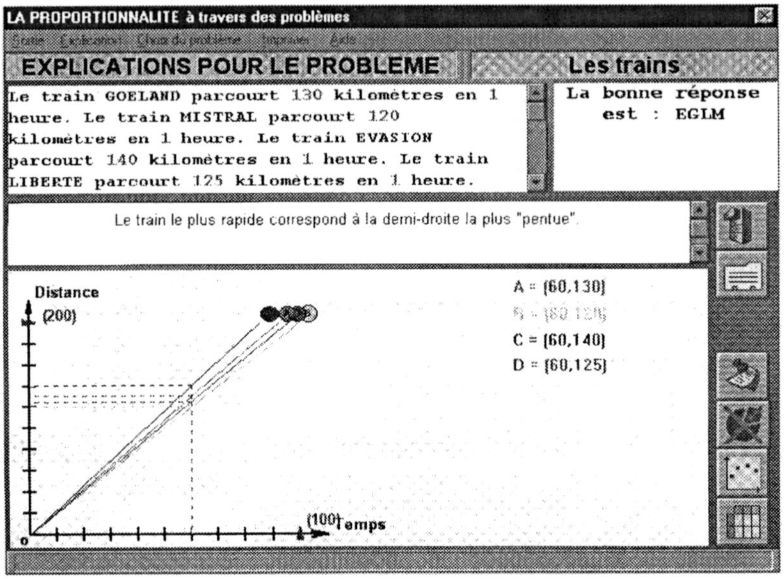

Figure 2. Last screen of the graphical explanation.

For each new screen, a new sentence is added and the graph is completed. The full text of the explanation is the following:

The horizontal axis represents the time in minutes. The vertical axis represents the distance in kilometers. The movement of the train is represented by a ray passing through the origin. The train G covers 180 km in 390 minutes. The train M covers 120 km in 60 min. The train L covers 225 km in 60 min. The fastest train corresponds to the ray with the steepest slope.

One screen at a time, the graphical representations of the line corresponding to a train is given. The graph in Figure 2 corresponds to the last graph given with the explanations.

APPENDIX III: GRADE 8, A SUMMARY OF LESSON 1

In the table, episodes concerning new knowledge are marked in bold; those aiming to assure the knowledge needed to express I as a function of D algebraically are marked in italics; they will play an essential role in the milieu for the new question. Among them, episodes 2f and 3c aim at consolidating this knowledge.

Episode	Title	Distribution of responsibility	Micro-contract	Meso-contract
2a	Value of the interest for 100F	T and students	Collective production	Institutionalization of the knowledge: decimal coefficient between the deposit and the interest
2b	Meaning of 5% interest: 5F for 100F	T and students	Collective production	
2c	*Calculation of the interest for 200F*	*T and Romain*	*Individual production*	
2d	*Decimal expression of the factor between S and I*	*T and two students*	*Agreement*	
2e	Continuation of the discussion of the second row (interest)	T and students	Collective production	
2f	*Calculation of the factor for 8% and 15% interest*	*Students*	*Collective production*	
2g	End of the discussion of the second row	T and students	Collective production	
2h	Pascal's intervention, unit value	T and Pascal	Individual production	
2i	*Discussion of the third row (total sum), additive procedure*	*T and students*	*Collective production*	
2j	*Expression of I as a function of D*	*Students*	*Collective production*	
3a	**Expression of T as a function of D**	**T, Damien, Romain and Julien**	**Agreement**	**Devolution of the new question**
3b	**Justification of good expressions for the factor and rejection of others**	**T, Romain and Aurore**	**Agreement**	
3c	**Calculation of the factor between the total sum and the deposit for 8% and 15% interest**	**T and students**	**Collective production**	**First institutionalization**

REFERENCES

Artigue, M.: 1992, 'Didactic engineering', in R. Douady and A. Mercier (eds.), *Research in Didactique of mathematics. Selected papers*, extra issue of Recherches en didactique des mathématiques, La Pensée sauvage, Grenoble, pp. 41–65.

Artigue, M. and Perrin-Glorian, M.J.: 1991, 'Didactic engineering, research and development tool: Some theoretical problems linked to this duality', *For the learning of Mathematics* 11(1), 13–17.

Balacheff, N.: 1987, 'Processus de preuve et situations de validation', *Educational Studies in Mathematics* 18, 147–176.

Ball, D.L., Lubienski, S.T. and Mewborn, D.S.: 2001, 'Research on teaching mathematics: The unsolved problem of teachers' mathematical knowledge', in V. Richardson (ed.), *Handbook of research on teaching*, 4th ed., American Educational Research Association, Washington, D.C., pp. 433–456.

Bartolini Bussi, M.G.: 1994, 'Theoretical and empirical approaches to classroom interaction', in Biehler, Scholz, Stasser and Winkelmann (eds.), *Didactic of Mathematics as a Scientific Discipline*, Kluwer Academic Publishers, Dordrecht, pp. 121–132.

Bauersfeld, H.: 1994, 'Theoretical perspectives on interactions in the mathematical classroom', in Biehler, Scholz, Stasser and Winkelmann (eds.), *Didactic of Mathematics as a Scientific Discipline*, Kluwer Academic Publishers, Dordrecht, pp. 133–146.

Brousseau, G.: 1996, 'L'enseignant dans la théorie des situations didactiques', in Perrin-Glorian, M.J. and Noirfalise, R. (eds.), *Actes de la 8ème Ecole d'Eté de didactique des mathématiques*, I.R.E.M. de Clermont-Ferrand, pp. 3–46; A longer version developed in a lecture in Montreal may be found online http://dipmat.math.unipa.it/ grim/homebrousseau.htm.

Brousseau, G.: 1997, *Theory of didactical situations in mathematics. Didactique des mathématiques 1970–1990*, Kluwer Academic Publishers, Dordrecht, 336p.

Brousseau, G. and Centeno, J.: 1991, 'La mémoire du systéme didactique', *Recherches en Didactique des Mathématiques*, 11(2–3), 167–210.

Chevallard, Y.: 1992, 'Concepts fondamentaux de la didactique: Perspectives apportées par une approche anthropologique', *Recherches en Didactique des Mathématiques* 12(1), 73–111.

Chevallard, Y.: 1999, 'Pratiques enseignantes en théorie anthropologique', *Recherches en Didactique des Mathématiques* 19(2), 221–266.

Comiti, C. and Grenier, D.: 1997, 'Régulations didactiques et changements de contrats', *Recherches en didactique des mathématiques* 17 (3), 81–102.

Douady, R.: 1987, 'Jeux de cadres et dialectique outil-objet', *Recherches en Didactique des Mathématiques* 7(2), 5–31.

Herbst, P. and Kilpatrick, J.: 1999, 'Pour lire Brousseau', *For the learning of mathematics* 19(1), 3–10.

Hersant, M.: 2001, *Interactions didactiques et pratiques d'enseignement, le cas de la proportionnalité au collège*, Thèse de l'Université Paris 7.

Hersant, M.: 2003, 'Des logiciels dans les classes: impact sur les connaissances des élèves et intégration à l'enseignement. Un exemple avec "La proportionnalité à travers des problèmes" ', *Petit x* 61, 35–60.

Hersant, M.: 2004, 'Caractérisation d'une pratique d'enseignement des mathématiques, le cours dialogué', *Revue canadienne de l'enseignement des sciences, des mathématiques et des technologies* 4(2), 241–258.

Krummheuer, G.: 2000, 'Mathematics learning in narrative classroom cultures: studies of argumentation in primary mathematics education', *For the learning of mathematics* 20(1), 22–32.

Margolinas, C.: 1995, 'La structuration du milieu et ses apports dans l'analyse a posteriori des situations', in Margolinas (ed.), *Les débats en didactique des mathématiques*, La pensée Sauvage, Grenoble, pp. 89–102.

Perrin-Glorian, M.J.: 2001, 'A study of teachers' practices. Organisation of contents and of students' work', in K. Krainer, F. Goffree and P. Berger (eds.), *European Research in Mathematics Education. On research in Mathematics Teacher Education*, Forschungsinstitut für Mathematikdidaktik, Osnabrück, pp. 171-186; available online http://www.fmd.uni-osnabrueck.de/ebooks/erme/cerme1-proceedings/cerme1-group3.pdf

Perrin-Glorian, M.J. and Hersant, M.: 2003, 'Milieu et contrat didactique, outils pour l'analyse de séquences ordinaires', *Recherches en didactique des mathématiques* 23(2), 217–276.

Salin, M.H.: 1999, 'Pratiques ostensives des enseignants', in *Le cognitif en didactique des mathématiques* sous la direction de Lemoyne et Conne, , Ed. Les presses de l'Université de Montréal, pp. 327–352.

Sierpinska, A.: 1997, 'Formats of interaction and models readers', *For the learning of mathematics* 17(2), 3–12.

Sfard, A., Nesher, P., Streefland, L., Cobb, P. and Mason, J.:1998, 'Learning mathematics through conversation: is it as good as they say?', *For the learning of mathematics* 18(1), 41–51.

Steinbring, H.: 2001, 'Chapter 5: Analyses of mathematical interaction in teaching processes', *Proceedings of PME 25*.

Steinbring, H., Bartolini Bussi, M.G. and Sierpinska, A.: 1998, *Language and communication in the mathematics classroom*. National Council of Teachers of Mathematics, Reston, VA.

Voigt: 1985, 'Patterns and routines in classroom interaction', *Recherches en Didactique des Mathématiques* 6(1), 69–118.

MAGALI HERSANT[1] and MARIE-JEANNE PERRIN-GLORIAN[2]
[1]*IUFM des Pays de la Loire et Centre de Recherche en Education Nantais, Université de Nantes, France*
[2]*IUFM Nord Pas-de-Calais et Equipe DIDIREM, Université Paris 7, France*

GÉRARD SENSEVY, MARIA-LUISA SCHUBAUER-LEONI, ALAIN MERCIER, FLORENCE LIGOZAT and GÉRARD PERROT

AN ATTEMPT TO MODEL THE TEACHER'S ACTION IN THE MATHEMATICS CLASS

ABSTRACT. This paper outlines some theoretical categories (i.e. the *meso-, topo-, chronogeneses*, the "milieu", the didactical contract and the learning games), providing a *model* to study mathematics teacher's action. In order to show what this model brings to the didactical analysis, we present the action of two teachers, on the same content, and we attempt a *threefold description*, covering different scales of analyses of the teaching processes. To amplify the phenomena that are to be observed, we suggested the teachers include the "Race to 20" situation in their teaching. We expect that implementing an unusual teaching device should lead the teachers to take decisions and explain them more easily than in everyday lessons.

KEY WORDS: chronogenesis, didactical contract, mathematic situation, mesogenesis, milieu, teacher's action; race to 20, threefold descriptive model, topogenesis

1. INTRODUCTION

We study mathematics teachers' actions within a model that attempts to connect and enhance several theoretical frameworks, borrowed mainly from TDS, the theory of didactic situations, and from ATD, the anthropological theory of didactics.

Our approach is resolutely nonprescriptive; it consists in describing the interaction of a teacher and his students in order to improve our understanding while respecting the complexity of the teaching process. In Section 2 we provide a brief description of the context of the research and we define the categories we find necessary to model the teacher's action. Section 3 is devoted to a description of the empirical set-up of the model. The conclusion gives possible ways of continuing this research and points out the new implications of this type of work for teacher training.

2. THE FRAMEWORK OF THE RESEARCH

2.1. *Present work*

The general aim is to describe and understand the teacher's action in the mathematics class from a didactical point of view. We consider that the

Educational Studies in Mathematics (2005) 59: 153–181
DOI: 10.1007/s10649-005-5887-1

pieces of knowledge at stake affect the teacher–students interactions, in such a way that the teaching and learning processes cannot be studied separately from the mathematical subject. Therefore, we assume that *looking at the interaction patterns through classroom observations* may enable us to see what is, in fact, taught and especially *how it is taught*. More precisely, we are looking at *teaching techniques* that could be specific of the teacher's action in the mathematics class. These techniques, that may be classified into categories, mainly born from TDS and ATD (see Section 2.2), are to be the core of our teacher's work model. However, we will not neglect the fact that the teacher's acts may also be related to more general groundings in education and learning theories. This aspect will lead us to *take into account the teacher's comments* about his own action during the class obtained in interviews (see Sections 2.1 and 2.3). Beyond the spotting of the techniques, the links between the different categories of action should be showed in order to reveal how our model is being elaborated.

In this paper, we shall consider the teacher's action while he[1] has to carry out a "Race to 20" lesson, suggested by the research team, adapted from Brousseau's situation (Brousseau, 1997, pp. 3–17). This situation is used as a paradigm for studying the didactical action based on the classroom interactions, and as a means to improve our theoretical framework. Indeed, the structure of the "Race to 20" situation, should enable a wide range of actions to occur—both on the teacher's side and the students' side. We assume that most of these actions are induced by the mathematical features of this situation.

2.1.1. *What is the "Race to 20" situation?*

The situation is based upon a game which opposes two players. The first player says a natural number X_1 that is less than 3 (1, for example). The second player says a natural number Y_1 obtained by adding 1 or 2 to X_1 (for example, he says 3, a number obtained by adding 2 to 1). The first player then says a natural number X_2, obtained by adding 1 or 2 to Y_1 (for example, he adds 1 and says 4), etc. The player who is the first to say 20 is the winner.

There are numbers that it is sufficient to say in order to win: 2, 5, 8, 11, 14, 17, 20. However most students do not spontaneously understand this. They quite quickly figure out that 17 is a winning number[2] but they need to play many rounds of the game to find out that so is 14. Therefore, the teacher's intervention is necessary to enable the students to discover the role of these numbers in the game. Brousseau (1997) shows how didactic engineering about the "Race to 20" led to the discovery of the didactic conditions, in which students come up with theorems such as "17 wins, so the 'Race to 20' equals to the Race to 17," then, "14 wins, ... " and so on, followed by the notion of "winning strategy" and that of "game equivalence." "Euclidean division" is a general model of this type of game, i.e. reaching a number $A(A = 20$ in the "standard" race to 20) in an integer number Q of steps of length $B(B = 3$ in the "standard" race to 20), giving the following equation $A = B * Q + R$, where R is the remainder in division of A by Q, and relatively to the game, the number to start with in order to win.[3] According to Brousseau (1997, p. 3) the "Race to 20" is a situation "to revisit division (in circumstances in which the 'meaning' of the operation did not conform to the one learned earlier)." In practice, however, this approach is rarely observed.[4]

In fact, Brousseau's situation has another aim: "foster the discovery and demonstration, by the children, of a sequence of theorems" (ibid., p. 4). For that purpose, the situation is divided in three phases in Brousseau's engineering: a phase of action (playing one-against-one); a phase of formulation (playing team-against-team), in which "the teacher nominates one child as the team representative for each round, naming her at random (ibid., p. 4)"; a phase of validation (the game of discovery): the students "have to put forward propositions and to prove to an opponent that they are either true or false" (ibid., p. 4).

It is the design and study of the Race to 20 situation, that led Brousseau to the general concept of didactic situations and their classification into situations of "action", "formulation" and "validation". This is a strong reason to choose this didactic setup as a paradigm for the studying of teacher's work. Indeed, this situation can be regarded as a very appropriate one to understand the teacher's action (or lack of action) through the different dialectics of action, formulation, validation (Brousseau, 1997, p. 9–11) that it requires.

2.1.2. *The research organization and discussion*

The research involved two teachers (T1 and T2) of grade V classes in an elementary school.

We first introduced the teachers to the "Race to 20" situation, presenting the main mathematical aspects of the game in a 2 hours training. Brousseau's complete text on this subject was handed out at the end of the training session, but the didactic engineering itself was not a training topic during this session. Using this engineering in the teaching process was not compulsory, nor was the reading of the text. It was given as an opportunity that the teachers could take into account, or not. As we will see below, they used this possibility in different ways.

- In the second phase, we asked each teacher to teach one or more lessons on the situation "Race to 20." The teachers were free to plan the lessons as they wished. Both decided to devote two lessons to the situation. We conducted interviews with the teachers before and after the lessons.
- The third phase consisted in the teachers' self-analyzing their first lesson, based on a video recording of the class.

All the lessons were audio- and videotaped and verbal interactions were transcribed from an audio source mainly but including details from the pictures, like words on the blackboard, when needed.

In this paper, our analysis focuses mainly on the first lesson, in order to concentrate our descriptions on the fundamental techniques the teachers used. In this research, we tried to obtain a synthesis of studies of "ordinary classes in everyday conditions" and those based on teaching experiments based on "didactical engineering." This synthesis of clinical and experimental approaches refers to the *ad hoc* theoretical and empirical approaches described in Leutenegger (1999) and Schubauer-Leoni and Leutenegger (2002).

In our research, the experimental perspective is involved in proposing a situation (here, "Race to 20") to teachers who had never tried it before. First of all, because Brousseau's situation assumes a large spectrum of teaching actions concerning the action, formulation and validation phases management, we expect the situation to reveal many of the fundamental teaching techniques at work in the didactic process. Furthermore, we think the novelty of this situation for the teachers should instigate or concentrate in a short time some usual techniques that normally occur in every didactic process.

The clinical dimension relies, first of all, upon the relative freedom the teachers had been given in organizing the lessons based on the "Race to 20" situation. Indeed, they could use their "ordinary" teaching techniques, to some extent. In this case, the clinical study of didactic systems is meant to follow the dynamics of the usual didactic processes at work during the lesson. Moreover, beside the meanings inferred from the observation of the classes, we also took into account those emerging from the teacher's self-analyses, while watching the video recordings of their lessons.

In this paper, we do not take into account all aspects of our research: our specific aim is only to demonstrate how some theoretical categories can be used in the description of teaching processes.

2.2. *The theoretical framework*

To analyze the teacher's action, we need to describe this action by using categories specific to didactic interaction.

2.2.1. *The didactic relationship*

The didactic relationship is a ternary relation between the teacher, the students, and the pieces of knowledge at stake. We assume that the teacher's work consists in initiating, establishing, and monitoring this relationship. Using the concept of didactic relationship is a way to emphasize the communicative nature of teaching techniques, and to focus on the fact that the core of the relationship between the teacher and the students is their sharing of this piece of knowledge. We consider that the didactic relationship is fundamentally threefold: understanding between the teacher and his students thus implies not only analyzing their respective positions but, especially, taking into account the knowledge that will be the focus of the lesson.

2.2.2. *The adidactic situation*

An adidactic situation is a learning environment designed by the teacher. We need two criteria to design and to understand such a situation.

First, the student should not be aware of the teacher's intentions about the knowledge underlying the situation.

Second, the student is engaged in a game, "this game being such that a given piece of knowledge will appear as *the* means of producing winning strategies" (Brousseau, 1997, p. 7). "The set of constraints and resources available in this game (situation) allows and directs students' adidactic action. This set is named *milieu*" (Brousseau, 1997, p. 248).

In our research, the milieu is made of the rules of the "Race to 20", which is an adidactic situation. The students' interactions with the milieu are supposed to be sufficiently "significant and adequate" to enable them, case by case, to gain knowledge, to formulate strategies of action or validate their understandings (second criterion). When students use the feedback coming from these milieus, their activity is not influenced by the necessity to satisfy the assumed expectations of the teacher (first criterion). One must note that the milieu, as a "set of constraints and resources" includes material objects (e.g. the writings on the board or the students' notebooks) as well as symbolic objects (e.g. the rules of the game, and also the successive "theorems" produced by the students).

2.2.3. *The didactic contract*

The teacher has to monitor the students' activity and the associated learning, by handling the evolution of the situations and of their milieus. By doing so, he defines the didactic contract that governs the didactic relationship and defines the conditions of its existence.

Depending on the point of view one adopts, the didactic contract appears to the observer as a set of reciprocal expectations between the teacher and the students. As Brousseau (1997, p. 54–58, 225) says: "[the] (specific) habits of the teacher are expected by the students and the behaviour of the student is expected by the teacher; this is the didactic contract." These expectations can be viewed as a set of largely implicit rules, of usual ways of acting (with regard to the subject being studied) that the teacher and the students find suitable in the context of the didactic relationship. Some of these habits are perennial (Mercier, 1988), and we consider these to be the basis of the didactic relationship. Others are specific to the concept being taught, and therefore depend on the evolution of the milieu during the lesson.

2.2.4. *Specific and generic techniques*

To understand the teacher's action, we have to describe the techniques that he produces. Some of these techniques are specific to each piece of knowledge. In the "Race to 20", for example, the teacher may involve the students in demonstrating that "saying 17" is a "winning theorem", or that the "Race to 20" is a "Race to 17". For doing so, he needs to interact with the students in a mathematical way, very specific to the "Race to 20" knowledge. His behavior would be "mathematically" different if the content of the lesson were a geometrical piece of knowledge.

On the other hand, we also postulate that the teacher has to use generic techniques: for example, at any time in the learning process, for any piece of knowledge, the students have to get involved in the tasks, they have to memorize the essential features of the pieces of knowledge being taught. By doing that, they have to assume the didactic contract. Some of the means that the teacher uses for that purpose are not specific at all, but pertain to the general teaching-learning process. Thus we argue that the teacher, in a constant dialectic, calls upon teaching techniques that are specific to the material being taught as well as upon generic educational techniques. In the Section 3.3 of this paper, we will show how the relations between these two types of techniques can be analyzed.

2.2.5. *Mesogenesis, topogenesis, chronogenesis*

In a broader view of the theory of didactic transposition (Chevallard, 1991, 1992; Mercier, 2002) we consider a triple dimension that describes the teacher's work, relative to starting and maintaining a didactic relationship (Sensevy et al., 2000).

Mesogenesis describes the process by which the teacher organizes a milieu, with which the students are intended to interact in order to learn. Without a specific mesogenesis action from the teacher, students play the "Race to 20" without studying any winning strategy. For example, when the teacher asks the students to write down in a table the process of their games, he brings some new constraints and resources into the learning environment, so he creates a new milieu. This is a mesogenetic action.

Topogenesis describes the process of the division of the activity between the teacher and the students, according to their potentialities. The teacher should define and occupy a position, informing students of tasks which will allow them, in turn, to occupy their positions in the didactic space. For example, in the beginning of the "Race to 20", the teacher himself may play against the students, or act as a referee, or simply observe the game. These different techniques are three different ways of dividing the didactic space in function of what each participant is supposed to know and, therefore, do. They refer to three different topogeneses. In the same vein, if a student claims that "saying 14" is a winning theorem (the student may say "it's a good number") the teacher may assume a high status position in validating this proposition (that means : "I know you are right and it is my task to tell you you are right"), or he may keep a low profile, asking the other students to react ("I know you are right, but the others have to acknowledge it, so it is not my task to validate now"). This is a topogenetic choice.

Chronogenesis describes the evolution of the knowledge proposed by the teacher and studied by the students. This progression produces, for the teacher and the students alike, a temporality that is unique to learning institutions, and that we define as the didactic time. The teacher has to monitor the knowledge process through a lesson or several lessons, in order to meet his didactical intentions. For example, in the "Race to 20", if a student claims that "saying 14" is a winning theorem, the teacher can decide not to take up this proposition, because this argument is brought too soon with respect to the "cognitive state" that the teacher infers from the other students' work. This is a chronogenetic action.

Our aim is to connect the categories we have presented in this theoretical framework, and, by doing so, to enhance their relevance in the description of the teaching-learning process. This vocabulary will be used in the analyses that follow in Section 3 of this paper. In the last part of this article, we will focus on what we consider as the background of these techniques. This means that we will try to identify, in the teachers' discourse about their teaching actions, some beliefs and values which can explain the didactic patterns they use.

2.3. *The methodology of the analysis*

In the following parts, we will describe the teacher's action using three levels of analysis:

- the interaction of mesogenesis, topogenesis and chronogenesis (3.1)
- the relationship between contract and milieu in learning games (3.2)
- the teacher's beliefs and usual ways (3.3).

In these sections, we will specify three types of monitoring the teacher can develop, differing from each other, in particular, by the scale appropriate to their description.

In the first type of monitoring (3.1), we consider that to get the students to learn, the teacher must constantly move the knowledge forward producing some didactical time (a chronogenetic constraint), then ensure a sharing of the tasks between those the teacher is responsible for and those devolved to the students (a topogenetic constraint) and manage the class's relationship with a material or cognitive milieu (a mesogenetic constraint). This triple constraint is inherent to the learning games, which the teacher must define and monitor as learning situations.

This first level of analysis is most of the time based on brief interactions, made of only a few speech-turns.

A second type of monitoring is identifiable (3.2), thanks to a medium-scaled analysis, when the teacher brings about the evolution of the learning game, as knowledge advances. By making the students connect with it, the teacher moves the didactic interaction to another goal, by another stake to the game, and thus creates another milieu and contract. In order to understand the teacher's action, we have to describe the way different learning games follow one another throughout a lesson.

This second level of analysis is grounded in interactions longer than the first one, in order to identify different goals and game structures in the learning–teaching process.

To these two types of monitoring we must add the system of beliefs among the determinants of the teacher's action. As pragmatists do, we consider beliefs as habits of action. However, if we want to identify some of the teachers' "teaching beliefs", and behaviors these could entail, we have to stop describing classroom interactions, and focus on other levels of analysis.

Thus, this third level of analysis is mainly based on interviews with the teachers interviews, and on the analyses they provide of their own actions.

3. A COMPARATIVE DESCRIPTION OF THE TWO TEACHERS' ACTIONS

For each teacher (T1 and T2) we will provide both a breakdown and a summary of the information concerning the lesson they delivered in their first classes. Then, we will focus on an episode that came from these breakdowns, and we will show examples of mesogenetic, chronogenetic and topogenetic techniques. The comparison of two "modes" of action will show how the model of the interaction between the teacher and the students is used.

3.1. *Analysis level 1: The interaction of mesogenesis, chronogenesis and topogenesis*

3.1.1. *Lesson 1 of T1*

In Table I, the bordered sections are those that will be studied further. The use of italics indicates what the teacher wrote on the board.

This lesson lasts 60 minutes, out of which 32 consist of trio or team work. First of all, the teacher asks the students to imagine what kind of game "The Race to 20" could be. Then he introduces the rules of the game (minute 6). When the list of the winning numbers comes up on the board, the teacher maintains a doubt. Michaël's statement and demonstration of the technique are given neither attention nor comments: Does his proposal come too soon in T1's lesson plan ? It is worth noticing that the teacher does not take part in the game with his students, but instead chooses to be a referee. Thus, he does not follow Brousseau's original scenario in its first phase: "The teacher explains the rules of the game and starts playing a round at the chalk-board against one of the children, then relinquishes her place to a second child" (Brousseau, 1997, p. 3). In a more general way, during his lessons, T1's instructional device is far from Brousseau's engineering. Indeed, he creates a referee function assigned to a student in the individual one-against-one game.

3.1.1.1. *Mesogenetic techniques.* From a mesogenetic point of view, the period of the lesson highlighted by the border in Table I (minutes 22–28, lines 165–192) represents a major change in the milieu. Indeed, the students were first confronted with the title "The Race to 20" and asked to imagine what it could mean. Then, they were introduced to the rules of the game and played it in groups of three, one student against another under the supervision of a third student acting as a referee. Around line 165, however, T1 moves into another stage of activity, consisting in eliciting "comments" that could be formulated in terms of mathematical features related to the

TABLE I
Breakdown of lesson 1 of teacher T1

Time in minutes	Speech turns (SpT)	Didactic episodes	Work modes
0–6	1–46	Lesson starts. Class discusses possible meanings of the title "The Race to 20". T1 explains rules of the game.	Entire class
6–11	46–113	Trial game begins, involving various students. (T1 doesn't play.)	
11–15	113–123	T1 instructs students to play in groups of 3 (1 against 1 + 1 referee). T1 explains referee's role: "Ensure that the rules are followed and keep notes of the results, possibly adding comments. Then take a turn as a player."	
15–22	124–164	Game starts in groups of three. T1 walks up and down the aisles.	Students in groups of 3
22–28	165–192	After playing, the students remark "The one who says 17 wins." T1 writes on the board: When you get to 14 you're sure to win. S1: "11." S2: "Even at 8 and also at 5 and at 2." S3: "No, he said 2 and he lost." Michaël: "That's because he didn't use the right technique. I say 2, Cédric has to say 3 or 4. Me, 5. He has to say 6 or 7 and then I'll say 8. He will have to say 9 or 10, I will say 11. He will have to say 13 or 12, I will say 14. He will have to say 16 or 15, I will say 17 and then it's over: He'll have to say 19 or 18." T1 reinforces the idea that some students who said 2 lost the game "anyway" and concludes "It would seem that some numbers are more important than others." T1, addressing Michaël, says "That was pretty clear; you got there pretty quickly."	Entire class
28–40	193–209	T1 asks to play again, being himself the referee. Following a comment from a student, T1 writes on the board: *The one who starts, wins?*	Students in groups of 3
40–48	210–327	A group plays in front of the rest of the class. Students remark: S4: "It's cheating if the same person always starts the game."	Entire class

(*Continued on next page*)

TABLE I
(*Continued*)

Time in minutes	Speech turns (SpT)	Didactic episodes	Work modes
		S5: "With the right technique, the one who starts always wins." S6: "I started but my opponent won." T1 writes on the board: *17, 14, 11, 8, 5, 2* T1 asks: "Are these numbers important? Can we say that whoever starts wins?" Students: "No." T1: "Not all of us seem to agree."	
48–60	327–357	T1 asks to play 2 against 2 + 1 referee. T1 joins a group as referee. T1 asks: "They won: was it just luck?" T1 collects what the students have written.	In groups of 2 + 2 + referee

"Race to 20." Here, the teacher takes on the work of formulation. What is at stake is no longer the game itself but the way it is played. Nonetheless, T1 remains open to the various comments the students make.

T1, minutes 22–28

165.	T1	"Did anybody write down any comments?"
166.	Student	"We had to restart the second round because the referee whispered something."
167.	T1	"So that's rather surprising, because the referee has to make sure that the rules of the game are respected, and yet it's the referee who's whispering. So it was necessary to have a referee, but there, the referee plays an unusual role, huh?"
168.	Student	"We noticed that if someone says 17, the other player can't win. The one who says 17 wins."
169.	T1	"The one who says 17 wins. Did anyone else notice anything like that?"
170.	Student	"We noticed the same thing. At one point, Arnaud didn't win because we had the same technique and we used it in each turn. When I played with Benjamin, it was sort of random, so we thought about each number, but since Arnaud didn't have the technique... "
171.	T1	"Is it a technique?"

Since the conjecture "17 wins" emerged rapidly, T1 decided not to evaluate it but to focus on empirical observations made by other students (speech turn 169). Some students indeed noticed "17 wins". Others put forward remarks on broken rules (e.g. adding 3 instead of 2 or 1) or on the fact that the games were very short. When T1 wrote "17" circled on the board, thus recalling the milieu represented by the object "17 wins", the students pointed out in only three speech turns that other numeric values such as 11, 14, 8, 5, and 2, also win. That "2 wins" is questioned by Quentin, who produces a practical argument.

180.	T1	"He added 3 also. Well. The rules, we'll continue to have a referee, because while playing sometimes we forget what's going on. You made a remark about the game itself,. . . that apparently 17, ... P. writes a circled 17 on the right-hand side of board. Yes, Perrine? You wanted to add. . . ?"
181.	Student	"There is also 11 and 14. When you get to 11, not when you get to 14, you're sure to win."
182.	T1	"When you get to 14, you're sure to win."
183.	Student	"Yes, but that's it. Even at 8."
184.	T1	"8 also."
185.	Student	"And also 5. Also 5 and 2."
186.	T1	"Quentin?"
187.	Quentin	"No, because I played against Hugo. He said 2 first and he lost."
188.	T1	"Ah"
189.	Michaël	"That's because he didn't use the right technique. I say 2, Cédric has to say 3 or 4. Me, 5. He has to say 6 or 7 and then I'll say 8. He will have to say 9 or 10, I will say 11. He will have to say 13 or 12, I will say 14. He will have to say 16 or 15, I will say 17 and then it's over: He'll have to say 19 or 18."
190.	T1	"Apparently, he played a round, he said 2, and he lost anyway. And if three of you speak at the same time, we'll have trouble hearing you. Right?"
191.	Student	"That's also why, there are lots of numbers that allow you to win. Not right away."
192.	T1	"Well, it seems there are some numbers that are a little more important than others, let's say. It's a difficult idea to express. That was pretty clear; you got there pretty quickly. Let's play again. But this time, the referee will be more of a secretary, taking notes on the games. The secretary will write down the numbers played. Since you've noticed some things, we can discuss them afterwards. There is the "Comments" section of your papers that you can use. Look at the paper together before starting. Keep it for now; you might need it. . . ." The students start a new game-playing phase.

T1's "Ah!" (188) could be the sign of the beginning of a debate on this question. But Michaël then proceeds (189) to outline the steps to take and explain why they're necessary. This object is apparently perceived by T1 to be "too important" to be the first formulation (see the chronogenetic dimensions, discussed next). So T1 reintroduces the debate started by Quentin and suggests to go back to the empirical milieu (190). He thus challenges Michaël's proposal. However, he asks for a referee to note down

TABLE II
Breakdown of lesson 1, teacher T2

Time in minutes	Speech turns	Didactic episodes	Work modes
0–4	1–23	T2 organizes the lesson. Lesson starts. T2 explains rules of the game and asks students to repeat the instructions.	Entire class
4–6	24–56	T2 plays a trial game with a student. T2 starts and says "1"; another student is called to take T2's place. T2: "You get the idea?"	
6–8	56–60	T2 organizes and instructs students to play in pairs, 1 against 1. T2 hands out the worksheets: T2: "You will note down who started... what numbers were played at each turn." T2 writes on the board: *How can I play well?* *What do I have to do to win?* T2: "You have seven minutes."	
8–17.30	60–62	game in pairs; T2 goes up and down the rows. T2: "Play as many games as possible."	In pairs, 1 against 1
17.30–19	62–82	Students remark: S1: "You have to get to 17 because then you're sure to win." S2: "You have to be careful." S3: "I do plus 2 plus 1 plus 2 plus 1." T2: "Put your worksheets aside for the moment."	Entire class
19–24	82–99	T2 organizes a game by creating a purple team and an orange team. T2 reminds students that team members need to advise their players.	
24–49	99–305	Members of the two teams play 12 games, and everyone has a turn. T2 reminds students that for each round the teams need to work together and advise their players.	
49–1h–10	305–448	T2: "We're going to think about the discoveries you've made." Students remark: S4: "One even number one odd number." T2: "How should I write that?" S5: "You have to try to get to 14." T2: "The one who says 17 is sure to win?" T2 writes what he hears on the board.	

(*Continued on next page*)

TABLE II
(*Contined*)

Time in minutes	Speech turns	Didactic episodes	Work modes
		T2: "The team that puts forward a decision everyone agrees on wins a point . . . but you have to prove it." S6: "You have to get 14." S7: "No, 15." T2: "Do we need to play a round for proof?" Round between 2 students: S8: "At 15 you're sure to lose" S9: "If you want to say 11 you have to say 8." S10: "If you want to say 8 you have to say 5." S11: "And if you want to say 5 you have to say 2." T2: "Let's verify this; who wants to play?" S12: "If you know the techniques and you start with 2 you've won." T2: "We'll write that in our notebooks." On the board: *17, 14, 11, 8, 5, 2*	

the rounds and he draws the students' attention to the "Comments" section of their papers (192): The game-playing action is thus coupled with formulation within the groups. This seems a relevant example of how mesogenetics techniques (e.g. introducing a new writing rule in the students' activity) allow the teacher to make positive "integrating choices" and negative "ignoring choices" among the students' proposals, with regard to the particular set of material and symbolic objects (the "milieu") he wants to set up.

3.1.1.2. *Chronogenetic techniques.* The same episode enables us to describe the teacher's way of managing the didactic time during such lessons. Between speech turns 182 and 186, time could suddenly have accelerated. If the teacher had instituted the proposals of the students who produced the "winning series" straight away, it would have been detrimental to the class's "cognitive state", because the key to the series is the theorem "17 wins", which had not yet been demonstrated. Introducing Quentin's contestation into the discussion enables the teacher to slow down the didactic time. T1 will ignore Michaël's contribution (192), cast doubt on it and maintain a high level of uncertainty in the class. However, the idea will be reused when "Michaël's technique" will provide the teacher with a relevant argument, at the right time, i.e. when the class is mature enough to take it. This is a good example of how a chronogenetic technique rests on the teacher, both with regard to the knowledge (he decides which proposals are interesting) and to the students (he decides whom to let speak according to the proposals he wishes to emphasize). The teacher's position is high enough to ensure he maintains control of the pace of the lesson.

3.1.1.3. *Topogenetic techniques.* This excerpt highlights the fact that T1 expresses his opinion rather indirectly: "It seems there are some numbers that are a little more important than others. . . ." The teacher does not want to emphasize the "good answers" when they are coming too soon. We are faced with a classical problem that is at the heart of a large number of research situations. The teacher wants to make the students aware of a type of knowledge (here the dialectic between mathematical necessity and the effective steps to meet this necessity), but he wants to do so without "giving the answer straight away". He therefore pretends ignorance, which can be perceived by the students as the opportunity to explore the cognitive environment (Greeno, 1991, 1994), without restricting themselves to a particular direction too quickly. This attitude will remain constant throughout the two lessons and may correspond to a didactic style that the students are used to.

In the students' and the teacher's comments we see Wittgenstein's distinction between a constraining process (terms relating, according to him, to physical determinisms) and a directing process characterizing, in particular, mathematics, in which one can always do something other than what is prescribed by rule. We can also note that when T1 speaks to Michaël to gently rebuff his argument, evoking time ("you got there rather quickly"), he undoubtedly means "too quickly for the rest of the class and therefore for the good development of the activity."

3.1.2. *Lesson 1 of T2*

During the 1 hour and 10 minutes first lesson T2 devotes 9.5 minutes to one-on-one work and the rest to group work. The stakes of the one-against-one game are announced and written on the board: From the start the students know that they have not only to win but also to explain how they did it. Very little time is devoted to observations about this one-against-one phase of the game, and, during the group game between the orange and purple teams, the team members are told they are expected to advise their players properly. This phase of the game, in teams, is the longest (25 minutes) and provides an opportunity for all the students to participate (in 12 rounds). The concluding phase (21 minutes) is also rather long: it is a time during which T2 emphasizes the students' "discoveries" and suggests them to "think about them." The conclusion is clearly to be instituted since T2 proposes that students write the final observation in their notebooks. Contrary to T1,

T2' choices are very close to Brousseau's engineering (introduction, game with two groups, game of discovery).

T2, minutes 49–1 hour 10 (bordered part of Table II)

308.	T2	"(...) What have you discovered? What enabled you to win? So, we're going to write down your suggestions here. So you'll give me your proposals–proposal of discoveries. Here we'll check that? it's true–thus, a discovery accepted and verified. So does the purple team have anything to say? Who wants to start? It doesn't matter; so, Joris."
309.	Joris	"You can give one even number and one odd number, so for example, 2, the other says another number, so even, odd, even, odd."
310.	T2	"How should I write that? You noticed it, so are you sure for now? What do you have to do to win? What you've discovered, you've discovered something specific, a long while ago, Maxime?"
311.	Maxime	"Try to have certain numbers."
312.	T2	"So what did you discover?"
313.	Maxime	"You have to try to get to 14."
314.	T2	"The number 14, you say, so what about the number 14?"
315.	Maxime	"With 14 or 17 they don't have the option of saying 2 or 1, the opponent is bound to win."
316.	T2	"So you win if, how am I going to write this?"
317.	Maxime	"If you get to a certain number."
318.	T2	"But which?"
319.	Maxime	"14."
320.	T2	"If you, if you say what? 17, what happens, you're sure to win? OK? If you say 17 you're sure to win. That's one proposal. Is it accepted by the orange team?"
321.	Orange team	"Yes."
322.	T2	"When a decision is declared to be accepted, you win a point, that's how we'll do it."
323.	Student	"Who won a point there?"
324.	T2	"The purple team."
325.	Students	"Oh no!!!"
326.	T2	"But then it'll be your turn, you may have discovered something else; if someone gives a false response and you show that it's wrong you get 3 points."
327.	Student	"3, oh yeah, that's OK!"
328.	T2	"You have to prove it, of course, you have to prove it! It's the orange team's turn now."
329.	Laura	"OK, when you get to 17 you win, but to get to 17 there has to be certain numbers beforehand."
330.	T2	"Does the orange team have any ideas about that; I'll only write it down when you agree. Go ahead, Laura."
331-332		inaudible
333	Laura	"To get to 17 you have to have 14."

334	T2	"If. . . ."
335	Laura	"If you want to get to 17."
336	T2	"If you want to say 17, is that it? 17"
337	Laura	"You first have to have 14."
338	Fanny	"No, it could be 15 !!!"
339	T2	"You have to say, Laura?"
340	Laura	"14."
341	T2	"You have to say 14. OK, we'll try to check that."
342	Orange team	"Or 15."
343	T2	"Make up your mind!"
344	Orange team	"No, we say 14."
345	T2	"Do you agree or do we have to play a round to prove it?"
346	Purple team	"It's OK!"
347	T2	"Who has another opinion? (. . .) If you want to say 17 you have to say 14, do you agree?"
348.	Student	"Yes, but not too much, because you can say 15."
349.	T2	"You can say 15, so your counterproposal is that you can say 15."
350.	Student	"But you can also say 14."
351.	Student	"You can say 14 or 15!!"
352.	T2	"OK, we're going to play a round, let's start. Let's have Laura and Fanny, and we'll look for the answer together."
. . . /. . .		
362.	T2	"You said you could say 15, can you say 15 here? You said earlier that you could say 15, you're the one who said it, huh! So are you keeping 14 or will you say 15?"
363-367.	Students	(the students say in turn) "16-17-18-20"
. . . /. . .		
368.	T2	"You said you could put 15 there. Is that proposal validated here?"
369.	Student	"Yes. . . "
. . . /. . .		
441.	T2	"(. . .) so who wants to tell me what you've just learned here? What do you need to do to win the "Race to 20"? Hurry up. Amélie, you want to tell us what you've just discovered?
442.	Amélie	"The first condition is to say 2.
443.	T2	"Is it enough to say 2? What's the technique? Go ahead, Amélie. Who wants to help her? Choose someone to help you.
444.	Amélie	"Jérôme."
445.	Jérôme	"If you say 2 and you know the techniques, when the opponent answers you can say 5, after you can say 8, you can say 11, you can say 14, you can say 17 and you can say 20."
446.	T2	"And you will have?"
447.	Student	"Won!!"
448.	T2	"We'll write that down in our notebooks, I think it's right, thanks. Class is over for today."

At the 49th minute, the students were confronted with a new milieu that they had been warned of, on the basis of comments they made while playing the game.

Indeed (see Table II), the students were first given the rules of the game, then they played a trial game, and they ended up playing in pairs (1 on 1). By doing so, they encountered a milieu dedicated to action (according to Brousseau's theory), while knowing that the goal was to formulate conjectures on "how to win". The debate on discoveries was at first very brief (speech turns 62–82) and then led to a game in two teams. At this point, the first milieu dedicated to action turned to a milieu dedicated to formulation. Indeed, within the team, a strategy has to be formulated and communicated to the player representing each team.

3.1.2.1. *Mesogenetic techniques.* In the phase devoted to the debate on conjectures (game of discovery, in Brousseau's engineering), T2 dismisses the proposal concerning even and odd numbers.[5] But the decision in favour of "14 wins" or "15 wins" is not obvious because the reasoning was not applied to the case of 17, which led T2 to reintroduce the idea (320).[6]

Then, nonmathematical features, which nonetheless constitute the groundings of the didactic situation, appear in the milieu. The team that makes a proposal accepted by the other team wins one point (322) while a team succeeding in proving that the other team's proposal is false, wins 3 points (326). But it did not work. T2 (352) goes back to the milieu dedicated to action (the one-against-one game), as T1 did. T2 did not accept Laura's statement, he called it an opinion (347); the same goes for Maxime (316). Eventually, the class seems to agree that the winning discoveries in the last game have to be the winning strategies of the first game (the one-against-one "Race to 20" game corresponding to the milieu dedicated to action). The "validated" milieu thus includes the entire numerical series as well as the principle of "who starts" in conjunction with the possibility of saying 2 and necessarily winning. To end up with the lesson, T2 "sums up" the milieu, to some extent, by asking students to reiterate "what they have to do in order to win the "Race to 20": Once the discovery has been written on the board, it can be put in the students' notebooks (448) and, thus, instituted. During this phase, the teacher and the students co-elaborate new symbolic objects (discovery, validation. . . , most of the time without naming these objects) and new material objects (e.g. the writings on the board). The mesogenetic techniques enable the teacher to introduce such objects.

3.1.2.2. *Chronogenetic techniques.* The milieu described earlier is built up through interactions between T2 and the students, but the organization of the objets emerging successively, relies mainly upon the teacher. T2 does not impose an unflagging pace. It is only at the end, during the final summary of discoveries, that T2 expresses a wish that the students state them "quickly", as a final verification and as a way to confirm what they have learned. We also notice T2's wish to end this session with this shared declaration attesting the achievement of a common knowledge during the session (437–448).

442.	Amélie	"The first condition is to say 2.
443.	T2	"Is it enough to say 2? What's the technique? Go ahead, Amélie. Who wants to help her? Choose someone to help you.

We can note that these T2's speech turns may allow for a chronogenetic description, because the teacher, in asking Amélie to go ahead, prepares the institutionalization, and speeds up the didactic time. But this action may also be characterized under a topogenetic description: this time regulation is possible only because the teacher assumes a high profile in the didactic relation.

3.1.2.3. *Topogenetic techniques.* In a similar way, T2 takes positions, makes choices, and does not hesitate to press students in order to move the situation forward. The chronogenetic techniques often need the teacher to be in high topogenetic positions, and reciprocally, the topogenetic techniques are linked to the pacing of the didactic time.

When Fanny questions the conjecture "If you want to play 17 you have to play 14" (338), T2 tries to press Laura for the "right" conjecture, but the members of the orange team do not agree at all. The choice of 14 or 15 as winning numbers continues to be debated (342 and 344). After a while, because the proposal is still being questioned, T2 finally decides to have two students to play another round, including Fanny. Playing first, Fanny finds herself in the position of choosing between saying 14 or 15. She says 14, a choice that T2 points out to her (362), reminding her of her previous argument and asking her to confirm her choice ("So, are you keeping 14 or will you say 15?"). After reminding Fanny's previous position for the last time, T2 validates the proposal that "14 wins."

This example shows a technique that seems to us to be central to the teacher's work: Teachers and students are engaged, in the beginning of the episode, in a learning "game" (as a situation on which the teacher plays) that supposes an evoked relation to the milieu. The situation consists in debating proposals, either in a logical manner or by referring to a certain part of the game that can serve as a milieu for argumentation. In such a situation, where the theorem cannot be stated and proved, in order to make progress, the teacher must return to the initial situation, asking the students to replay the game, so that they and the class as a whole, can be faced again with the practical difficulty. They play, but with the precise goal of proving a conjecture through the feedback provided by an "actual" relationship to the game. We can note, here, that this mesogenetic technique (generating a milieu), consists, in fact, in actually changing the type of playing: from then on, it becomes a learning situation.

3.2. *Analysis level 2: The relationship between contract and milieu*

The chronogenetic or topogenetic techniques that we have identified, even if they can be described on relatively large scales, were shown essentially in the midst of short interactions, and they only slightly modify the general framework of the didactic relationship. Mesogenetic techniques can also often be described in the context of a brief interaction, yet a modification in the milieu can sometimes change the nature of the interaction. This means that, in the context of an apparently stable situation with a single object, the teacher changes the stakes of the situation by proposing new forms of activity. For instance, in Brousseau's engineering, beyond the first phase (called the "dialectic of action") there comes a "dialectic of formulation" when the stake is the production and diffusion of winning strategies. Then, a "dialectic of validation" appears, when the proof of the efficiency of these strategies and the study of their consistency with the piece of knowledge already acquired and instituted are proposed to be at stake.

In the two lessons we studied, we can find such changes, in which the teacher attempts to establish a new learning game in order to move the didactic time forward.

T1, during the second lesson, for example, has the students replay rounds that they kept notes on from their first lesson, in front of the whole class, in order to study them publicly.

In this case the new learning game has a specific goal. The students have to evaluate the former rounds. By taking into account the acquired knowledge, they have to determine, for example, the error, which is supposed to lead to proving or disproving the conjectures.

This new learning game requires a particular milieu. The didactic setting is constituted by rounds replayed with new knowledge.

In addition, a didactic contract is linked to this goal and this milieu. For example, the students make an attempt at providing a critical evaluation of the moves from the previous rounds.

For the teacher, these are new ventures. It is no longer a question of acting on the local activity of the students but of modifying, in the didactic plan, the very form of their interactions with the game.[7]

With this example we can identify the main features of the link between contract and milieu. The milieu can be considered as a set of objects. The students' activity is focused on these particular objects that are the "old rounds" of which the students have kept a written record. These objects are conceptual (some "theorems") as well as material (the sheets of paper used to keep note of the rounds). But we can consider these objects in another way, if we focus on the rules of action one needs to follow in order to act in the classroom. For example, a round noted on the paper can be used as an object of reminding the students of a strategy. The same round can be used as an object of evaluation, thanks to which it is possible to identify "good" and "bad" moves. So, a certain set of objects can be viewed under two different descriptions. Under the "contract" description, one is focused on the rules of actions that objects elicit and the expectations that arise from these objects in a given situation. Under the "milieu" description, one is focused on the very objects which pertain to a given situation. But the two descriptions are completing each other.

To bring the contract-milieu relationship into focus, we can observe how the expectations of the teacher T2 in the round played by Laura and Fanny (330–349 in the excerpt quoted) are not the same as the ones he had had earlier in the lesson. This example shows a technique that seems to us to be crucial to the teacher's work. Teachers and students are engaged, at the beginning of the episode, in a learning game that consists in debating proposals, either in a logical way or by using examples. The previous rounds of the game can serve as a milieu for argumentation. But this choice has its limits, which this specifc episode enables us to notice. The teacher cannot stop using examples and counterexamples. He does not succeed in bringing out and devolving a dialectic of theoretical validation.[8] So he has to go back to the starting point, having students play a round in order for them, and the class as a whole, to be faced again with the practical difficulty. He is trying to produce the demonstration of a conjecture, but with the feedback of an "actual" relationship to the milieu, in an action situation centered on *playing the game*. This is what Brousseau calls "pragmatic proof". By changing the milieu (an actual relationship rather that an evoked relationship), the teacher changes the rules of action too. He wants the students to test the pragmatic consequences of their claims (for example, when Fanny says that she can win by playing 15). The expectations embedded in the "actual" milieu generate a new contract. The students know that when playing the game, the teacher expects them to give the proof of what they have declared.

3.3. *Analysis level 3: Beliefs and usual ways of the teachers*

The third level of our analysis is devoted to studying the teachers' beliefs. Here we will sketch out some of the characteristics of the universe of beliefs evoked by T1 and T2 during their self-analysis interviews. We will emphasize aspects that prove to be directly linked to certain actions that we have identified in the two preceding levels of analysis, to show how the didactic techniques are produced on a background of beliefs whose description belongs to anthropological studies. Thus, for each teacher, some beliefs directly organize the teaching process. This points out to the need in didactic analyses to demonstrate the permeability between content specific objects and more generic educational objects across the various levels in analyzing the didactic contract.

In his management of the "Race to 20", only T1 to used the technique which consisted in having the students imagine the goal and possible rules of a game called "Race to 20." This teacher is also set apart by having placed himself in the "lowest" topogenetic position possible and by his quasisystematic failure to institute any piece of knowledge. During the self-analysis, here is what the teacher said about his techniques. The significance he attributes to these techniques led us to believe they were a deliberate choice on his part:

(...) So from a word or an expression or a sentence, the setup, in fact, there, it was to try to find out how it could, what ideas that could give us about the game (...) I'm used to doing it that way, trying to establish a link between the greatest number of activities that apparently don't have anything to do with each other, so as to try to find out a coherence in particular areas, which is also at the origin of the idea of working from plans. It's true that in maths it's a little more difficult (...) The difficulty with regard to that, that's something that I frequently try to do because in my own experience, I, myself, didn't experience that coherence until very late in my studies and we worked in a 'compartmentalized' way, so I undoubtedly discovered a lot thanks to my profession. But the difficulty, in fact, we see that there are many students who aren't really in the game (...), who aren't interested in the questions. So... you can really feel there, that there's a group (...) indeed, there's a certain number of students who, for the moment, don't seem concerned, don't seem motivated. And the idea of questioning, it's also to motivate them a bit (...)

In this "explanation", we find the effect of an effort to convey coherence, rooted in the personal educational history of T1, who wishes to involve the students who are slow to get interested in the work. What this teacher says is a good demonstration of how difficult it is to associate general motivational processes with topogenetic techniques of devolution. It is likely that T1's choice of having a student act as a referee for each round of the game also comes from his wish to involve the students:

(...)It was... so there was observation, in fact, by the referee, especially with regard to the rules of the game. So, indeed, I had to check that, yes, they really added 1 or 2 because that (...) so we had to understand each other, so I did not want the situation to get out of hand, we had to avoid misunderstandings because it went too fast, for example. (...) We do that often. For assessments, for example, they're often associated with... either when we're doing some research on something... (*gives an example in Physical Education*) (...)

This type of thinking undoubtedly played a major role in the use of the arbitration device, since it is a frequently used organizing principle to which one can certainly attribute, beyond its particularities, a function of coherence in the varied learning of elementary education. Arbitration thus serves a double function of linking present activity to the past activities and of distancing the students from their activity through taking on "different roles." One can say that T1 seeks to involve the students in the management of the educational relationship by attributing to them circumstantial roles that allow him to balance the interaction between people: "(...) I'm certainly thinking about the taking of power, let's say of the teacher, I really want to avoid that, undoubtedly because of a past experience." We note that this seeking of symmetry is thought to be, fundamentally, a distinctive feature of all relationships. It is not due to an interaction between agents, named by an institution, which interaction would be mediated by the sharing of knowledge that one (the agent in the position of teacher) has before the other (the students in the position of those taught). This view is consistent with the low profile systematically kept by T1 throughout the lesson:

(...) It's something that I do, to play the innocent or I pretend I do not understand (...) Maybe I tend to do this too much and too often, I don't know, I like doing it (...) It sometimes causes trouble... I have students who wait for me outside the classroom (...) who don't understand the game, who are going to say to themselves, 'The teacher messed it up again'. (...) It's a big risk I take, to play with them and say stupid things and pretend not to understand or to make a student say something that seems obvious (...)

The coherence in the system of beliefs is clearly stated, including its "risky" effects. Familiar with this type of thinking, the reader will not be surprised to know that T1 "[doesn't] much like the competition aspect" and that from there comes the fact that consequently he "didn't focus much on who won." We feel that the teacher's explanations can be interpreted as follows. One of T1's fundamental beliefs lies in the need to "try to establish a link between a maximum of activities that don't, on the face of it, have anything to do with each other, so as to try and find a coherence between particular areas." From this perspective, we can interpret a large number of this teacher's techniques: We see that the technique of drawing out questions from the students at the beginning of the lesson is what we can call a devolution premise that goes beyond "the bare minimum" of all didactic interaction, in an attempt to attain larger and more ambitious objectives for the rational appropriation of knowledge. Undoubtedly this focus has biographical origins that we would need to look into more deeply. Here, the teacher exposes a general view of the use of a topogenetic technique ("playing the innocent") about which he points out its communicational complexity. We see how this technique can create didactic necessities, in which space must be given to the students in order for the didactic time to move forward as well as to the cognitive values of the teacher, for whom affecting ignorance constitutes a way of separating knowledge from the effects of imposition. Some didactic techniques (that we note in Sections 3.1 and 3.2) are truly rooted in educational beliefs.

The words of T2 are less characterized by traces of his "professional philosophy" to some extend, but his comments—made while watching the videorecording of the lesson—on certain topogenetic or chronogenetic actions revealed a lot about his didactic and educational beliefs. In relation to the beginning of the activity, T2 says, significantly, "(...) There I am, waiting for everyone's attention, and that's an instituted code. After a while everyone knows that you have to, you have to prepare yourself (...) It's something I should probably do instinctively but it is quite frequent (...)." With regard to the organization of the class into working groups, the teacher specifies "(...)I make it so that in each group there is a dynamic element, a leader who will carry the group". The teacher thus secures a possible didactic tool for moving forward the lesson by differentiating the students' positions. In particular, with regard to letting the class speak up, T2 explains how the spokespeople are selected:

(...)I must have intended to choose a certain student so I think I let the others raise their hands and have the intention of speaking (...) I try to make them aware that when you raise your hand you've formulated your words in your head and you intend to communicate them (...) You see one hand, two, three, and little by little you do see that the students need some time (...)

Among the practices declared as "frequent" during this lesson, there is also the return to the instructions written on the board "so that those who don't remember after a while can refer to it if they're really autonomous and are used to doing so (...) I prefer to ensure that it's very visible and that it's a point of reference for the kids (...)." In the same monitoring spirit, in devolution of organizational habits, T2 says "(...) I also always give them a specific time during which they can try to achieve the objective so that they learn to manage their own time," and he adds "In groups, in our ritual, there is always someone who watches the time." We see that the words of T2, as opposed to those of T1, remain directly linked to the "Race to 20", while simultaneously revealing methods, and thus traces of a teaching style, inherent to the didactic contract established in the class.

(...) something I do a lot; that is to say, I let them answer, I write everything on the board, and then... So, everyone has a turn... There's no direct validation, you see. It's afterward that we take a second look at things and give counterexamples, and all the students get a chance to talk. . . , the kids who ask questions and then, in general, it's always the same ones who ask the really interesting questions (. . .)

The implication of all students in the lesson also emerges in the organization of enough games at the board to enable every student to represent his or her team ("My concern was that each student should come and represent the team).

4. CONCLUSION

This paper is an attempt to take into account the complexity of the teacher's action. In order to do that, we have tried to characterize this action under three different descriptions.

The first and second type of descriptions mainly allow us to determine the effects of the didactic constraints on the teacher's behavior in the classroom. These two types of descriptions are only differentiated by the scale of the analysis. The third type of description consists in bringing the educational background of these techniques into focus.

It is crucial to point out that these three levels are interconnected. We think that the quality of description of the teacher's action depends on how we manage to show the three levels interweaving.

Let us try to manage such a threefold description on an example.

As we saw above, T1, during his first lesson, has to deal with an early declaration of a student (Michaël, in 189).

182.	T1	"When you get to 14, you're sure to win."
183.	Student	"Yes, but that's it. Even at 8."
184.	T1	"8 also."
185.	Student	"And also 5. Also 5 and 2."
186.	T1	"Quentin?"
187.	Quentin	"No, because I played against Hugo. He said 2 first and he lost."
188.	T1	"Ah"
189.	Michaël	"That's because he didn't use the right technique. If I say 2, Cédric has to say 3 or 4. Me, 5. He has to say 6 or 7 and then I'll say 8. He will have to say 9 or 10, I will say 11. He will have to say 13 or 12, I will say 14. He will have to say 16 or 15, I will say 17 and then it's over: He'll have to say 19 or 18."
190.	T1	"Apparently, he played a game, he said 2, and he lost anyway. And if the three of you speak at the same time, we'll have trouble hearing you. Right?"

At the first level, the sharpest, we can point out how the three meso, chrono-, topo- genesis techniques work together to sow a doubt in the class.

When T1 interacts through speech turn 190, there are already many objets brought into the milieu by the students. Then, the mesogenetic technique consists in selecting and indicating one object (put forward by Quentin) in order to enlarge the milieu, dangerously shrunk by Michaël's proposal. By choosing Quentin's objection, the teacher emphasizes a contradictory proposal and behaves as if he didn't know more than the students.

Therefore, at this moment, he positions himself within the same didactical space as the students. This is a topogenetic technique, that enables doubt to be sustained.

Furthermore, the teacher uses a chronogenetic technique, intended to slow down the didactical time by jumping backward on a previous proposal (187), prior to *Michaël's* one. Through this example, the narrow intertwining of the first level techniques is revealed.

Enlarging the description scale from minutes 15 to 28, at the second level of description, we understand that the increased uncertainty in the class leads to a modification of the learning game. After the speech turn 191, continuing the same discussion could prove counterproductive. Indeed, there would be no means to validate or invalidate the proposals. So, a new round has to be suggested in order to reduce uncertainty (in which "The secretary will write down the numbers played, since you've noticed some things, we can discuss them afterward"; see above 192). This turns out to be a new game with slightly modified rules. Therefore it can be described as a new milieu and a new contract. In this type of description, we can consider the teacher's utterance as a means to create uncertainty, and by doing so, to create the necessity to reduce this uncertainty by introducing a new learning game.

One should note that these two types of descriptions are "effect descriptions". In describing the teacher's action with a peculiar vocabulary (mesogenesis, chronogenesis, topogenesis, milieu, contract...) we show how the teacher's action is constrained by the knowledge. The linguistic form of these descriptions could be generally paraphrased as follows: in order to teach pieces of knowledge, the teacher has to respect didactic constraints, and make use of related techniques.

At the third level of description, the teacher's utterance appears to be based on a general belief. As we saw above, for T1, making use of the "playing the innocent" technique is in harmony with one of his educational aims. It seems to be crucial, to T1, to give some didactic space to the students to avoid effects of authority. Hence, this third type of description is no longer an "effect-description": the teacher's behavior is studied in the frame of more general habits, which are not directly produced by the didactic constraints. It does not mean that these general habits are disconnected from the "didactic" ones. One could suppose, for example, that the "playing the innocent" technique is an important one for this teacher because it has both didactic effects and educational (in a broad sense) effects. The relevance of the techniques should be evaluated under these two ranges of descriptions.

According to us, this example seems to constitute a fundamental dialectic of the teacher's work. One part of the determinants of the teacher's action is to be found in the inherent constraints of the structures of the didactic relationship; another part is identified in the beliefs developed through the contact with these constraints, beliefs converted into habits of action, into pragmatic matrices, like the "playing the innocent" technique. These two systems of determination are in constant interaction: habits of action are continuously redefined under the constraints of the didactic processes, which can themselves be displaced as the action unfolds.

The attempt to produce such threefold descriptions has some methodological consequences.

First, the researchers have to collect data both on the teaching process itself and about the teacher's beliefs. In order to do that, the self-analysis, by the teacher, of his performance is very useful.

Secondly, as we did in this research, it seems relevant to elaborate methodological devices in which both the experimental and the clinical dimensions are included. We conjecture that the "Race to 20" experimentation elicited some essential techniques that it would be more difficult to isolate in ordinary lessons. The "Race to 20" seems to be relevant in showing a range of techniques linked with the moves from one phase to another ("action" to "formulation" then to "validation") through the appropriate games. At the same time, it seems to us that the teaching conditions were ordinary enough to preserve the "ecological validity" of our findings. However, this will have to be confirmed through further work, on other mathematical situations, involving different types of knowledge.

The research described in this article is furthered in the following two directions which seem very important for us to get a deeper understanding of the dynamics of the teaching-learning process:

- Enabling us to apply the same system of categories to the description of an ever-growing number of situations. By doing so, one can hope to make useful comparisons between the teachers' techniques in almost every ordinary situation, the components of which are determined by an a priori analysis.
- Developing a teacher training unit about the teacher's action in the so-called "investigative activities" in mathematics classes. It is a question of documenting the extremely technical nature of the actions a teacher must undertake in an adidactic situation. Our aim is to create thus the

conditions to transform general teaching techniques (so often remarkably well mastered by teachers) into specific mathematics teaching techniques and, in doing so, to grant them their true scope.

NOTES

1. The gender of the teacher was not a variable in our research and we will not disclose it here. We will use the generic masculine pronouns instead of compounds such as "s/he" merely for the sake of alleviating the text.
2. The player who says 17 can control the progress of the game, saying 2 if the opponent says 1 and 1 if the opponent says 2. So a turn in the game makes 3.
3. The game "Race to A" is therefore equivalent to the game "Race to R".
4. Most teachers who tried the situation never connected it with the Euclidean division during the teaching process. It requires to then play the Race to 30 by adding 1, 2 or 3 . . . etc. Even in this research, where the teachers were trained with the mathematical meaning of this situation, the Euclidean division did not emerge during the teaching process. The expected model is obtained from the perspective of equivalence of games conceived as mathematical structures, not from the perspective of finding a strategy of winning in real games of a certain type, which is most certainly the perspective of the player, and to a certain extent, of the teacher.
5. During the self-analysis interview, P2 will say about this "Yes, I eliminate that because (. . .) I think that it could have made us lose track of what we were observing".
6. In the post-lesson interview P2 clarifies, "She (a student) picks up on isolated pieces of information, without taking other information into account. . . ."
7. The effort fails, however. The teacher does not succeed in installing a new dialectic that will meet the requirements necessary to the evolution of the students' problem. At the end of the study, the students were nonetheless able to play a trial game (on the whole correctly) of the "Race to 30", with a common difference of 3.
8. The teacher has not instituted a strong demonstration of the proposal "17 wins against all defence." He could have reused this demonstration, which would have formed the basis of/for the notion of a "winning strategy."

REFERENCES

Brousseau, G.: 1997, *Theory of Didactical Situations in Mathematics*, Kluwer Academic Publishers, Dordrecht.

Chevallard, Y.: 1991, *La transposition didactique*, La Pensée Sauvage, Grenoble.

Chevallard, Y.: 1992, 'Fundamental concepts in didactics: perspectives provided by an anthropological approach', in R. Douady and A. Mercier (eds.), *Research in Didactique of Mathematics*, Selected Papers, Grenoble: La Pensée Sauvage, pp. 131–168.

Greeno, J.: 1991, 'Number sense as situated knowing in a conceptual domain', *Journal for Research in Mathematics Education* 22, 170–218.

Greeno, J.: 1994, 'Some further observations of the environment/model metaphor', *Journal for Research in Mathematics Education* 25, 94–99.

Leutenegger, F.: 1999, Construction d'une clinique pour le didactique. *Une étude des phénomènes temporels de l'enseignement. Recherches en didactique des mathématiques* 20(2), 209–250.

Mercier, A.: 1988, Didactical contract: permanent clauses and local breaches, *Proceedings of ICME 6*, Budapest.

Mercier, A., Sensevy, G. and Schubauer-Leoni, M-L.: 2000, 'How social interactions within a class depend on the teacher's assessment of the various pupils' mathematical capabilities, a case study', *Zentralblatt fur Didaktik der Mathematik, International Review of Mathematics Education* 32, 126–130.

Mercier, A.: 2002, 'La transposition des objets d'enseignement et la définition de l'espace didactique, en mathématiques', Note de synthèse, *Revue Française de pédagogie* 141, 135–171.

Mercier, A., Schubauer-Leoni, M-L. and Sensevy, G. (eds.): 2002, 'Vers une didactique comparée', *Revue Française de pédagogie* 141, 5–16.

Schubauer-Leoni, M.L. and Leutenegger, F.: 2002, Expliquer et comprendre dans une approche clinique/expérimentale du didactique ordinaire. In Leutenegger, F. and Saada-Robert, M. (eds.), *Expliquer et Comprendre en Sciences de l'Education*, Paris, Bruxelles: De Boeck, pp. 227–251.

Sensevy, G., Mercier, A. and Schubauer-Leoni, M-L.: 2000, 'Vers un modèle de l'action didactique du professeur. A propos de la Course à 20', *Recherches en Didactique des Mathématiques* 20, 263–304.

Sensevy, G., Mercier, A. and Schubauer-Leoni, M-L.: 2002, 'A model for examining teachers' didactic action in mathematics, the case of the game "Race to 20"', *Proceedings of European Research in Mathematics Education II*, Vol. II, Marienbad, February 24–27, 2001, 420–433.

GÉRARD SENSEVY
CREAD Rennes 2/IUFM de Bretagne
E-mail: gerard.sensevy@bretagne.iufm.fr

MARIA-LUISA SCHUBAUER-LEONI
FPSE Genève
E-mail: schubauer.leoni@bluewin.ch

ALAIN MERCIER
INRP
E-mail: mercier@inrp.fr

FLORENCE LIGOZAT
FPSE Genève
E-mail: florence.ligozat@pse.unige.ch

GÉRARD PERROT
IUFM de Bretagne
E-mail: mogeper@free.fr

TERESA ASSUDE

TIME MANAGEMENT IN THE WORK ECONOMY OF A CLASS, A CASE STUDY: INTEGRATION OF CABRI IN PRIMARY SCHOOL MATHEMATICS TEACHING

ABSTRACT. Time is a constraint but also a condition of operating within a didactic system. Don't we need to distinguish several kinds of times? Our study will focus on teachers' time management strategies. We will identify these strategies by taking into account two temporal dimensions – didactic time and time capital – and the rate at which the former advances relative to the latter, called the pace of an activity, a lesson or a teaching sequence. Those strategies have been identified in the specific context of the integration of the Cabri-geometry dynamic geometry software in the daily work of a French primary school.

KEY WORDS: Cabri, didactic time, geometry, pace, time capital, time management

"I have to cover the whole syllabus," "I haven't got the time to let pupils resolve the problems," "it takes too long," "I can't waste my time with this kind of activities"; such complaints are frequently heard from teachers. These complaints can even be considered as an argument against teaching based on problem solving because "it requires a lot of time and gets in the way of covering the prescribed material." The recurrent theme of "lack of time" points to time as one of the main problems in classroom management.

We will approach this problem by looking at strategies developed by teachers to manage the different kinds of time at work in a classroom. We will limit our study to some of these "times," which we will define in the theoretical part of the paper. Then we will analyze teachers' management strategies observed in a research on the integration of Cabri-geometry in a French primary school. This analysis will bring to light time management strategies, which may appear also in other situations. We postulate the generality of these strategies but do not provide evidence for it in this paper, which is focused on Cabri-geometry.

1. Theoretical Framework and the Problematique of Time in Teaching

The starting point of our study is the integration of new technologies ("ICT" in the following) in mathematics teaching at the primary school,

Educational Studies in Mathematics (2005) 59: 183–203
DOI: 10.1007/s10649-005-5888-0

particularly the integration of Cabri, a dynamic geometry software, in the geometry practices of students. One of the issues, which led to this research, is

> What are the conditions and constraints that make possible the integration of Cabri in teaching without creating a clash between tradition (i.e., what is usually done) and innovation?

One of the conditions, developed in a previous paper (Assude and Gélis, 2002), is to find the "right distance" between the old and the new as far as the functioning principles of the class, as well as the types of tasks and available techniques, are concerned. In the present paper, we want to stress other conditions, related to different kinds of time and teachers' management strategies in relation to these kinds of time.

The question of time management in integrating ICT is a key issue, which, so far, has not received sufficient attention in research. There are several reasons why this issue should be addressed. One of them has to do with the subject matter to be taught: How to divide it up over a period of time when using the ICT? This is not explained in syllabuses and can create a difficulty for teachers who start using them because they do not know for sure what to do and how to do it. Another reason why it can be hard on teachers concerns the time devoted to becoming familiar with the software and to learn how to operate it. Time has to be spent (sometimes too much) to familiarize students with the software before any time can be gained in the learning process. How to shorten the familiarization process yet give students sufficient control over the software (so that they do not feel awkward with it) to make its use more "economical"? Time is both a constraint and a condition in implementing this new way of doing mathematics, as it is in every case, which makes it a relevant variable when studying the functioning of didactic systems. As Lemke (2000) says: "Every human action, all human activity takes place on one or more characteristic timescales." What timescales do we need to identify to answer our questions?

The existence of different kinds of time in a classroom has been noticed and studied by several researchers. Chevallard and Mercier (1987) and Leutenegger (2000) showed how the didactic time is important in textualizing knowledge and regulating the didactic contract. Brousseau and Centeno (1991) and Matheron (2001) studied the importance of didactic memory to remember past events when pupils are learning something new. Following Varela's works (1999), Arzarello et al. (2002) distinguished two times: "physical time" or clock time, and "inner time," which emphasize actors' time, especially pupils' learning time. Pupils' learning time was also studied by means of didactic biographies by Mercier (1995), or other ways such as the "fractions diary" (Sensevy, 1996), or a multiple device that construes remembering as a process encompassing memory, attention and waiting (Assude and Paquelier, 2004). What kinds of time are we speaking about? Physical time, didactic time, teaching time and learning time. These kinds of time can be displayed on a smaller scale and then we could speak about kinds of time such as, for example, the time of a mathematical narration or the time of a mathematical discussion (Arzarello et al., 2002).

In this paper, since we are interested in teachers' time-saving strategies, we will concentrate on the following concepts: didactic time, time capital, pace, which we are now going to define and explain.

Didactic time is defined par Chevallard and Mercier (1987) as related to scheduling the teaching of some knowledge. Didactic time is produced by the textualization of knowledge, which results from the process of didactic transposition, i.e., the process of transformation of a body of knowledge into a knowledge that can be taught (Chevallard, 1985). Didactic time is linear (the objects of knowledge follow one another in linear order) and sequential (knowledge is divided in syllabus items, which are then arranged in tasks, lessons, courses of lessons). Didactic time is used as a gauge of the advancement of knowledge, and, in this sense, it is a framework, which regulates the activity of the teacher. In France, syllabuses prescribe the portion of knowledge to be taught but the teacher is relatively free to divide it in lessons and sequences of lessons according to his pupils′ progress. So our question is: How does the teacher manage this didactic time when the divisions have not been previously made at all, or not made to fit with the use of ICT in teaching?

A second temporal element is the *time capital*, i.e., the "objective" time counted down by the clock and available for the classroom work: the year, the month, the day, the hour, and the minute. Such time cannot be compressed but represents a capital, i.e., the value attributed to each time interval depends on what can be done within it. Management of this time capital by the teacher takes into account the estimated temporal cost of each activity and the global time of all activities put together. For example, a teacher may deem a particular problem-solving activity a waste of time, since the temporal cost of implementing it is very high compared to all situations put together. Another activity may be seen as time saving, since it does not require spending too much of the time capital. As Pronovost (1996) says: "most anthropologists, who had studied this issue, emphasized that social time is also structured according to meaningful activities of which it is made. There is a meaningful relation between symbolically constituted temporal periods and the content of activities."[1]

By dividing his time capital into activities, syntheses, assessments, the teacher organizes the work in the classroom. A smaller time scale can be taken into account; thus activity time is time capital devoted to classroom activities, synthesis time is time capital devoted to whole class syntheses of previous work, and assessment time is time capital devoted to tests or other kinds of examinations.

We can go further in dividing the time capital. For example, activity time can be seen as a more general category including times such as tool time (time needed for becoming familiar with and learning how to use a tool), time for individual work, time for working in small groups, time for whole class discussion, etc. Those times are interconnected and partly inclusive. For example, in a situation where the pupil is required to use Cabri, the tool time of the software is included in the pupil's working time; but, in some situations, the pupil does not use the tool but only describes what the teacher had been doing with the software. These activity times may also be explained in more specific terms with respect to given mathematical content and/or type of activity. As we will see in Section 4, the time of a figure construction activity is not the same as the time spent on analyzing a previously constructed figure.

How does the teacher manage these different times in the classroom? We propose that the teacher views the time he[2] has available indeed as a certain capital, which can be used (invested) in different ways. He makes decisions based on his assessment of the costs of relations such as the relation between tool time and didactic time, the relation between activity time and didactic time, the relation between tool time and activity time, etc., in view of saving as much time capital as possible, whilst promoting didactic time and pupils' learning time.

Investment in time capital does not necessarily yield a gain in didactic time. We can suppose it does or wish for it, but does it really happen?

We will call *the pace* of a course or a part of it, the rate at which didactic time advances relative to the time capital allotted to it. The pace can be slow, fast or moderate. Diagrams 1–3 illustrate what we mean by these different possibilities.

Diagram 1 represents a fast-paced course: didactic time makes rapid progress compared to the time capital that is being spent. We may expect that, in this situation, pupils will not be able to follow the pace, and the teacher will have to return to the subject later in the course.

Diagram 2 represents a slow-paced course: knowledge is not moving very much forward while a large amount of time capital is being spent. We can expect difficulties with covering the material; the teacher must monitor his actions if progress in didactic time is to be achieved.

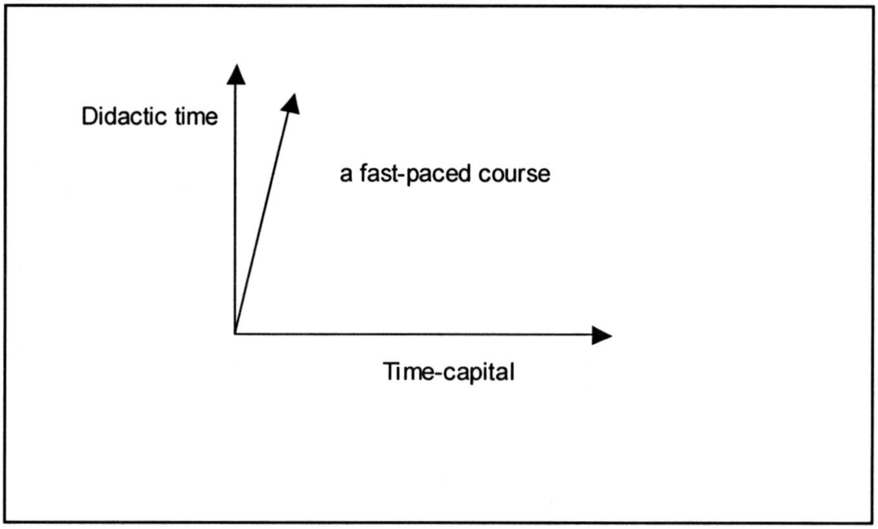

Diagram 1. A fast-paced course.

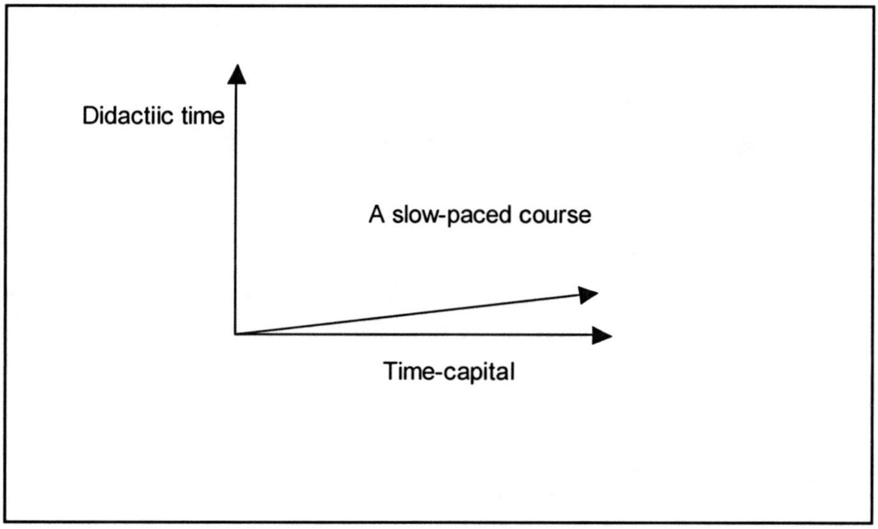

Diagram 2. A slow-paced course.

Diagram 3 illustrates a moderate-paced course, where a reasonable amount of time capital is spent to achieve progress in the didactic time.

In a moderate-paced course, the teacher can still try to save more time capital, by making use of time saving strategies. We will give some examples of these strategies later on.

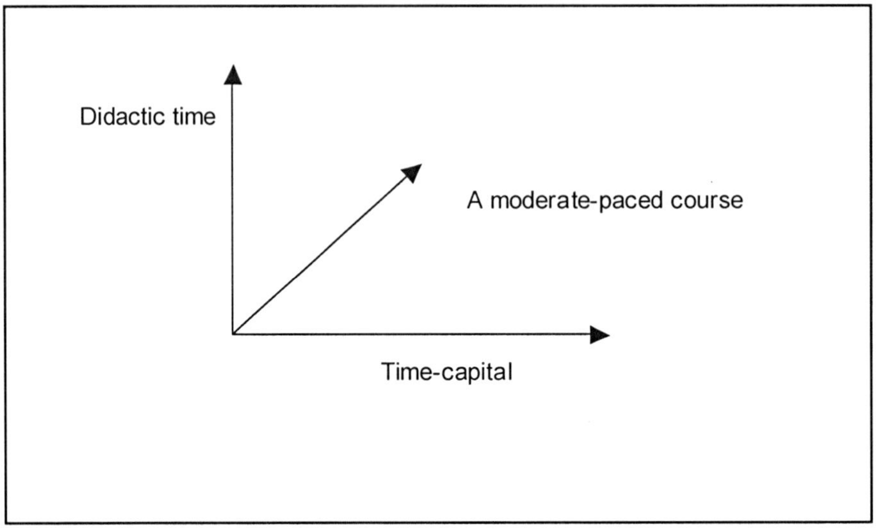

Diagram 3. A moderate-paced course.

Based of these theoretical elements, our research questions are the following:

- How does the teacher manage his didactic time?
- How is the pace of activities taken into account, if it is taken into account at all?
- What time management strategies should the teacher adopt to use his time capital in the most efficient way?

Before answering these questions, we briefly outline the context and the methodology of our research in the next section.

2. The context and methodology of research[3]

Research into the integration of Cabri software was carried out in two CM2[4] classes during the school years 2000/2001 and 2001/2002. Our team of two researchers and two teachers shared their tasks as follows: the teachers were responsible for choosing and implementing classroom situations, the researchers served as resource providers through the production of a "task box"[5] and they observed the classes; researchers and teachers analyzed the data together.

The body of data was constructed on the basis of teachers' lesson preparation notes, pupils' exercise books, classroom observation (video- and audio-recordings and researchers' field notes). The analyses that follow concern all geometry-related activities, and more particularly the subject of quadrilaterals. This subject was chosen because it was included in the curriculum for this grade level.

We will use the comparison between the different years to highlight the different types of time in presence, the relationships between these times, and what this tells us about teachers' time management strategies. The comparison between the outlines of topics and activities covered in 1999/2000 (before the integration of Cabri) and in 2000/2001 (first year of integration) will allow us to analyze the changes brought about by the presence of the software in the production of didactic time. The comparison of such outlines for the years 2000/2001 and 2001/2002 (second year of integration), will allow us to identify different strategies adopted by the teacher in the aim of saving time when working with Cabri.

3. Achieving control over didactic time

The production and management of didactic time are a challenge for a teacher, no matter whether he is a beginner or an expert, whenever he has a new subject to teach or a new technology to integrate. The teacher often does not know what to do, what is essential to do, or how to do it.

We will now compare the two teachers' management of didactic time, before the integration of Cabri and in the first year of this integration (see Table I). We will be looking for essential differences.

The outline of the geometry course for the year 2000/2001, compared to that of 1999/2000, preserved the reproduction of plane figures (item number 1), on the description and reproduction of solids (item 3), and reflection about a line. Some differences can be observed: familiarization with the software was added in 2000/2001, as expected; and there was a change in the construction activities. In the year 1999/2000, pupils had to construct particular quadrilaterals using a construction program and, during these constructions, they encountered drawing requirements such as constructing parallel or perpendicular lines. Thus, the concepts of parallelism and perpendicularity of two straight lines appeared to pupils as answers to construction problems. In 2000/2001, pupils using Cabri primitives did not encounter these concepts as construction tools, and therefore their teachers added a unit on parallelism and perpendicularity, where such lines were drawn using standard physical geometric instruments. The teachers viewed this addition as problematic. In their ordinary lessons, they would try to find situations, in which concepts would appear as tools for solving encountered problems. This unit was not justified this way, and therefore adding it appeared to the teachers as a breach of their standards of practice. However, in the first year of integration of the software they felt they had

TABLE I
Comparing didactic time without and with ICT

Year 1999/2000 (without software)	Year 2000/2001 (with software)
(1) Reproduction of plane figures (5 hours) • reproduction of complex figures • use of geometrical instruments (ruler, compass, set-square, etc.) • use of precise geometrical language (line segment, ray, straight line, arc of a circle, centre, radius, right-angle, etc.) • recognition of certain figures • reproduction of polygons • classification of different polygons • identification of geometrical properties of figures	(1) Reproduction of plane figures (without software) (5 hours) • reproduction of complex figures • use of geometrical instruments (ruler, compass, set-square, etc.) • use of precise geometrical language (line segment, ray, straight line, arc of a circle, center, radius, right-angle, etc.) • recognition of certain figures • reproduction of polygons • classification of different polygons • identification of geometrical properties of figures

(2) Construction of geometrical figures using a construction programme[6] (5 hours)
• construction of a parallelogram
  ○ notion of parallel lines
• construction of a square
  ○ notion of symmetry
  ○ perpendicular straight lines
• construction of a rectangle
  ○ notion of circle, diameter
• construction of a rhombus
  ○ distance between two points
• construction of a tangram
  ○ recognition of different polygons
• elaborating a construction program for a figure
• recapitulation of the properties of different parallelograms
	(2) Familiarization with Cabri (3 hours)

(3) Description, representation and construction (5 hours) • description of solids	(3) Construction of geometrical figures, in particular quadrilaterals (9 hours) • construction of irregular quadrilaterals using Cabri and standard instruments

(*Continued on next page*)

TABLE I
(Continued)

Year 1999/2000 (without software)	Year 2000/2001 (with software)
• acquisition of a vocabulary (cube, cylinder, cone, sphere, pyramid, parallelepiped, rectangle, face, vertex, edge, etc.) • plane representation of solids • classification of solids: polyhedrons – nonpolyhedrons • investigation of different nets of a single polyhedron • construction of solids • study of the properties of the cube, of the parallelepiped	• construction of particular quadrilaterals based on their diagonals • analysis of figures already produced • identification of the properties of particular quadrilaterals • connections between different quadrilaterals • notion of figure • elaborating a construction programme for a figure • construction of a square using a construction program
(4) Construction and transformation (3 hours) • enlargement dilation • reduction dilation • reflection about a line	(4) drawing perpendicular and parallel lines (1 hour) • use of standard instruments
(5) Taking bearings on a squaring or a map (2 hours)	(5) Description, representation and construction (5 hours) • description of solids • acquisition of a vocabulary (cube, cylinder, cone, sphere, pyramid, parallelepiped, rectangle, face, vertex, edge, etc.) • plane representation of solids • classification of solids: polyhedrons – nonpolyhedrons • investigation of different nets of a single polyhedron • construction of solids • study of the properties of the cube, the parallelepiped
	(6) Reflection about a line (3 hours)

no choice but to add this isolated unit because it was explicitly part of the curriculum.

Another difference is that, in the year 2000/2001, the teachers did not have the time to include the units on dilation or location in the plane. It is worth noting that 13 hours were spent on items 2, 3 and 4 of the 2000/2001 outline, while 5 hours were spent in 1999/2000 on the corresponding item 2. Time capital cannot be stretched, and if a decision is made to use some of it on integrating Cabri, one should be aware that certain items on the outline will have to be dropped.

The integration of Cabri in the two CM2 classes did not cause any major changes with regard to the broad types of tasks: construction, description and property identification. However, there were local changes in the tasks and the types of techniques used (for more details see Assude and Gélis, 2002). For example, constructing a particular quadrilateral from diagonals is a new task for pupils. This task was first assigned in the pencil and paper environment, and then with Cabri, and it involved working on the properties of diagonals of particular quadrilaterals. It became an essential point of the entire general unit on quadrilaterals, since it allowed work on the concept of property as a construction tool and as a means of characterizing a particular quadrilateral.

One comment can be made regarding the *old/new dialectic*: integration does not mean a complete revolution with respect to former practices. While some practices remain, for example, preparing a construction program or constructing a square by means of a construction program, new ones appear, for example, constructing quadrilaterals starting from their diagonals. One factor of integration appears to us to be the "*right distance*" between the old and the new with regards to the proposed types of tasks.

This condition of integration – even though essential from our point of view – was not identified as such by the teachers, who repeatedly mentioned the discomfort of working in a hurry, without knowing what was coming next, the taking of risks, the radical change in how they planned and conducted their lessons. Yet a degree of consistency was built up by interconnecting the old and the new in terms of principles, tasks and techniques. So why did the teachers experience discomfort?

To incorporate the work with Cabri into their geometry course outlines, the teachers adapted the outline of the previous year's course. But this adaptation did not necessarily allow the teachers, in the first year of integration of Cabri, to have a global vision of the new didactic time. It was necessary for them to accept this temporal instability in order to step into this "adventure," in which they took many risks (as they themselves admitted). The shift from "day-to-day" management to a more global control over didactic time appears to us to be another essential condition of integration. Control over didactic time implies tracing out a temporal division of knowledge (an order) that is linear and fits logically into the overall work of the class. Such control gives teachers a general view of how knowledge unfolds over time, even though it is possible to change things afterward, and makes it possible to anticipate pupils' difficulties during this progress. Didactic time appears here as a framework and as a means of regulating the teacher's activity; this role of didactic time was overlooked in the first year of integration.

In comparing the first and second years of our research, we noticed an increase in teachers' control over didactic time. During the year 2001/2002 (third year of our research project), teachers started from the previous year's outline, changing some items that appeared problematic, and trying to provide for time-saving actions (this point is taken up in more detail below). For example, systematic revision work was carried out on the circle and on the diagonals of a polygon prior to Cabri lessons, because a lot of time was wasted in the previous year on making these revisions during Cabri lessons. In addition, this second year proved to be an easier experience, since teachers could anticipate what pupils would be doing, and therefore they were able to manage their time more economically: they were saying, "we know where we're going," "we don't need to rush," "things can be planned in advance." Without the necessary landmarks, as in the first year of integration, teachers did not know when to speed up and when to slow down. The difference between the first and the second year could be explained, largely, by the teachers' greater command of both the software and the didactic time in the second year. It made it possible for them to anticipate the time things would take and how they would be interrelated in the mixed paper and pencil and ITC environment, and thus to establish a tentative temporal order, which, even if not kept, was open to absorb any unexpected irregularities or gaps.

We shall now see how, when teachers have control over didactic time, they can manage various times in a different way in order to save time in their work with Cabri. To do this, we will analyze the unit on quadrilaterals.

4. TIME ECONOMY AND THE PACE OF SITUATIONS

In each of the two years of integration of Cabri, the unit on quadrilaterals was organized in three stages each of which occupied two to three one-hour classroom periods:

- The first stage focused on the *construction* of quadrilaterals, with a paper and pencil phase and a Cabri phase. In the former, pupils had to construct quadrilaterals starting from diagonals; in the latter, they had to construct quadrilaterals using the software. In both phases, pupils could construct particular quadrilaterals of their own choice.
- The second stage was devoted to *analyzing* figures. The analysis of paper and pencil figures aimed at the identification of different quadrilaterals and their properties (for example, identification of a square and its properties). The analysis of Cabri figures was meant to draw students' attention to the distinction between a drawing and a figure[7] and to the relationships between different quadrilaterals (for example, the fact that a square is a particular type of rectangle[8]).
- The third stage focused on the *construction* of quadrilaterals with paper and pencil and with Cabri, in particular the construction of a square starting from the diagonals, or starting from a side, using a construction program written on the basis of the "construction replay" feature[9] of Cabri. This third stage closed with an assessment phase: pupils were asked to draw a square following a construction program.

These three stages in the unit on quadrilaterals were determined by variously paced activities. A construction activity was much slower than an activity in which previously constructed figures were analyzed, and a very open-ended construction situation such as those in the first stage was much slower than a construction activity with standard instruments following a construction program. When we say that the pace is slower, we mean that the relation between the pupils' working time and didactic time is not economical relative to time capital, because didactic time hardly moves forward; pupils must have sufficient command of the software in order to do the constructions, which is not necessarily the case for analysis; moreover, since the situation can be very open-ended, collective time is also more important.

Each activity proceeds, therefore, at a different pace, and becoming aware of that allowed teachers to make decisions and choices as to the types of activities. Pace is therefore one factor of selection of activities amongst others, such as the knowledge to be taught. For example, during the first year of our research and the first lesson of the first stage, pupils were working individually, with half of the class constructing quadrilaterals using Cabri, and the other half constructing them in paper and pencil environment starting from specific diagonals; the situation was so open that most pupils were lost. A decision regarding what to do next was necessary; the situation was closed down by means of a pencil and paper activity requiring the identification of the properties of diagonals based on an analysis of figures, and, in the second stage, a faster paced activity was chosen, namely, the analysis of figures previously constructed with Cabri.

However, the teachers thought that the pace obtained in the first year was globally convenient and they reintroduced it in the second year with respect to the succession of construction, analysis and construction activities, since it allowed pupils to master the software and, at the same time, to advance the didactic time in terms of new mathematical content. The pacing of the sequence of activities is all the more noticeable as there is a main axis of focus for the three stages. This axis is the notion of properties of the quadrilaterals, since the properties are used both to characterize and construct particular quadrilaterals.

Besides this notion of pace of a sequence of activities, every activity also has its own pace, which can be a factor in choosing it, for the teachers. In the first year of our research, the teachers dared not intervene orally in the whole classroom except when giving their final synthesis of each activity. They had noticed that in some cases an intermediate synthesis was a way to boost the work of individuals who were lagging behind; however, they would not risk doing this in the first year, thinking that they ought not intervene in the pupil's working time. In the second year, the pace of some activities was changed. We could observe several strategies.

One of the strategies was to cut down on secondary activities and go directly to the heart of the matter. For example, the first lesson required the production of many different quadrilaterals, which were then classified. At that time classification was not an essential element and was very costly in time capital, since the pupils had not constructed very varied quadrilaterals, and it would have been necessary for the teachers to start working again, making it clear at this time what they wished to obtain. In the second year, classification was no longer taken up as a target objective in the activity.

Another strategy was to familiarize pupils with mathematical objects that appeared problematic for students in the first year, prior to the lessons with Cabri. In the first year, some pupils no longer knew what a diagonal was, and the teachers spent much of the available time capital on necessary revision work thus somewhat disrupting the planned pace of the activity. In the second year, they chose to revise the concept of diagonal earlier in the course, when dealing with the reproduction of plane figures, particularly polygons.

Another strategy was to change pupils' relationship to an object of knowledge in view of its prospective use in the Cabri environment. For example, pupils' relationship to the compass in the first year did not include the use of the instrument to transfer lengths. But when teachers wanted pupils to use the "compass" primitive in Cabri, they had to revise the use of the compass and the notion of circle. In the second year, pupils were asked to reproduce figures including circles using the compass to transfer lengths, which made it possible to save time.

Yet another strategy, introduced in the second year, involved giving intermediate syntheses; in those brief syntheses one or two pupils would be asked to show how far their work had progressed. The teachers would make sure beforehand that these pupils' input would advance the work of others.

Finally, we observed a strategy consisting in teachers' making "small" authoritative contributions. In the first year, they dared not make comments during individual work done by pupils; however, in the second year they believed that certain contributions could help save time. For example, some pupils were unable, in Cabri, to draw a circle passing through a point, because they did not know that it is necessary to specify to the software that the circle should be "passing through this point." In the first year, teachers discovered pupils' usage patterns, such as the one just mentioned. Without experience in using technology with pupils, teachers would not be able to anticipate those patterns. In the second year, the teachers already aware of this difficulty, showed pupils what needs to be done. This input of information was beneficial to some pupils, who, without it, could not understand why their circle was not passing through the required point.

Those different strategies allowed to change the pace of some activities and to save time capital. Let us see now other time-saving strategies.

5. Time economy and the individual/collective relation

In a discussion with the teachers at the end of the first year, one of the teachers said, with reference to the unit on quadrilaterals: "we spent too much time and the pupils did not have enough time." This means that too much time capital was spent, while didactic time did not advance and the "tool time" for getting familiar with the software was insufficient. She thought that pupils had not had enough usage time overall. How can didactic time be made to advance more rapidly while increasing the tool time? Several strategies were implemented during the second year of experimentation. Here we give two examples, one in this section and the other in the next. The first example is concerned with a strategy that makes use of the relation between the individual and the collective activity.

One of the sessions on quadrilaterals was based on a series of activities based on questions about preconstructed Cabri figures. Figure 1 shows an

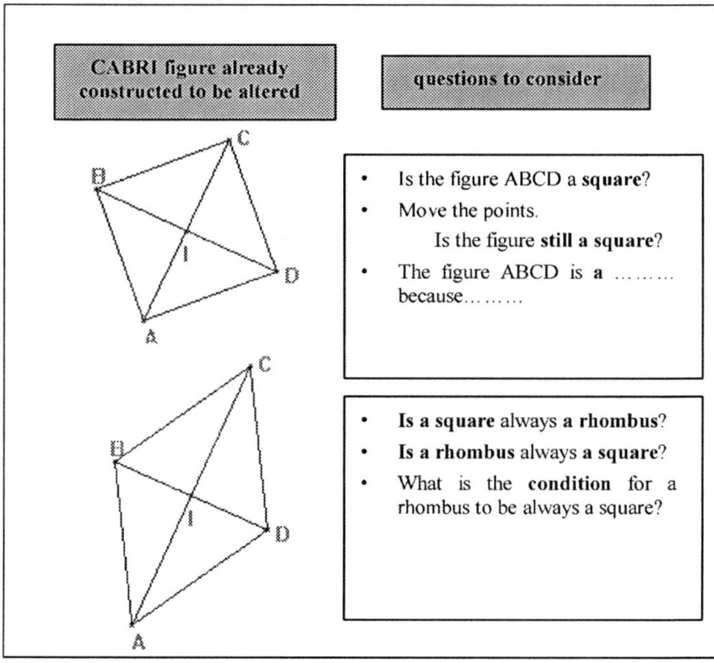

Figure 1. Example of type of task proposed.

example of such activity. In this activity, a Cabri file opened on a square that had to be put out of shape. The drawings obtained from this Cabri figure were all rhombuses, due to the geometrical properties governing the construction of the initial square. The objectives of the proposed task were both instrumental and mathematical. In instrumental terms, it focused on the distinction between figure and drawing and on the invariance of the properties used in the construction of the figure under the change of its shape. In mathematical terms, this task provided an opportunity to revise the properties of different quadrilaterals (showing that a quadrilateral is a rhombus by measuring the lengths of its sides with Cabri) and to work on inclusions between classes of particular quadrilaterals (every square is a rhombus, but not all rhombuses are squares). The other activities were similar and concerned other pairings of particular quadrilaterals (such as rectangles and squares, or parallelograms and rectangles).

There were five exercises of the same type, where the pairings of quadrilaterals changed. During the year 2000/2001, all pupils had to complete all five exercises, and the teachers did not have enough time to make a synthesis at the end of the class. They were compelled to allow the pupils to finish the exercises during another class, where there was also a synthesis and institutionalization phase for the mathematical and instrumental knowledge involved in these tasks. In this case, much time capital was spent whereas pupils did not have much tool time, and no progress was made in didactic time.

In the second year of the research, a different choice was made concerning the same activities. Each pair of pupils was presented with only three of the activities, but the whole class was presented with all of them. Thus, the teachers added tool time and, in the phase of synthesis, all pupils became acquainted with all of the activities, either first hand or through the reports of other pupils. The teachers achieved a gain in time capital and the attention time of pupils did not flag, as in the first year, because the pupils had to give an account of their activity, and they realized they had the same drawing but not the same figure. For example, two pairs started with a square but those squares had not been constructed in the same way: one was constructed as a rectangle, the other as a rhombus. Here, too, there is a gain in time capital (instead of two class periods, one and a half was enough), in tool time and in pupils' working time; pupils paid more attention to the mathematical content of the activity, partly because their curiosity was aroused by noticing that they had the same drawing but not the same figure.

6. Time economy and material or symbolic means

We shall presently show how, in relation to the synthesis made during these activities, the teachers adopted a strategy that enabled them to save time by using a poster that freezes the dynamic properties of Cabri.

As mentioned, in the second year, individual pupils got acquainted with three activities only, but collectively they saw all of them. The pupils worked in pairs, and afterward a collective synthesis was organized as follows. Each pair presented their work using the software, and after the presentation and discussion, the teacher displayed the information provided by each pair in a poster (Figure 2).

By freezing the dynamic properties, the poster condenses the essential information of the activity. Thus, each pair presented the cases they had worked on or placed them in the context of the cases already presented. The use of this poster made it possible to leave a trace of the work of each pair, and the teachers used it to institutionalize the mathematical knowledge concerned, in particular, the relations between different quadrilaterals and their properties, as well as the difference between drawing and figure. In addition, this poster allowed pupils to understand the purpose of the situations, as some affirmed in the synthesis afterward when the teacher gave a recapitulation of the session. For example, one pupil said: "The

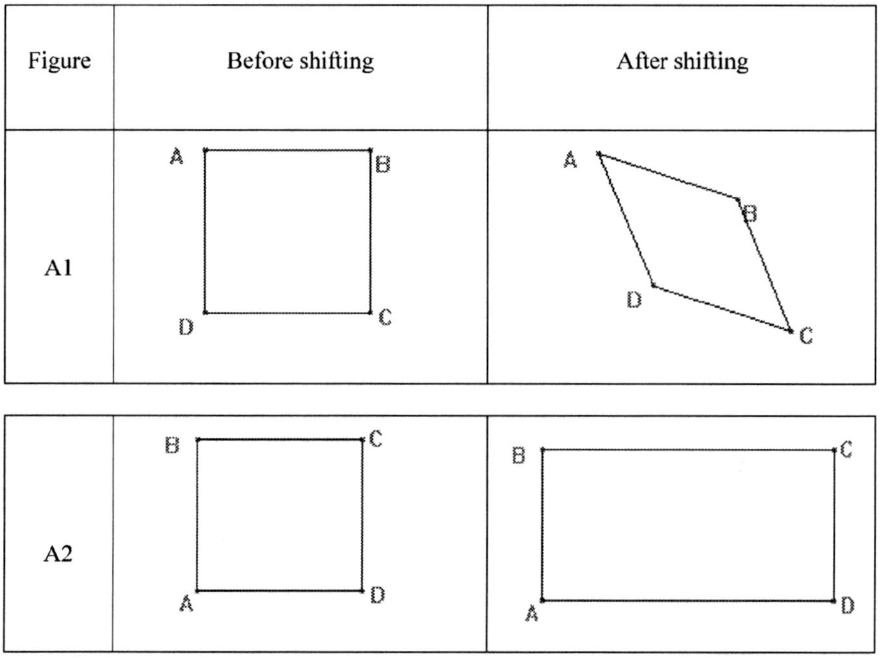

Figure 2. Information processing poster.

square has all the properties of the rhombus, but it also has other properties," while another pupil said: "Had we constructed a square, it would remain a square, it would have right angles."

This poster is a means of processing information that saves time, since pupils represent and then visualize the different activities, instead of talking about all of the activities without representing them. This way of representing activities is also a means of freezing the dynamic properties of Cabri. The dialectic between static representation and the dynamic properties of Cabri figures provides pupils with a better understanding of cases they did not study compared to cases they studied themselves, and of the purpose of the situation. Taking a snapshot of a Cabri activity also served as a means of classroom management by the teacher, since it enabled him to process information from pupils' work in relation to her objectives. One of the management difficulties expressed by the teachers was that of getting information from the pupils' work that they wished to later use in the synthesis: "everything happens very quickly," "there isn't enough time to see what they are doing and suddenly they are on to something else." By freezing the dynamic properties of Cabri, by leaving a static trace of an activity in a poster, the teachers partly resolved the problem of splintered observations and information regarding pupils' individual work, since they channeled the type of information that the pupils would be presenting during the synthesis. The information that interested the teachers was primarily concerned with the mathematical and instrumental knowledge that had to be taught.

The poster shows the purpose of the activity, and the institutionalized content that pupils are supposed to know.

7. Conclusion

Teachers' feelings about time expressed in the introduction to this paper point to the difficulties of time management in the classroom. What can be done to manage time capital while taking into account the different times and bearing in mind that didactic time must move forward ("the syllabus has to be covered")? In our research the comparison between work carried out before the integration of Cabri and the first year of integration, and subsequently the comparison between the first and second year of integration, shows several conditions of integration linked to the problem of time.

One of the conditions of integration is the teacher's command of the didactic time, which allows the teacher to have a global view of how the teaching of certain content is progressing, and to have an idea of what has to come after an activity. This condition allows teachers to know where they are and where they are going. This condition cannot necessarily be satisfied in the first year of integration of new technologies. We think that even experienced teachers (the teachers participating in our research had each been teaching for more than 15 years) are not necessarily ready to face time management difficulties when the way of working with the class changes and when ready-made outlines are not available. Indeed, we have to work hard to create and to distribute scenarios of Cabri integration in primary school to assist teachers who are starting on this "adventurous road." But it is probable that teachers who invest themselves more readily in this process of integration are those who accept a degree of temporal instability due to an initial lack of control over didactic time. Didactic time has an important functional role in classroom management, because it allows the teacher to anticipate pupils' difficulties and to work on the pacing of particular situations or their sequences. Several strategies are used by teachers to achieve saving time of working with Cabri (the software has to be mastered and has to provide gains in terms of the quality of pupils' learning without spending much time capital).

Another condition of integration is therefore that of saving as much time capital as possible by manipulating the relationships between the different times using multiple strategies. The strategies observed in our work were as follows:

- fine-tuning of the individual/collective relationship;
- using material or symbolic means such as posters that make it possible to condense pupils' work information by freezing the dynamic properties of Cabri;
- knowing when to go to the heart of the matter, which is not unconnected with control over didactic time;
- changing the order in which material is taught, either to revise aspects where difficulties persist (for example, diagonals) or to change the relationship to an object (for example, use of the compass to transfer lengths); and
- making intermediate syntheses or "small" authoritative contributions.

Such strategies allowed the teachers to save their time capital in working with Cabri and, as a result, this software could be integrated in the day-to-day work of the class.

As we said in our introduction, we think that time management strategies observed in our research are not specific to this particular situation. They can be found in other situations. Identifying generic strategies of time management is not the purpose of this paper but through this particular example we can already show that time is a key point in classroom management. We believe research in mathematics education should to take it up as a worthwhile issue.

NOTES

1. The translation is mine.
2. Generic masculine form is used to alleviate the text.
3. For a complete presentation and analysis of the context and methodology of this research, see Assude and Gélis (2002) and Gélis and Assude (2002). This research also uses the contributions of the following papers: Artigue (1998, 2001), Artigue and Lagrange (1999), Chevallard (1999), Lagrange (2001), Lagrange et al. (2003) and Rabatel (1999).
4. "CM" stands for "cours moyen". The pupils are 10-year old; it is the last year of French primary school. Teachers teach the same grade level during several years and their pupils change every year.
5. The "task box" is a series of activities on quadrilaterals that is not structured according to a period of time.
6. A construction program is a list of instruction for building a geometric figure.
7. The drawing is what one sees, while the figure is the class of drawings having the same geometrical invariants (see for example, Parzysz (1988), Laborde and Capponi (1994), Laborde (1998), Fischbein (1993)).
8. See an analysis of this type of activity in Jones (2000).

9. In the Cabri software, the "historical" or "return to construction" command in the most recent versions allows the user to go over the different stages of the construction of a figure by displaying consecutive representations of the figure and naming the objects constructed in succession.

REFERENCES

Artigue, M.: 1998, 'Rapports entre la dimension technique et conceptuelle dans l'activité mathématique avec des systèmes de mathématiques symboliques', *Actes de l'Université d'été 1996 "Des outils informatiques dans la classe..."*, IREM de Rennes, pp. 19–40.

Artigue, M.: 2001, 'Learning mathematics in a CAS environment : The genesis of a reflection about instrumentation and the dialectics between technical and conceptual work', *Journal of Computers for Mathematical Learning* 7(3), 245–274.

Artigue, M. and Lagrange, J.-B.: 1999, 'Instrumentation et écologie didactique de calculatrices complexes: éléments d'analyse à partir d'une expérimentation en classe de Première S', in D. Guin (ed.), *Actes du congrès "Calculatrices symboliques et géométriques dans l'enseignement des mathématiques"*, IREM de Montpellier, pp. 15–38.

Arzarello, F., Bartolini-Bussi, M. G. and Robutti, O.: 2002, 'Time(s) in Didactics of Mathematics. A Methodological Challenge', in L. English and alii (eds.), *Handbook of International Research in Mathematics Education*, Lawrence Erlbaum Associates Publishers, Mahwah, pp. 525–552.

Assude, T. and Gélis, J.M.: 2002, 'Dialectique ancien-nouveau dans l'intégration de Cabri-géomètre à l'école primaire', *Educational Studies in Mathematics* 50, 259–287.

Assude, T. and Paquelier, Y.: 2004, 'Acte de souvenir et approche temporelle des apprentissages mathématiques', *Revue Canadienne de l'Enseignement des Sciences, des Mathématiques et des Technologies* (in press).

Brousseau, G. and Centeno, J.: 1991, 'Rôle de la mémoire didactique de l'enseignant', *Recherches en Didactique des Mathématiques* 11(2/3), 167–210.

Chevallard, Y.: 1985, *La transposition didactique. Du savoir savant au savoir enseigné*, La Pensée Sauvage, Grenoble.

Chevallard, Y.: 1997, 'Familière et problématique, la figure du professeur', *Recherches en didactique des mathématiques* 17(3), 17–54.

Chevallard, Y.: 1999, 'L'analyse des pratiques enseignantes en théorie anthropologique du didactique', *Recherches en didactique des mathématiques* 19(2), 221–266.

Chevallard, Y. and Mercier, A.: 1987, *Sur la formation historique du temps didactique*, Publication de l'IREM d'Aix-Marseille, n°8, Marseille.

Fischbein, E.: 1993, 'The theory of figural concepts', *Educational Studies in Mathematics* 24(2), 139–162.

Gélis, J.-M. and Assude, T.: 2002, 'Indicateurs et modes d'intégration du logiciel Cabri en CM2', *Sciences et Techniques Educatives* 9(3.4), 457–490.

Jones, K.: 2000, 'Providing a foundation for deductive reasoning : Students' interpretations when using dynamic geometry software and their evolving mathematical explanations', *Educational Studies in Mathematics* 44(1–3), 55–85.

Laborde, C.: 1998, 'Visual phenomena in the teaching/learning of geometry in a computer-based environment', in C. Mammana and V. Villani (eds.), *Perspectives on the teaching of geometry for the 21st Century*, Kluwer Academic Publishers, Dordrecht, pp. 113–121.

Laborde, C. and Capponi, B.: 1994, 'Cabri-géomètre constituant d'un milieu pour l'apprentissage de la notion de figure géométrique', *Recherches en Didactique des Mathématiques* 14(1.2), 165–210.

Lagrange, J.-B.: 2001, 'L'intégration d'instruments informatiques dans l'enseignement: une approche par les techniques', *Educational Studies in Mathematics* 43, 1–30.

Lagrange, J.-B., Artigue, M., Laborde, C. and Trouche, L.: 2001, 'A meta study on IC technologies in education', *PME* 25(1), 111–125.

Lagrange, J.-B., Artigue, M., Laborde, C. and Trouche, L.: 2003, 'Technology and mathematics education: A multidimensional study of the evolution of research and innovation', in A. Bishop and alii (eds.), *Second International Handbook of Research in Mathematics Education*, Kluwer Academic Publishers, Dordrecht, pp. 239–271.

Lemke, J.L.: 2000, 'Across the scales of time: Artifacts, activities, and meanings in ecosocial systems', *Mind Culture and Activity* 7, 273–290.

Leutenegger, F.: 2000, 'Construction d'une "clinique" pour le didactique. Une etude des phénomènes temporels de l'enseignement', *Recherches en Didactique des Mathématiques* 20(2), 209–250.

Matheron, Y.: 2001, 'Une modélisation pour l'étude de la mémoire', *Recherches en Didactique des Mathématiques* 21(3), 207–246.

Mercier, A.: 1995, 'La biographie didactique d'un élève et les contraintes de l'enseignement', *Recherches en didactique des mathématiques* 15(1), 97–142.

Parzysz, B.: 1988, 'Knowing vs seeing, problems of the plane representation of space geometry figures', *Educational Studies in Mathematics* 19(1), 79–92.

Pronovost, G.: 1996, *Sociologie du temps*, De Boeck Université, Bruxelles.

Rabardel, P.: 1999, 'Eléments pour une approche instrumentale en didactique des mathématiques', *Actes de la Xème Ecole d'Eté de Didactique des Mathématiques*, Houlgate, Vol. I, pp. 203–213.

Sensevy, G.: 1996, 'Le temps didactique et la durée de l'élève. Etude d'un cas au cours moyen : le journal des fractions', *Recherches en didactique des mathématiques* 16(1), 7–46.

Varela, F.J.: 1999, 'The specious present: A neurophenomenology of time consciousness', in J. Petitot (ed.), *Naturalizing Phenomenology*, Stanford University Press, Stanford, pp. 266–314.

TERESA ASSUDE
UMR "Apprentissage, Didactiques, Evaluation, Formation"
IUFM d'Aix-Marseille, Université de Provence, INRP
2 av Jules Isaac
13626 Aix-en-Provence cedex 1
France
Fax: 33-494090468
E-mail: t.assude@aix-mrs.iufm.fr

CLAIRE MARGOLINAS, LALINA COULANGE and ANNIE BESSOT

WHAT CAN THE TEACHER LEARN IN THE CLASSROOM?

ABSTRACT. Our research is concerned with teacher's knowledge, and especially with teacher's processes of learning, in the classroom, from observing and interacting with students' work. In the first part of the paper, we outline the theoretical framework of our study and distinguish it from some other perspectives. We argue for the importance of distinguishing a kind of teacher's knowledge, which we call *didactic knowledge*. In this paper, we concentrate on a subcategory of this knowledge, namely *observational didactic knowledge*, which grows from teacher's observation and reflection upon students' mathematical activity in the classroom. In modeling the processes of evolution of this particular knowledge in teachers, we are inspired, among others, by some general aspects of the theory of didactic situations. In the second part of the paper, the model is applied in two case studies of teachers conducting ordinary lessons. In conclusion, we will discuss what seems to be taken into account by teachers as they observe students' activity, and how in-service teacher training can play a role in modifying their knowledge about students' ways of dealing with mathematical problems.

KEY WORDS: case study, didactic knowledge, didactique of mathematics, ordinary mathematics lesson, theory of didactic situations, teacher's activity, teacher's knowledge

INTRODUCTION

Teacher's knowledge is a very important topic for mathematics education, and has been developed in many publications. Our approach to this question has been built in the context of research in "*didactique* of mathematics," developed mainly in France around the basic notions of the theory of didactic situations, which, so far, remains in some way an independent field of research. Therefore, it is only after our research that we discovered many links with other publications. This provided us with quite a new point of view on our problematique and research results.

In this paper, we concentrate on a special part of teacher's knowledge, called "didactic knowledge". We start by explaining this term and position it with respect to other theoretical approaches to teacher's knowledge. We then outline a model of teacher's activity (Margolinas, 2002), which is our main theoretical framework in this study. We focus on a particular level of this model: the level of observation of students' mathematical activity when interacting with a problem. We discuss the interactions that are, *a priori*, possible between this level and the teacher's project for the lesson. This leads to some methodological issues, which justify our use of two

Educational Studies in Mathematics (2005) 59: 205–234
DOI: 10.1007/s10649-005-3135-3

case studies, in a biographic approach. In one of these studies Serge, an experienced teacher, devotes a lesson to algebra; in the other, Beatrice, a novice teacher, teaches for the first time translations and rotations. In our conclusions, we will try to link our findings in these specific case studies with those presented in other publications, in the aim of deepening our understanding of the phenomenon of teacher's learning from classroom experience.

TEACHER'S DIDACTIC KNOWLEDGE

Shulman (1986) has identified the following components of the professional knowledge of teachers: content knowledge, pedagogical knowledge and pedagogical content knowledge. There has been a lot of debate about these distinctions. Bromme (1994) and Kahan et al. (2003) adapted Shulman's general term to mathematics – "mathematics content knowledge" – and focused on the specificity of mathematics.

Steinbring (1998) stressed the fact that the distinction between content knowledge and pedagogical knowledge is not independent of the model of the teaching/learning process. In a linear model of this process, "mathematical content knowledge is primarily needed during the first step in this process, whereas pedagogical content knowledge is necessary for the conditions and forms of the transmission of school mathematics" (p. 158). But if we see teaching and learning mathematics as an autonomous system, "pedagogical content knowledge does not primarily serve to organize the transmission of mathematical content knowledge" (p. 159). Therefore, he states that "a new type of professional knowledge for mathematics teachers is needed – a kind of a mixture between mathematical content knowledge and pedagogical knowledge" (p. 159).

As stated earlier in this paper, the field of research in *didactique* of mathematics has developed a somewhat independent perspective. We now explain what led us to the notion of didactic knowledge as a part of teachers' professional knowledge, and why we will prefer this term to Shulman's "pedagogical content knowledge," even if they seem very close.

In general, in our field, the adjective "didactic" qualifies a more general concept, e.g., "contract" in general pedagogy, as content specific. Thus, "didactic contract" refers to pedagogical contracts that are subject matter or content specific. We are aware of the fact that "didactic," in other theoretical perspectives, also seems related to the representation of subject matter (Ponte et al., 1994), but in didactique of mathematics, its use follows a regular pattern of content specification that seems very useful.

Therefore, from our perspective, the teacher's *didactic knowledge* refers to the part of this knowledge, which is related to the mathematical knowledge to be taught. In this sense, *knowing that* (something is so) and *knowing why* (it is so) (Shulman, 1986) are part of didactical knowledge if they are related to some mathematical content. In the following section, we develop a model aimed at specifying some elements of this particular knowledge.

A MODEL OF TEACHER'S ACTIVITY

In earlier studies (Margolinas and Steinbring, 1993; Margolinas, 2002, 2004), we have developed a model, which was first intended as a model of the teacher's *milieu*, but which can also serve, in a somewhat weaker version, as a model of the teacher's activity. This model was designed to better take into account the complexity of the teacher's activity, and in particular to capture the elements the teacher is dealing with. The first model was based on a modification of Brousseau's model of the structure of the milieu (Brousseau, 1990, 1997; see Margolinas, 1995 for a complete vision of what this modification was meant for). In the first part of this section, we briefly outline the model (Figure 1).

In a first interpretation, this model can be understood in a linear way (from +3 to −1), but, as Steinbring (1998) pointed out about Shulman's model, it is not a good model for teacher's activity, which is more complicated. In fact, the "linear" way of understanding this model represents more the researcher's way of planning classroom experimentation than the teacher's actual work (Margolinas, 2004).

- +3 Values and conceptions about learning and teaching
  - Educational project: educational values, conceptions of learning, conceptions of teaching
- +2 The global didactic project
  - The global didactic project, of which the planned sequence of lessons is a part: notions to study and knowledge to acquire
- +1 The local didactic project
  - The specific didactic project in the planned sequence of lessons: objectives, organization of work
- 0 Didactic action
  - Interactions with pupils, decisions during action
- -1 Observation of pupils' activity
  - Perception of pupils' activity, regulation of pupils' work

Figure 1. Levels of teacher's activity.

We have to imagine that at every level, the teacher has to deal with at least two of the components (Perrin-Glorian, 1999): the upper component and the lower component. This creates a sort of "tension" for the teacher.

For instance, when the teacher is interacting with the students (0, didactic action), he[1] is always bound up with his didactic project (+1, local didactic project) but he also has to consider and deal with what he understands of the actual difficulties of the students (−1, observation of pupil's activity) (Figure 2).

With this focus on didactic action (0), we can understand why the 'linear' interpretation of teacher's work is not accurate. Even if some elements of the local didactic project (+1) are set up before the actual interaction, some are modified, on the spot, during the lesson, and sometimes the teacher is planning the local project of a future lesson during the actual interaction, in view of the results. Similarly, what the teacher is able to observe (−1) and interpret during the lesson relies on what he has planned for the interaction (+1) and what he has anticipated, as stated very clearly in Ponte et al. (1994): "Views and attitudes act as a sort of filter. They are indispensable in forming and organizing the meaning of things, but on the other hand they can block the perception of new realities and the identification of new problems" (p. 347).

Our research into the teacher's activity during the past years (Coulange, 2000, 2001, 2002; Margolinas, 1997, 2000, 2002) enabled us to explore numerous possibilities offered by this model. In this paper, we will focus on the part of teacher's didactical knowledge related to the observation level (−1), which we will label with the acronym "ODK" (observational didactic knowledge).

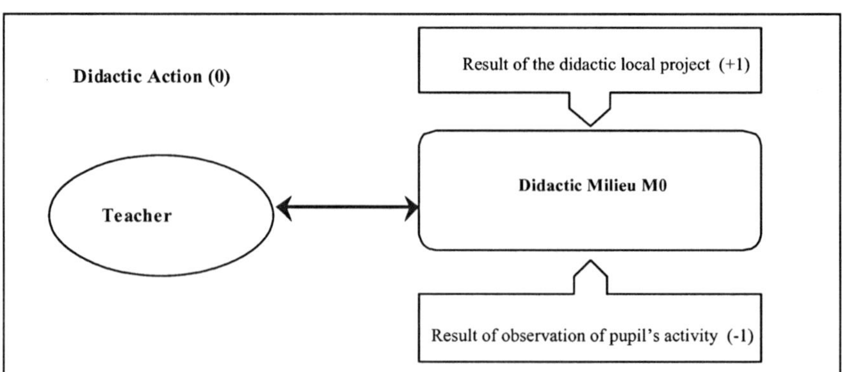

Figure 2. Didactic milieu.

But before we enter this subject, we want to briefly discuss the similarities and differences between our model and the one independently developed by Kahan et al. (2003), which we have discovered only recently. Kahan's framework (p. 227) "provides a way of examining and categorizing current research in [the field of the role of teacher's mathematical content knowledge], with an eye toward articulating what is known and determining where there are gaps to be addressed". The six elements of teaching Kahan et al. considered are, in some way, a more detailed consideration of our (+2, +1, 0) levels, whereas the crossing with processes of teaching is lacking in our model. On the other hand, according to the focus on mathematical content knowledge (which refer to Shulman's category), levels +3 and −1 are lacking in Kahan et al.'s model. Our earlier work (Goigoux et al., 2004) has shown the importance, even during instruction, of the relations between level +3 and lower levels. In this paper, we will discuss the relations between the observation level (−1) and upper levels. Therefore, we find it useful, from our point of view (which has not exactly the same scope as Kahan et al.'s), to have the whole set of levels included in one framework. On the other hand, we find ourselves very much in accordance with Kahan et al. when they discuss their framework, in particular, when they say,

> We prefer to identify teaching process rather than phases to indicate that the process may be ongoing and overlapping. For example, assessment and reflection happen not only after instruction, but also during it as well, as when teachers use questions to assess what the class understands or circulate to supervise student work. (Kahan et al., 2003, p. 228).

FOCUS ON THE OBSERVATIONAL LEVEL

It is generally agreed that it may be important for the teacher to take into account the student's conception in mathematics: "A teacher who pays attention to where the students are conceptually can challenge and extend student thinking and modify or develop appropriate activities for students" (Even and Tirosh, 1995). But, on the other hand, the same authors stressed, in the conclusion of the same paper, that "many of the teachers made no attempts at understanding the sources of students' responses. [...] Therefore, we suggest that teacher's awareness of sources of student's responses be developed".

Other doubts related to the actual teacher's knowledge about student's mathematical knowledge, or "sensitivity to students" (Jaworski, 2002) is stressed in other publications. For example,

Although research findings in cognitive psychology have provided a rich knowledge base about how children learn, it is not yet clear how these ideas might inform systematic changes in teacher's instructional practice. Even less understood are the kinds of concomitant changes in teachers' cognitions that are likely to occur with a student-centered focus in classroom practice. (Artzt and Armour-Thomas, 1999)

Steinbring raised a major point:

The teacher has to be able to diagnose and analyze students' constructions of mathematical knowledge and has to compare those constructions to what was intended to be learned in order to vary the learning offers accordingly. (Steinbring, 1998)

If we interpret this quotation within our model, we can stress the importance of the interdependence between the local didactical level (+1): "what was intended to be learned" and the observation level (−1) that enabled the teacher to "diagnose and analyze," but also how the observation level (−1) can enable the teacher to "vary the learning offer" (level +1). There should exist a certain interplay between these levels, which can be very important for changing teaching so that students' reactions are taken into account.

We have not found, in the literature, any studies of teacher's observational knowledge. In the conclusion of their paper, Tirosh et al. (1998) wrote: "rather than to familiarize teachers with an 'inventory' of student misconceptions, a main objective may be to raise their general sensitivity to students' ways of making sense of the subject matter and the instruction" (p. 62). We can interpret this suggestion by saying that, if teachers were more conscious of the importance of students' ways of making sense of mathematics, they would be able, in any situation, to learn from students' reactions and to reconstruct students' answers. To our knowledge, this hypothesis has not been sufficiently studied in mathematics education research. But this is exactly our focus in this paper.

The two case studies we will present now are an attempt to gain some insight into this question: how can the teacher acquire some ODK about students' ways of solving problems during classroom interaction? This question seems important because it is quite obvious that nobody can learn an extensive 'inventory' of student misconceptions, and even less an inventory of different problem solving strategies, or an inventory of the different difficulties that can arise in all kinds of particular problem settings. Therefore, the classroom interaction itself must be, for some part of teacher's knowledge (at least the observation level −1), the very site of its learning, even if nobody has organized the classroom situation to teach something to the teacher!

THE GUIDING PRINCIPLES OF OUR STUDY

We needed some guidelines to orientate our study of the conditions of teacher's learning. Even if this choice is debatable, we have decided to draw upon assumptions that could be made about any learning, in any situation, and not only about teacher's learning in the classroom. It might have been better to develop or use a framework specific of teacher's learning, or, more precisely, of teacher's ODK. But, as we will see in the following, the rather rough principles, adapted from more general theories of learning in situations turned out to be quite useful in the interpretation of our case studies.

Thus, the first principles were adapted from the basic assumptions of the theory of didactic situations (Brousseau, 1997); they are well-known and don't need any further elaboration:

1) *The antagonistic milieu principle.* Teacher's learning occurs in interaction with an antagonistic milieu, i.e, the teacher must interact with his milieu and recognize feedback coming from this milieu.
2) *The reflection principle.* Teacher's learning requires reflection upon his own actions.
3) *The usefulness principle.* Conne (1992) has pointed to the importance of the recognition of the "usefulness" of one's knowledge in building permanent knowledge. This inspired us to assume that teacher's acquires an element of ODK if he becomes aware of its usefulness.
4) *The awareness of ignorance principle.* Finally, Mercier (1995), and certainly many other authors (e.g., such classics as Dewey) have stressed the importance of the recognition of one's own ignorance in any learning process. Therefore, we assumed: The teacher has a chance to learn if he becomes aware of his own ignorance about something related to students' reactions to a problem.

These assumptions have focused our observations of teachers' actions on certain aspects, and guided our interpretation of these actions.

METHODOLOGICAL ISSUES

Our study focused on the ODK that the teacher can learn during classroom interaction. The facts that were of interest to us are very difficult to observe; actual situations, in which the teacher needs to recall or construct some new ODK to be able to interpret what students are dealing with, are not frequent, and are not easily recognizable.

A priori, we have rejected as unpromising, for the purposes of our study, the cases of teachers who were not interested in students' ways of dealing with problems. Learning new ODK was not likely to occur in these situations.

We have thus selected, for our case studies, two teachers, Serge and Beatrice, who were both convinced of the importance of dealing with students' conceptions, and who had chosen to devote a significant amount of classroom time to student problem solving, individually or in small groups. These problem-solving situations, by their unpredictability, put the teacher's knowledge at risk:

> Teaching for problem solving is a risky business because it invites the unpredictable and raises the question as to how many perturbable events a typical teacher can accommodate without fear of losing control of the class. (Cooney, 1999, p. 169)

But this is exactly why these situations were interesting for us: they require the teacher's public "accommodation," and therefore learning, which then becomes observable by the researcher. In more 'frontal teaching' situations, the teacher's learning, even if it were taking place, could be quite difficult to observe.

TWO CASE STUDIES OF OPPORTUNITIES FOR TEACHERS TO LEARN

For our case studies here, we use some of the data collected in the aim of studying the teacher's situation as a whole (and not only the teacher's learning). We have selected two episodes, which seemed relevant for our specific purposes. These examples may be considered somewhat isolated, but it was our aim in this paper to show precisely in detail what can and cannot happen for the teacher during classroom interaction, trying to describe the facts very accurately and to link them to our theoretical framework.

Our model of teacher's activity led us to constructing a large database with information not only about actual class interactions but also about the school institution and more generally about the environment of teachers' didactic information.

Concerning classroom interactions, we collected two types of data:

- Inside data: audio or video recordings, copies of pupils' work, etc.
- Outside data: teachers' written preparation, interviews with the teacher before or after the lesson, etc.

Concerning the school system, we collected present and past information about programs, textbooks, and teachers' journals. This information allowed us to characterize the possible state of knowledge of a given teacher, which was relevant for understanding his particular mathematical teaching project in the observed lessons.

We have selected two of the observed lessons, which seemed relevant from the point of view of studying teacher's learning in the classroom. In both lessons, the teachers' ODK was not sufficient for dealing with pupils' solutions, and the teachers became aware of some difficulty. The outcomes of the situations in the two lessons was different: the first was a case of short-lived or *ephemeral* learning (without future); the second a case of local learning (restricted to the given problem but more permanent).

Ephemeral learning

This first study is drawn from Coulange's doctoral research (2000, see also Coulange, 2001, 2002). It is centered on the teaching of systems of equations in grade 9.[2] The teacher, Serge, who is being observed, is an experienced teacher. He has worked with researchers in the didactics of mathematics[3] for a long time. Our analysis is based on the data system represented in Figure 3.

Serge chose nine concrete problems[4] to develop his lesson planning on systems of equations. Here we present only the first three problems.

1. Here are two heaps of stones.

   x indicates the number of stones in the first heap,
   y indicates the number of stones in the second heap,

   The second heap has 19 more stones than the first.

a) Write an expression for y using x
b) There are 133 stones in all. Write an equality that is verified by x and y.
c) Find x and y.

<table>
<tr><th colspan="2">Classroom interaction data</th><th colspan="2">School institution data</th></tr>
<tr><th>Inside data</th><th>Outside data</th><th>Past
From the beginning of the twentieth century</th><th>Present
1998</th></tr>
<tr><td rowspan="2">Observation of 3 ninety-minute classroom sessions on systems of equations; audio recording of Serge and his pupils.</td><td rowspan="2">- Serge's interview before the lessons: his teaching project, lesson planning on systems of equations and the setting up of equations.
- Written preparation.
- Interviews after, during and between the observed sessions: Serge's opinion about previous sessions and about changes in the development of his project.</td><td colspan="2">Curriculum, mathematical syllabuses, schoolbooks, papers in didactic journals on systems of equations and the setting up of equations.</td></tr>
<tr><td colspan="2">Curriculum, mathematical syllabuses, schoolbooks, and papers in didactic journals on concrete problems: arithmetic and algebra.</td></tr>
</table>

Figure 3. Data system – Serge's teaching of systems of equations and the setting up of equations.

2. Same problem as 1 with the following data:

   – The second heap has 7 times more than the first.
   – There are 56 stones in all.

3. Same problem as 1 with the following data:

   – The second heap has 26 stones less than the first.
   – There are 88 stones in all.

Serge's plan is to confront algebraic and non-algebraic strategies. This confrontation is meant to give meaning to the algebraic tool as a better way to solve word problems. This conception of teaching concerning the meaning of mathematical knowledge (activity level +3) is conform with the contemporary ideology of the secondary school establishment in France, as represented, for instance, in the general introductions to mathematics curricula. The opposition between algebraic and non-algebraic strategies for solving concrete problems as a way to introduce the systems of linear equations (activity level +2) also seems conform to the contemporary textbooks. What we call here "non-algebraic strategies" represent types of solutions that are not taught at secondary level (some such basic problem-solving methods are taught at the elementary school). Secondary teachers often refer to these methods as "trial and error" strategies. Serge himself calls them "arithmetical strategies," which is in accordance with his didactic knowledge (in didactic journals such as *Petit x* there have been papers published about the duality between algebra and arithmetic; see e.g., Chevallard, 1985).

We will now try to determine what a contemporary teacher could (but not necessarily does) know about arithmetical ways of solving the "stones" problems chosen by Serge.

In the French mathematics curricula, arithmetical ways of solving problems were systematically taught in the first part of the 20th century, but this disappeared from curricula and teaching by the end of the sixties. The teaching included a systematic classification of problems into types and an exposition of exemplary solutions of problems of each type. The "stones" problems would fall mostly into the "unequal share problems". For instance, here is an extract from a 1932 textbook (Delfault and Millet, 1932).

Unequal share – 2 parts – sum and difference known
279. Typical problem – Paul and Charles share £28; Paul has £4 more than Charles. Find each part.
1st Solution. – If I take £4 from Paul's part, I obtain Charles'. But the sum of the parts is reduced by these £4, making $28 - 4 = £24$ and is twice as much as Charles' part.

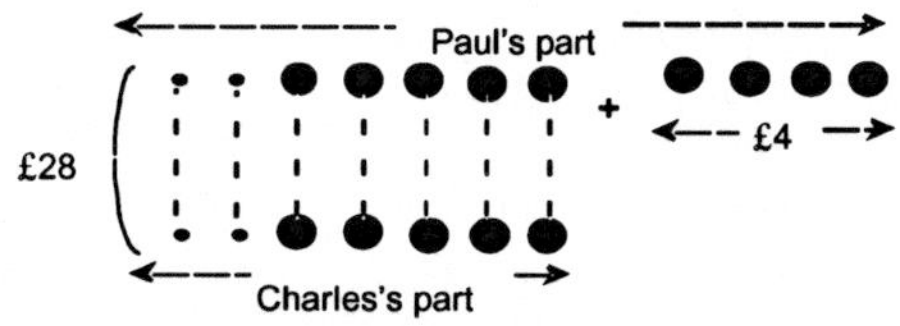

Figure 4. Illustration for the 1st solution (Delfault and Millet, 1932).

Charles' part: 24 : 2 = £24
Paul's part: 12 + 4 = £16
[see Figure 4]

2nd Solution. – By adding £4 to Charles' part, I obtain Paul's, the sum of the parts is increased by these £4; therefore, it is equal to 28 + £4 = 32£ and is twice as much as Paul's part.

Therefore, an arithmetical solution of the third "stones" problem could be one of the following:

Solution 1 (see Figure 5). If I take 26 stones from the first heap, I obtain the number of stones in the second heap. But the total number of stones is reduced by these 26 stones, making 88 – 26 = 62 and is twice as much as the number of stones in the second heap.

The number of stones in the 2nd heap: 62 : 2 = 31 stones
The number of stones in the 1st heap: 31 + 26 = 57 stones

Solution 2 (*see Figure 6*). By adding 26 stones to the second heap, I obtain the number of stones in the first heap. But the total number of stones will be increased by these 26 stones. Therefore, it is equal to 88 + 26 = 114 and is twice as much as the number of stones in the first heap.

The number of stones in the 1st heap: 114 : 2 = 57 stones
The number of stones in the 2nd heap: 57 – 26 = 31 stones.

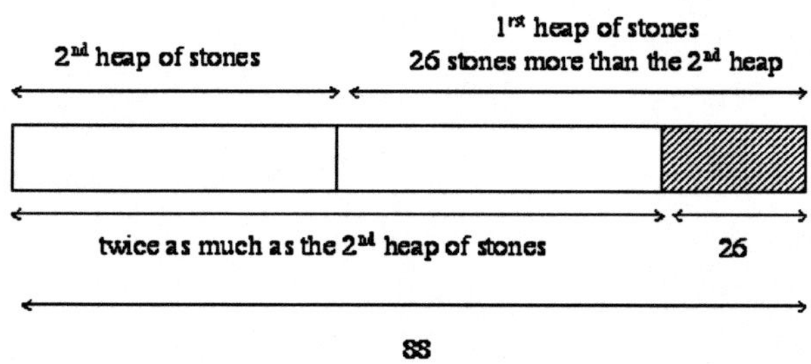

Figure 5. Diagram for solution 1 of the third "stones" problem.

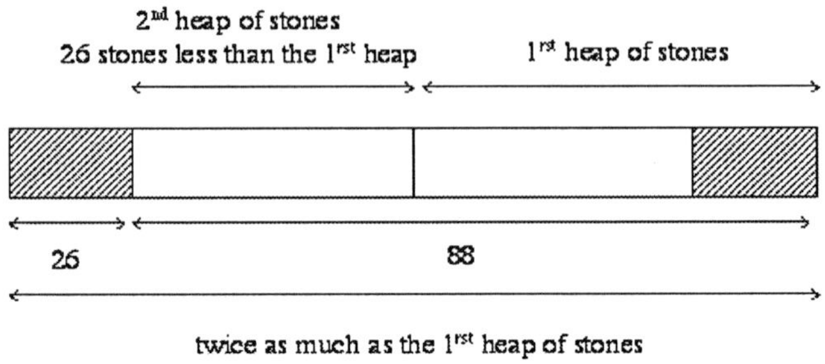

Figure 6. Diagram for solution 2 of the third "stones" problem.

We can see, therefore, that it is possible to solve the 'stones problems' with standard arithmetic procedures.

We will now try to understand Serge's knowledge of arithmetical strategies and the pupils' attempts to solve the "stones" problems using arithmetical approaches.

During the interviews, Serge talked in detail about algebraic strategies, and it is possible to link his comments with specific algebraic knowledge. However, Serge's comments about arithmetical strategies were few and imprecise. In class, in his interaction with a student (Thibault) who chose an arithmetical approach, it is clear that he is a bit lost and does not know how to react. We present below an excerpt from this interaction.

In answer to the 3rd 'stone' problem, Thibault wrote

$$
\begin{aligned}
x + y &= 88 \\
y &= x - 26 \\
88 + 26 &= 114 \\
\frac{114}{2} &= x \\
x &= 57 \quad y = 57 - 26 = 31
\end{aligned}
$$

Then he called the teacher. Here is a transcript of their conversation:

Thibault: Sir. Can you explain to me why I added 26?

Serge: Why you added 26?

Thibault: These 26, here... first I did 88 plus 26, and I found it, I did 88 plus 26, and I found it.

Serge: So, how did you get that? Here you have 88 in all; here you say 26.

Thibault: Because x is minus 26. y is x minus 26.

Serge: Why are you doing? Wait... No, here it's correct here.

Thibault: These are just the terms of the problem.

Serge: Yes. OK. Therefore 88

Thibault: I add on 26, that gives me 114, so then I divide by 2 to find the two heaps, in fact it's as if the two heaps were the size of x, I did it as if the two heaps were of the size of x.

Serge: Wait... Let me have a look.

Thibault: It's like there were two big heaps; therefore, we have to add more, on the right.

Serge: Ah... OK you reverse; therefore, you add the difference.,

Thibault: That's it.

Serge: That is you add the 26.

Thibault: That's it.

Serge: And after you divide by 2. Is that it?

Thibault: I divide by 2 and take off 26.

The first two lines of Thibault's solution are similar to the beginning of an algebraic solution. But they are only Thibault's answer to: "Write an expression of y using x". The lines that follow are similar to the second arithmetical solution. Thibault's technique for solving the problem lacks an explanation and that's what he asks his teacher to help him with.

At first, Serge is totally confused by Thibault's solution, which gives the correct numerical answer, and he is unable to give the explanation required. In fact, what he says in the end "Ah OK you reverse... therefore you add the difference," doesn't correspond to Thibault's procedure, because there is no straightforward algebraic link between $y = x - 26$ and $88 + 26 = 114$.[5]

Besides Serge's difficulty in understanding arithmetical strategies, this episode also reveals that, even if arithmetical strategies are no longer taught at school, students seem to be able to construct such reasoning by themselves, in didactic situations.

Using our model, we can say that Serge is in equilibrium with the upper components of the milieu, that include the contemporary establishment's relationship to algebra and arithmetic and the absence of arithmetic as a body of knowledge within the secondary school curricula. But Serge is destabilized by the pupil's ability to create sophisticated arithmetical reasoning, instead of mere 'trial and error' strategies. Thus, Serge is not in equilibrium with the lower components of the milieu, and therefore there is a possibility for him to learn by adapting to these components.

Indeed, Thibault's explanations offer Serge an opportunity for learning. Immediately after the Serge-Thibault episode, Serge rushes to inform an observer (A. Bessot), present in the classroom but who was not, at the time, aware of the interaction.

Serge [to L. Coulange beside him]: Hey, this is a good one. Take a look at this. Wait... I want to show it to Annie [who is sitting further away].

Serge [to A. Bessot]: He reverses, that is he adds 26, that is he adds on the two heaps at maximum; therefore, at the beginning 88; hence, he adds on 26 in order to have two equivalent heaps and he divides by 2 [. . .] and therefore he has the big heap. Afterwards he takes off 26.

We will now try to highlight the general findings of this case study related to our focus on the evolution of Serge's ODK.

During his interaction with Thibault, Serge is clearly dealing with an antagonistic milieu, which contains Thibault's arithmetical reasoning (level −1 of teacher's activity). The conditions for the existence of this antagonistic milieu (first principle) are linked to the upper components of the milieu:

- Serge wants to problematize mathematical concepts in general (level +3) and particularly the algebraic way of solving problems as opposed to non-algebraic ways (level +2).
- He establishes the didactic situation so that pupils enjoy a significant autonomy, which is related to his general conception of teaching (level +3) and his conception of this particular lesson (level +1).
- He considers the mathematical attempts of the pupils as important and potentially meaningful (level +3), which leads to precise observations of their written productions (level −1).

Therefore, we propose as a well founded hypothesis, to be confirmed by further study, that:

(1) the teacher can encounter an antagonistic milieu when he deals with the observation of pupil's mathematical activity (level −1), but,
(2) this can occur only if this pupil's mathematical activity is really observed by the teacher, which is linked to the global and local conceptions of teaching (positive levels).

The following questions deal with learning as a process that can lead to some stable knowledge.

Serge is an accomplished teacher, which is manifest not only in the number of years of teaching, but also in local observation, because he is successful in dealing with the class in a very open situation. Therefore, he certainly doesn't *a priori* consider the teaching situation as an opportunity for him to lean something new. However, in this particular case, the existence of researchers in the classroom allows him to initiate a process of reflection (second principle). During the action (interaction with Thibault) Serge is not able to answer Thibault's question, which in fact refers to the validity of his strategy. But when he explains Thibault's strategy to one of the researchers, the situation of formulation thus created allows him to express this arithmetic solution successfully. There is no trace, in our data, however, of a further reflection; no attempt at validating or generalizing of Thibault's procedure (which was in fact the pupil's initial question); no expressed interest for the arithmetical body of mathematical knowledge.

We do not see any institution where this particular knowledge could be transformed into professional knowledge. The contemporary mathematical teaching institution (in France) does not consider arithmetical methods of solving problems as a body of mathematical knowledge worthy of teaching. This kind of knowledge is not presently considered as professional knowledge and, therefore, it is not seen as useful (third principle).

But even from a personal point of view, Serge is interested not so much in the pupil's strategies as in his own teaching project, which is to establish algebra as a better way to solve problems. The exact wording, which we translated by "it's a good one," was in French, "*elle est pas mal celle-là*" which often introduces a good joke, something not so serious, but nice or funny. Serge neither realizes his own ignorance about arithmetical solution, nor considers his new understanding as useful (fourth principle).

To conclude on this example, we can say that Serge's situation satisfies all the conditions for becoming aware of the pupil's strategies, and therefore for transforming his knowledge through an interaction with the antagonistic lower milieu. But the understanding of these strategies is not supported by his local or global didactic projects (levels +1 and +2) which would be conform to the institutional project for algebra. Therefore, we cannot expect this learning to leave any permanent traces in his didactic knowledge.

Local learning

The second case study is based on Margolinas' research data (see Margolinas, 1997, 2000, 2002). Four mathematics teachers from the same school have decided[6] to prepare all the lessons for their grade 8[7] classes together (as was the case in Ponte et al., 1994). The data are based on observations of the same lesson taught by three teachers in the group (Beatrice, Marie-Paule and Danièle) and an observation of the working group before and after the lessons. The lesson observed is the first of a very short unit devoted to the introduction of two new transformations: translation and rotation (symmetries are already known to the pupils). In this paper, we will focus on Beatrice's lesson. Our analysis is based on a data system represented in Figure 7.

Beatrice is a student teacher[8] at the I.U.F.M.;[9] she teaches only the observed 8th grade class. During the period of the observation (June 97'), she had already been validated as a teacher, and had received brilliant grades

<table>
<tr><th colspan="2">Classroom interaction data</th><th colspan="2">School institution data</th></tr>
<tr><th colspan="2">Beatrice's lesson</th><th colspan="2"></th></tr>
<tr><th>Inside data</th><th>Outside data</th><th>Past
From 1970</th><th>Present
1997</th></tr>
<tr><td>Observation of 1 session (50 min) concerning the transformations:
- Video and audio recording of Beatrice and her pupils,
- Copy of pupils' written productions</td><td>- Beatrice's short interview before the lesson: her teaching project, lesson planning.
- Written preparation.
- Long interview one week after the lesson, which allows us to collect data about all levels of activity.
- Pupils' written questionnaire at the end of the unit.</td><td colspan="2" rowspan="9">Curriculum, mathematical syllabuses, schoolbooks on transformations</td></tr>
<tr><th colspan="2">Working group (Beatrice, Danièle, Marie-Paule)</th></tr>
<tr><td colspan="2">Danièle's lesson
Marie-Paule's lesson</td></tr>
<tr><th>Inside data</th><th>Outside data</th></tr>
<tr><td>cf. Beatrice's lesson</td><td>cf. Beatrice's lesson</td></tr>
<tr><th colspan="2">Working group sessions</th></tr>
<tr><th>Inside data</th><th>Outside data</th></tr>
<tr><td>Observation of 2 sessions (1h each): one session before the unit concerning transformation and one session after the unit: audio recording
copy of written material</td><td>Documents used by the present and past working groups relative to transformations.
History of the working group and relationships between teachers.</td></tr>
</table>

Figure 7. Data system – Beatrice's teaching of geometric transformations.

at the Institute. Marie-Paule, who is the leader of the working group, is also a teacher at the I.U.F.M. and Beatrice's tutor.[10] C. Margolinas, who is the observer of the working group, is a teacher at the I.U.F.M., where she is head of the mathematics department for the secondary school student teachers. Therefore, even if Beatrice is no longer a student teacher at the time of the observation, she has considered Marie-Paule and Claire[11] during the year as her teachers. Right before she taught her class, she observed Marie-Paule teaching the same material, and she has done regularly so during the year.

The lesson was based on the study of the "fish" problem (see Annex 1). The problem was given on a sheet of paper with 10 pairs of figures made of line segments, triangles and semi-circles and looking a little bit like fish. One figure in the pair was always labeled A, and the other B. The question was (in English):

"Here are 10 situations. For each of the 10 cases, how can figure B be obtained from figure A? Group analogous situations."

The lesson was organized as follows.

First, pupils individually answer the questions (study of the ten situations). Afterwards, they study the problem in small groups of 3 to 4 pupils; in Beatrice's class, each group had to study two situations only. After that, a representative from each group has to present some of the answers with the help of a transparency prepared by the group. During the presentations, the teacher stresses the important points and writes them on the board. A synthesis concerning translations and rotations is planned for the next lesson.

Pupils have already studied symmetries (reflections in grade 6 and in elementary school; central symmetry in grade 7). But here, for the first time, they are faced with rotations and translations. The curriculum does not prescribe an exhaustive study of these transformations, but rather an introduction to their properties that are to be studied in more detail in the next year (grade 9).

The teaching project of the teachers' working group is to study all the transformations recommended by the curriculum together, and to compare their effects when applied to the same figure. This project, rarely present in the textbooks but conform to the curriculum (activity level +2), allows a comparison of the transformations. We can notice the conformity of the working group's project with the contemporary ideology of the secondary school establishment: pupils are to work together and develop their own conceptions of the mathematical concept (activity level +3 and +1).

To determine the mathematical environment of the 1997 curricula, we have to take into account the changes that have occurred since 1970. In the 70's (during the "new math" reform movement), the transformations (isometries) played an important part in a theoretical study of transformations of the plane into itself. An intuitive approach to the study of transformations was not envisaged in the curriculum. In 1997, the current program, published in 1985, states "Translation and rotation should never be presented as mappings[12] of the plane into itself. In each case, they should appear by means of their actions on the figure or by their leaving the figure unchanged". The first sentence makes reference to the previous curricular instructions, where the emphasis was on the classification of isometries and modes of their generation (based on translation and rotation). In 1985, the program states: "Activities will first consist in experimental work [...] the properties will therefore appear progressively."

Personally, Beatrice has only known the 1985 curriculum, even as a pupil. The first point of our study will focus on Beatrice's understanding of the curriculum. We will study an incident, which occurred during the collective planning of the lesson. The formulation of the problem used before by the previous working group led by Marie-Paule was:

"Here are ten situations. Classify them according to the transformation that allows us to go from figure A to figure B."

The statement of the problem in these terms immediately provoked strong reactions from Beatrice and Danièle:

Danièle: I think in the curriculum they say.

Beatrice[in chorus with Danièle]: Never refer to a transformation of the plane into itself. You will say that we haven't said transformation of the plane into itself but all the same.

Beatrice is very careful to conform to the curricular instructions (levels +2 and +3). She studies the official texts seriously. Her interpretation of these instructions is that it is forbidden to refer to, and even to use the term "transformation". The curriculum does not rule out the use of the term, and only states, "Translation and rotation should never be presented as mapping of the plane into itself," which is difficult to understand if one doesn't know the history of the curricula (it refers to the axiomatic presentation of transformation as mapping of the plane into itself). Beatrice's knowledge of this part of the program (level +2) is therefore not in agreement with the official instructions. She is not in equilibrium with this upper component of the milieu. Even the fact that she has heard her tutor, Marie-Paule, use the term "transformation" in her own lesson doesn't have any impact on Beatrice's interpretation. Beatrice refrains from using the term "transformation," which she struggles to replace by all kinds of other expressions, as in the following examples of her communication with students:

Beatrice: I told you that there is a figure which is the image of another by a certain er. . ., based on a certain construction [. . .]

Beatrice: So I'm going to give each of you two figures, you will give a name to these, to these, situations [. . .]

The pupils react by calling all the transformations "symmetries". It is logical on their part, since the transformations they have learned about up till now were named "axial symmetry" and "central symmetry". The pupils have tried to find new adjectives to add to the noun "symmetry" to characterize the new transformations.

At the end of the unit, the researchers asked the pupils to write a few lines "to explain, to someone who was absent from school Monday and Tuesday, what you have learned in the new unit". The written answers of some pupils clearly show the coherent use of this *ad hoc* vocabulary:

Adeline: We have learned several sorts of symmetry: rotation translation and vectors.

Vanessa: There are several sorts of symmetry; I have learned 2 more, symmetry by translation and rotation.

During the lesson, Beatrice noticed the pupils' use of the term "symmetry". She interpreted this fact as a failure of her teaching project. When interviewed, one week after the lesson, she declared:

Beatrice: [. . .] and then someone said symmetry, it was Geraldine, and from that moment on everyone thought that all the figures were er. . ., cases of symmetry [see below for the following]

She also noticed that the pupils were drawing segments between homologous points, even in the rotation situation. This strategy allowed the pupils to answer the questions correctly for all the situations except for rotation. We can find a strong coherence in the pupils' actions. In fact, an accurate observation of the pupils' work on the rotation situation shows that frequently pupils who managed to find the right strategy (drawing arcs between homologous points) have previously made some attempts with segments. But Beatrice, who had not anticipated this strategy, was unable to interpret its pertinence and reacted rather aggressively at the 13th minute of individual work:

Beatrice: Try to use your brains a bit. You see a figure there. I told you that there is a figure which is the image of another by a certain er. . ., based on a certain construction, so when you draw a construction on your paper, try to use your brains a bit, try to show how you go from one figure to another, there is no point in using straight lines, er. . ., if it doesn't mean anything for you.

Immediately after the lesson, she expressed her disappointment with her pupils' reactions:

Beatrice: My pupils didn't do anything at the beginning. I said to myself I just couldn't understand it.

One week after the lesson, during an interview, she still insisted on this problem:

Beatrice: [the rest of a previous quotation] They would draw axes of symmetry everywhere.

Claire: That's what you saw when they were working alone.

Beatrice: Yeah, right at the start, and so then I. I tried to explain a bit better, to tell them how? So there I gave them a bit more guidance, and then I saw that it didn't work, and er. . ., to show them that they were not just symmetries, but new things too, new situations, and then er. . ., so then it got going a bit better, but it was still not really great and then I looked at my watch and it was half past nine.

At this point of our study of Beatrice's case, we can already draw some conclusions. Beatrice perceives scrupulous respect of curricular instructions as an important part of her role (Level +3 of teacher's activity). Therefore, her interpretation of the instructions strongly determines her global and local projects for the unit (Level +2 and +1) and even the details of her interactions with pupils (Level 0). Beatrice's linguistic struggles show the importance of the upper component of the milieu for her, not only before the lesson but also during the interactions. On the other hand, she also deals with what she is able to notice of the pupils' work (Level -1). The timing of the lesson is based on her planned projects (upper components) but the regulation of the actual interactions depends also on her judgment about her pupils' activity (lower component). The tension between these two components is particularly obvious in Beatrice's case.

Beatrice has a professional problem, specific to the 'fish' activity: she is unable to interpret the pupils' activity, which interferes with the timing of the various elements of the lesson, and gives her a feeling of failure. At the beginning of the interview, she expresses this feeling but doesn't find any precise explanation for the difficulty:

Beatrice: Yes in fact I had seen Marie-Paule's class and in mine things didn't get going. [...] the explanation I found for this was that the terms [of the problem] were not very clear. [...] they hadn't understood what they had to do and then someone said symmetry [see previous quotation]

We will show that, during the interview, Beatrice's knowledge about pupils' strategies has changed. In the following extract, we have italicized the phrases that show her evolution.

Claire: Yes, and otherwise do you think that those who made mistakes do you think they understood their mistakes that they corrected them alone or rather you ended up guiding them.

Beatrice: Goodness me. *For example there was one girl who was really obsessed by symmetry and she kept drawing those things there, sorts of axes of symmetry*,

[Claire takes out a black and white copy of the transparency of the group[13] (Aurélie, Emeline, Gérardine, and Laure), see Annex 2. On the group's original transparency the 'arcs' appear in green and the segments in black.],

Beatrice: What do they do? Now, there it was all right apparently, the rotation er...; we didn't see these[14],

Claire: I can't remember that.,

Beatrice: But no because *she did the construction lines,* in fact, as in the..., as in the..., *as in the translation,* and *it's only there that the arcs of circle appear*/so perhaps she had not had the time to, to draw it or, and *what's the meaning of the segments she had drawn.*,

Claire: Yes.,

Beatrice: Géraldine er..., she always wants to bring things back to something she knows, and that's why there she had some ideas er..., I remember her first drawing very well, where she said it was a central symmetry, and here I get the feeling that she had done the same, she had seen that, *she had seen that there we join C and C' and D and D' and they wanted to do the same, they joined it, the points and their images, but that doesn't really show the rotation.*,

Claire: Yeah,

Beatrice: So, why have they done that? *It was stupid of me not to have exploited that.*,

Claire: Yes, but you have not, you could not do it all because you have seen. I don't know what it shows in color either, because sometimes [Claire shows the original transparency] yes, it's quite the same idea.,

Beatrice: At last there is an idea, an idea, a wrong idea anyway, basically mmm..., but all the same, so anyway it's written 61° rotation, yeah, from my point of view *they don't see that when the figure rotates a point is on the circle, a point is on a circle and not on a straight line.*,

Claire: Here, there is something funny that is, have you seen they didn't use the same color, have you seen, she does the segments in black and then, every code and this kind of circle, it's not very clear but...,

Beatrice: Yeah,

Claire: and on the other, because there it was Géraldine.,

Beatrice: So there the same question doesn't apply[15],

Claire: There it was Heloïse and the others.,

Beatrice: The same question doesn't apply about er..., because here it's only segments that are needed each time, in fact *it's strange,* that takes away just as many,[16]*for the rotations, one doesn't draw the segment one draws the circle's arcs and er..., I didn't think about it in fact.*,

Claire: And one can't draw the segments in fact, that,

Beatrice: Yes, one can draw a segment and its image but one can't draw, say that when one applies the transformation of the points we wrote they are not segments, and they are not lines.

This excerpt from the dialogue between Claire and Beatrice shows a slow transformation of Beatrice's understanding of the pupils' strategy. At first she considers that it is Géraldine's fault if the word symmetry has spread in the class, and she interprets the segments drawn between homologous points as an attempt to reduce every situation to symmetry. She slowly realizes that there is some logic to this procedure, and its nature.

Using our model, we can say that Beatrice is in equilibrium neither with the upper components of the milieu, because her interpretation of the curricular instructions is erroneous, nor with the lower components, because her pupils' activity does not correspond to her planning. But only the pupils' general reactions to the activity constitute the feedback of the milieu during the lesson. During the interview, Beatrice is confronted with the pupils' written productions. Therefore, in the situation that includes both the lesson and the interview, a possibility exists for Beatrice to learn by adaptation to the lower components of the milieu. In fact, we have seen that there is an evolution of Beatrice's knowledge about the coherence of her pupils' strategy in solving the fish problem.

We will try to highlight the general findings of this case study related to our focus on the evolution of Beatrice's ODK.

As mentioned above, Beatrice is not in equilibrium with the upper components of the milieu because she has an erroneous interpretation of the curriculum. But there is no feedback from this milieu:

- Beatrice does not go back to the official texts;
- Marie-Paule, who knows all the official instructions very well (and is present as an observer during Beatrice's lesson) has not detected the problem (but uses the word 'transformation' with her own class), and has not reacted to Beatrice quotation from the program.

She is not in equilibrium with the lower components of the milieu either, and during the lesson, she is not pleased with her interaction with the class and has the feeling of a struggle with the pupils in order to guide them to the resolution of the problem. Since the -1 component of the milieu includes only what the teacher perceives of the pupils' activity, we will try to describe this perception.

- Beatrice wants to problematize mathematical concepts in general (level $+3$) but on the other hand she considers the precise and accurate constructions in geometry very important. Therefore, she does not consider the drawing itself as part of a possible work in progress that can lead to the problematization of geometry (Chevallard and Jullien, 1990), but only as a final production.
- She is part of a group of teachers who consider it important to grant pupils significant autonomy, and constructed the lesson according to this conception (during the first three quarters of the lesson, pupils work autonomously) but when her pupils' reactions do not correspond to what she expects, she feels the need to reassert her authority rather than get involved in a further devolution of the problem.
- She considers, as she was taught at the IUFM, that the mathematical attempts of the pupils are important but during the actual interactions she has too much to deal with to be really attentive to their procedures.

Therefore, the conditions for dealing with an antagonistic milieu (first principle) during the lesson are not reached. In fact, what she is able to infer from the class interaction is very general, but not accurate. She states only that the pupils have not fully understood the problem and that the organization of the activity was not very good. What gives Beatrice the opportunity to deal with an antagonistic milieu are the particular conditions of the interview, because she is asked to consider her pupils' activity and she is faced with their productions, without any other things to deal with. The conditions for the existence of this antagonistic milieu are: Beatrice's initial global dissatisfaction, which results from classroom interaction, and the confrontation with the pupils' actual work. As it was the case with Serge, global and local conceptions of teaching are a condition to consider pupils' mathematical activity, but some conditions linked to the tasks of the teacher have to be added. If the teacher has too much to deal with during the interactions in class, he may be unable to observe pupils' activity.

C. Margolinas plays a mixed role between research and training here, since Beatrice has been her student for some months before the observation. Claire's intention and Beatrice's reading of the interview situation is something between a research interview and a moment of training after the observation of a lesson. An intention to teach (on the part of Claire) and to learn (on the part of Beatrice) from this interaction may exist. But the interview situation also creates a necessity, for Beatrice, to express herself and to justify her actions in front of Claire, which gives her as opportunity to reflect on her previous action (second principle). Were Claire present not only as an observer but also as a trainer, she would be able to refer to Beatrice's problem during a training session at the IUFM and to provide Beatrice's knowledge with a didactic environment. On the other hand, the working group could have played the role of an institution but, during the working group's session after the unit, nobody talked about the pupils' strategy when confronted with the 'fish' problem. Therefore, the working group as a professional institution did not give Beatrice's fresh knowledge a chance to become institutionalized. We can foresee that this knowledge, which is only personal and local, may be rather fragile.

Since there is no institution to deal with Beatrice's knowledge, it cannot be linked to any more general knowledge. Had Claire played the role of a trainer instead of a researcher, it could have been possible to establish this connection. For instance, Géraldine's tendency to rely on something already known could be related to the dialectics between old and new or based on the hypothesis of coherence on the pupils' part. Therefore, the utility of the bit of knowledge (the similarity between the constructions in the cases of symmetries and rotation) acquired by Beatrice is strongly linked to the particular mathematical activity observed, which she would revisit only in the following year, at best. The "usefulness principle" for the stabilization of knowledge is not fully respected. To be more precise, we can say that if Beatrice chooses to give the fish problem to her class again in the future, she may recollect and capitalize on her knowledge about the pupils' attempt to link homologous points with segments rather than arcs in the rotation situation.

The same reflection is at stake with regard to the awareness of "ignorance principle". Indeed, Beatrice faces her ignorance, and she states explicitly: "I didn't think about it, in fact" but this realization is related to a very particular fact, which is not linked to some more extensive knowledge.

To conclude, we can say that Beatrice was not in the position to learn from the classroom situation, but the interaction with Claire has created the conditions for the observations of pupils' strategies and therefore for an antagonistic milieu to exists. The conditions for stable general knowledge that could be applied to other classroom situations are not present: the link to cultural or professional knowledge, that would have made it possible to understand the importance of pupils' construction strategies in geometry, is lacking. This condition might be satisfied if this case study was used as a basis for a session at the IUFM. On the other hand, the conditions are fulfilled for a rather stable local knowledge about the pupils' strategy for the "fish" problem.

Conclusion

We will now try to understand how our case study can lead to some new views about ODK, and how these views can be related to earlier studies in this field. We will stress only two points: what is the teacher noticing about the students' strategy? What can be the relative roles of in-service training and teaching experience?

In our case studies, both Serge and Beatrice seem to have a rather blurred vision of students' ways of dealing with the problem. Serge is rather happy with the lesson, and Beatrice extremely unhappy, but in both cases, the reasons seems not so much related with what the students actually did in the lesson, but more with a general feeling. The results are thus very similar to those obtained by Tirosh et al. (1998, cases of Benny, page 55, and Drora, page 56). In another study (Artzt and Armour-Thomas, 1999, p. 222–223), we have found some similarities between Beatrice and Ellen: "Ellen seemed bewildered by the students' incorrect responses and explained it by saying, 'I think he wasn't thinking', 'They don't think.' More generally, we also found, in our teachers, "inaccurate post lesson judgments that their lesson went well or that their students understood" (p. 228), but not for the same reasons that these authors have stated: "providing little room for the expression of student ideas". In our case studies, although the students were granted a lot of opportunities to express their ideas, and the teacher did listen to them, insufficient ODK prevented Serge and Beatrice to have an accurate post lesson judgment. Therefore, we can make the assumption that it is not only "a teacher directed style of teaching

[that] can serve as a mask for teacher" (idem, p. 229) but, more generally, the local didactical project of the teacher (level +1) can serve as a mask as well. The general advancement of the project can serve as a rule for validating the adequacy of the teacher's teaching project but not what students actually learn from it. These hypotheses can highlight differently the discrepancy of some studies about the influence (or not) of teaching on teacher's knowledge (as discussed in Cooney, 1999 p. 169). But we found another rather distressing similarity with Cooney's paper (pp. 182–183) when he explains his uneasiness with a very well performed lesson without any mathematical content. Perhaps with more pedagogical (as opposed to didactic) experience, Beatrice would have been able to get her lesson going, with exactly the same setting and difficulties for the students, as it is the case for Serge? The lack of didactical knowledge and in particular of ODK for experienced teachers should become an object of further inquiry.

The community of research in mathematics education generally agrees on the importance of "enabling teachers to reflect on their practice from a cognitive perspective" (Artzt and Armour-Thomas, 1999, p. 211). But the conditions for this reflection do not seem to be satisfied in the ordinary practice of teaching, even when teachers are working as a group. We reach, in fact, a similar conclusion as Ponte et al. (1994):

> Quite significantly, the views and attitudes that underwent the most significant changes had to do with issues that were specifically addressed in the training activities and meetings. On the other hand, the views and attitudes that proved to be more resilient were related to some hidden cultural and professional dimensions which had not been addressed on those occasions. This suggests that significant change may be brought about by external influences when teachers interact in groups with the potential for strong internal dynamics. (Ponte et al., 1994, p. 357)

These findings may appear as obvious to some readers: how can the teacher learn without any external interaction? But in some countries, the in-service training does not seem important, because it is supposed that teachers will learn everything by mere 'experience'. The case study of Serge may lead to a different perspective: How can an external intervention lead an experienced teacher to 'see with fresh eyes' what is really happening with his students?

ACKNOWLEDGMENTS

We appreciate the discerning comments of the two anonymous reviewers of an earlier version of this paper, and should like to thank them for the help they provided that allowed us to link our research with other results.

We would like to thank Judith Barnoin, Equipe PAEDI, IUFM d'Auvergne, for her linguistic help in the establishment of the first version of this paper.

NOTES

1. The generic masculine pronouns will be used to alleviate the text.
2. Aged 14–15, fourth and last grade of French "collège": 3e in the French system.
3. He participates in a working group in an IREM (Institute for Research in Mathematical Education) and reads a lot of publications related to research in mathematical education.
4. These problems were taken from "Petit x," a journal for secondary level mathematics teachers published by IREM de Grenoble.
5. This link would be the following: $x + y = 88$ and $y = x - 26$, which is equivalent to $x + y = 88$ and $y + 26 = x$, which is equivalent to $x + y + 26 = 88 + 26$ and $y + 26 = x$, which is equivalent to $2x = 114$ and $y = 26 - x$; this leads to the exact sequence of arithmetical computing that conserves only the numerical part (and is based on another mode of reasoning).
6. Without any suggestion from C. Margolinas; this working group is not linked to any research but only to the desire of these teachers to share their experiences.
7. Aged 13-14, third grade of French "collège": 4e in the French system.
8. Student–teachers have passed the Ministry of Education's teaching examination. During the first year after the examination, they are in probation and training. They work part time (1/3) as teachers, which for mathematics teachers means taking charge of one class. For the pupils the student–teacher is an absolutely ordinary teacher, who gives all the lessons, marks, etc. The rest of the time is devoted to training: sessions on didactic and general subjects at the I.U.F.M. (see below), writing of a professional dissertation. At the end of the academic year, they receive marks from the Institute and they must receive the Ministry's validation to become full teachers. For more description of the system of "formation" in the I.U.F.M., see Britton et al., 2003.
9. Institut Universitaire de Formation des Maîtres: University Teacher Training Institute.
10. "Conseiller pédagogique": student teachers work with experienced colleagues in the schools where they teach. Tutors observe student teachers in the classroom and give advice on the planning of the lessons. They are involved in the final evaluation of student teachers.
11. We would write "C. Margolinas" when we name her as the researcher involved in this study and Claire when she is implicated as a person in interactions with Béatrice.
12. In French: "Applications".
13. See Annex 3. N.B. each group was assigned two situations, for this group, these were figures 9 (axial symmetry) and 10 (rotation).
14. That means, we didn't see these during the collective phase of the presentation of transparencies.
15. This group didn't have any rotation in their situations.
16. What Béatrice wants to say (in French: "ça en enlève autant") is not very clear, which is rather frequent when someone thinks aloud, as it is the case here.

ANNEX 1: THE "FISH" PROBLEM

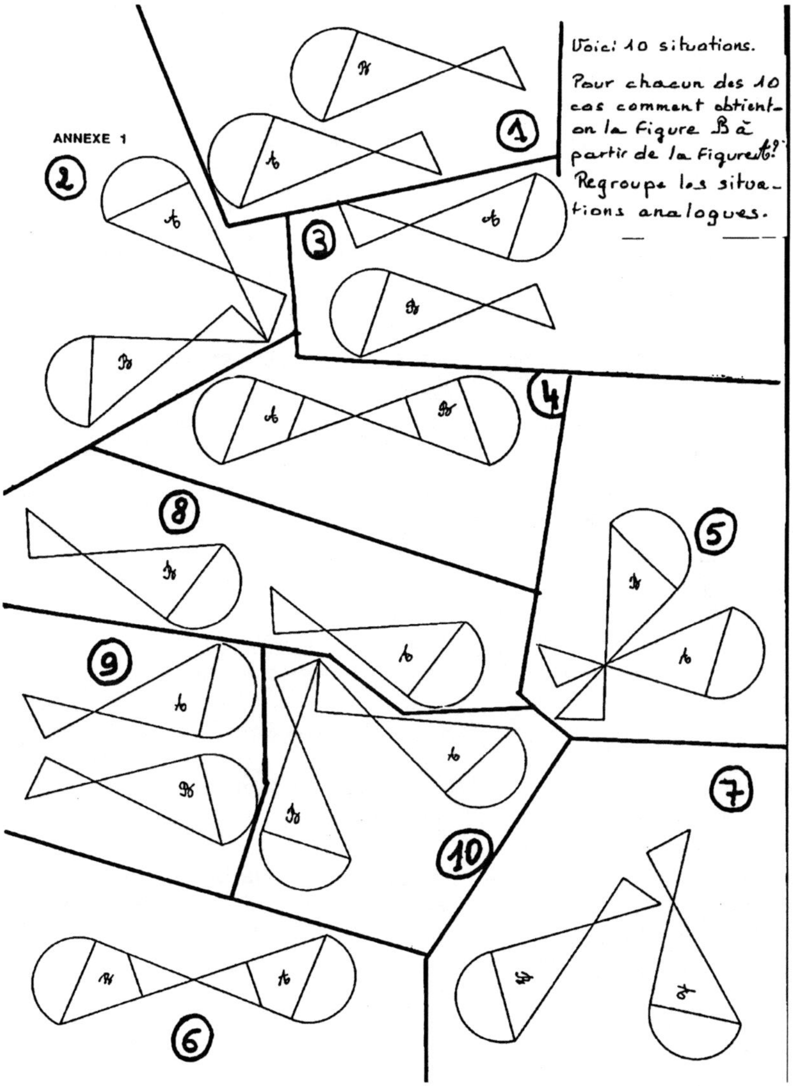

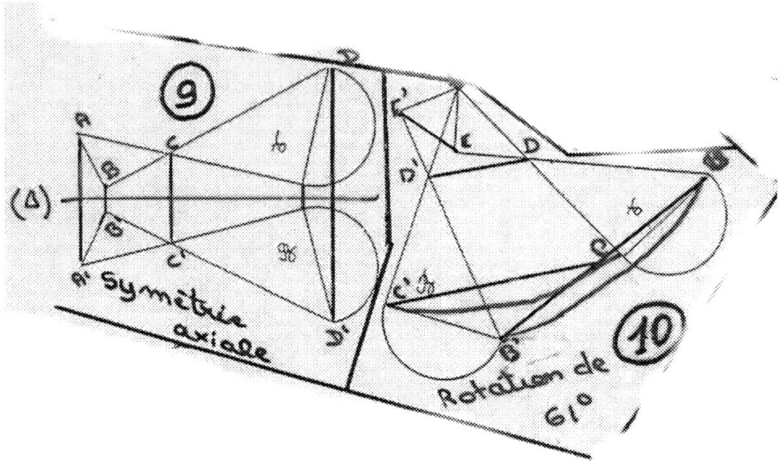

Annex 2: Géraldine's group transparency

References

Artzt, F.A. and Armour-Thomas, E.: 1999, 'A cognitive model for examining teachers' instructional practice in mathematics: A guide for facilitating teacher reflection', *Educational Studies in Mathematics* 40, 211–235.

Britton, E., Paine, L., Pimm, D. and Raizen, S.: 2003, *Comprehensive Teacher Induction. Systems for Early Career Learning*', Kluwer.

Bromme, R.: 1994, Beyond subject matter: A psychological topology of teacher's professional knowledge, in R. Biehler et al. (eds.), *Didactics of Mathematics as a Scientific Discipline*, Kluwer.

Brousseau, G.: 1990, 'Le contrat didactique: le milieu', *Recherches en Didactique des Mathématiques* 9/3, 309–336.

Brousseau, G.: 1997, *Theory of Didactical Situations in Mathematics*, Kluwer.

Chevallard, Y.: 1985, 'Le passage de l'arithmétique à l'algèbre dans l'enseignement des mathématiques au collège (1e partie)', *Petit x* 5, 51–94.

Chevallard, Y. and Jullien, M.: 1990, 'Autour de l'enseignement de la géométrie au Collège, Première partie, A- La géométrie et son enseignement comme problèmes, B- La notion de construction géométrique comme problème', *Petit x* 27, 41–76.

Conne, F.: 1992, 'Savoir et connaissance dans la perspective de la transposition didactique', *Recherches en Didactique des Mathématiques* 12(2/3), 221–270.

Cooney, J.T.: 1999, 'Conceptualizing teachers' ways of knowing', *Educational Studies in Mathematics* 38: 163–187.

Coulange, L.: 2000, 'Étude des pratiques du professeur du double point de vue écologique et économique. Cas de l'enseignement des systèmes d'équations et de la mise en équations en classe de troisième', Thèse d'Université, Université Joseph Fourier, Grenoble I.

Coulange, L.: 2001, 'Enseigner les systèmes d'équations en troisième, une étude écologique et économique', *Recherches en Didactique des Mathématiques* 21(3), 305–354.

Coulange, L.: 2002, 'Analyse de l'activité du professeur dans l'enseignement des systèmes d'équations en classe de troisième', *Actes de la 11ème Ecole d'Eté de Didactique des Mathématiques*, 197–206, La pensée sauvage.

Delfault, M. and Millet, A.: 1932, *Arithmétique cours moyen et supérieur – certificat d'études*, Hachette.

Even, R. and Tirosh, D.: 1995, 'Subject-matter knowledge and knowledge about students as sources of teacher presentation of the subject matter', *Educational Studies in Mathematics* 29, 1–20.

Goigoux, R., Margolinas, C. and Thomazet, S.: 2004, 'Controverses et malentendus entre enseignants expérimentés confrontés à l'image de leur activité professionnelle', *Bulletin de psychologie* 57/1, 65–69.

Jaworski, B.: 2002, 'Sensitivity and Challenge in University mathematics tutorial teaching', *Educational Studies in Mathematics* 51, 71–94.

Kahan, A.J., Cooper, A.D. and Bethea, A.K.: 2003, 'The role of mathematics teachers' content knowledge in their teaching: A framework for research applied to a study of student teacher', *Journal of Mathematics Teacher Education* 6, 223–252.

Margolinas, C.: 1995, La structuration du milieu et ses apports dans l'analyse a posteriori des situations, in Margolinas, C.: *Les débats de didactique des mathématiques*, La Pensée Sauvage.

Margolinas, C.: 1997, 'Etude de situations didactiques "ordinaires" à l'aide du concept de milieu: détermination d'une situation du professeur', *Actes de la 9ème Ecole d'Eté de Didactique des Mathématiques*, 35–43, ARDM.

Margolinas, C.: 2000, 'La production des faits en didactique des mathématiques', *Actes du séminaire du LIREST*, 33–55, ENS Cachan.

Margolinas, C.: 2002, 'Situations, milieux, connaissances – analyse de l'activité du professeur', *Actes de la 11ème Ecole d'Eté de Didactique des Mathématiques* 141–156.

Margolinas, C.: 2004, 'Modelling the Teacher's Situation in the Classroom, Regular Lecture', *Proceedings of the 9th International Congress on Mathematical Education*, Kluwer.

Margolinas, C. and Steinbring, H., 1993, Double analyse d'un épisode: Cercle épistémologique et structuration du milieu, in M. Artigue (ed.), 1993, *Vingt ans de didactique des mathématiques en France*, pp. 250–257, La pensée sauvage.

Mercier, A.: 1995, 'La biographie didactique d'un élève et les contraintes temporelles de l'enseignement', *Recherches en Didactique des Mathématiques* 15(1), 97–142.

Perrin-Glorian, M.J.: 1999, 'Problèmes d'articulation de cadres théoriques: l'exemple du concept de milieu', *Recherches en Didactique des Mathématiques* 19(3), 279–322.

Ponte, P.J., Matos, F.J., Guimarães, M.H., Leal, C.L. and Canavarro, P.A., 1994, 'Teachers' and students' views and attitudes towards a new mathematics curriculum: A case study', *Educational Studies in Mathematics* 26, 347–365.

Shulman, L.S.: 1986, 'Those who understand: Knowledge growth in teaching', *Educational Researcher* 15(2), 4–14.

Steinbring, H.: 1998, 'Elements of epistemological knowledge for mathematics teachers', *Journal for Mathematics Teacher Education* 1, 157–189.

Tirosh, D., Even, R. and Robinson, N.: 1998, 'Simplifying algebraic expressions: Teacher awareness and teaching approaches', *Educational Studies in Mathematics* 35, 51–64.

CLAIRE MARGOLINAS
INRP, UMR ADEF INRP
Université de Provence
IUFM d'Aix-Marseille

6 rue Barnier
63000 Clermont-Ferrand, France
E-mail: claire.margolinas@wanadoo.fr

LALINA COULANGE
Équipe DIDIREM, Université de Paris 7 et IUFM de Créteil

ANNIE BESSOT
Équipe DDM Laboratoire LEIBNIZ, UMR5522 CNRS/UJF/INPG

JOAQUIM BARBÉ, MARIANNA BOSCH, LORENA ESPINOZA
and JOSEP GASCÓN

DIDACTIC RESTRICTIONS ON THE TEACHER'S PRACTICE: THE CASE OF LIMITS OF FUNCTIONS IN SPANISH HIGH SCHOOLS

ABSTRACT. The Anthropological Theory of Didactics describes mathematical activity in terms of *mathematical organisations* or *praxeologies* and considers the teacher as the *director of the didactic process* the students carry out, a process that is structured along six dimensions or *didactic moments*. This paper begins with an outline of this epistemological and didactic model, which appears as a useful tool for the analysis of mathematical and teaching practices. It is used to identify the main characteristics of the mathematical organisation around the *limits of functions* as it is *proposed to be taught* at high school level. The observation of an *empirical didactic process* will finally show how the internal dynamics of the didactic process is affected by certain mathematical and didactic constraints that significantly determine the teacher's practice and ultimately the mathematical organisation *actually taught*.

KEY WORDS: Anthropological Theory of Didactics, Epistemological Approachs, mathematical organisation, praxeology, didactic moments, didactic transposition, limit of functions

1. Introduction

The main purpose of this paper is to show how teachers' practices are strongly conditioned by different restrictions, of mathematical origin, related to the particularities of the considered content, and of didactic origin, implied by the organisation of mathematics teaching. The case of the teaching of limits of functions in Spanish high schools will highlight these restrictions. Some of them – maybe the most well known (see for instance Artigue, 1998; Ferrini-Mundi and Graham, 1994; Williams, 1991) – refer to the particularities of the notion of limit and to the difficulties of its introduction as a functional tool to enhance students' mathematical problem solving ability. Other restrictions come from the mathematical knowledge as it is proposed to be taught in official syllabi and textbooks, related to, for instance, the difficulty of giving sense to the teaching of limits of functions when these are presented as a tool to study the continuity of functions. There are, moreover, didactic restrictions, which affect the teacher's practice at a more general level and can be linked to the atomisation of mathematical curriculum and to the limited scope for action traditionally assigned to the teacher.

Educational Studies in Mathematics (2005) 59: 235–268
DOI: 10.1007/s10649-005-5889-z

In the first section we present the main elements of the Anthropological Theory of Didactics in accordance with the recent works of Yves Chevallard (1997, 1999, 2002a and 2002b), which constitutes the theoretical basis of our research. The problem of teaching 'limits of functions' is then presented, in the second section, in terms of the three steps of the process of didactic transposition: the 'scholarly' mathematical knowledge, the mathematical knowledge as it is designed to be taught and the way it is actually taught by a concrete teacher in a concrete classroom. The third section presents this last component from the observation of an empirical didactic process that took place during 14 sessions in a Spanish high school class (15 to 16-year-old students). The particular way the observed teacher directs his students' practice is described in Section 4 referring to the dynamics of the *didactic moments* as proposed by Chevallard (1999). This brings us finally, in Section 5, to a paradigmatic example of some visible didactic restrictions, which affect the teacher's practice at the different levels of generalisation.

2. FUNDAMENTAL ELEMENTS OF THE ANTHROPOLOGICAL THEORY OF DIDACTICS

2.1. *Mathematical organisations*

What we call the *Epistemological Program* in didactics of mathematics – to be distinguished from the *Cognitive Program* (Gascón, 1998 and 2003b) – is the program of research which stems from the work of Guy Brousseau[1], and is prompted by the conviction that the construction of models of mathematical activity to study phenomena related to the diffusion of mathematics in social institutions constitutes the first step in mathematics education research. Within the Epistemological Program, the Anthropological Theory of Didactics proposed by Chevallard (1997, 1999, 2002a and 2002b) offers a general epistemological model of mathematical knowledge where mathematics is seen as a human activity of *study of types of problems*. Two inseparable aspects of mathematical activity are identified. On the one hand, there is the *practical block* (or know-how) formed by *types of problems* or *problematic tasks* and by the *techniques* used to solve them. Doing mathematics consists in studying (in order to solve) some problems of a given type. For instance, in upper secondary school, possible types of problems related to limits of functions are: to calculate the limit of a function, to demonstrate the existence of a limit, to define the notion of limit of a function, to check the validity of a proof, etc. The term 'technique' is used here in a very broad sense to refer to what is done to deal with a problematic task. There are different techniques to calculate the limit of a function (depending on the kind of function and on the way it is given), to do a proof, to propose a definition, etc. Some techniques are of algorithmical nature, but most are not; some are well known and easy to characterise, while others are not. The anthropological approach assumes that any 'way of working', the accomplishment of any task or the resolution of any problem requires the existence of a technique, even if this technique can be difficult to describe or show to others (even to ourselves).

A second anthropological assumption is that human practices rarely exist without a *discursive environment,* the aim of which is to describe, explain and justify what is done. Consequently, on the other hand, there is the *knowledge block* of mathematical activity that provides the mathematical discourse necessary to justify and interpret the practical block. This discourse is structured in two levels: the *technology* ('logos' – discourse – about the 'techne'), which refers directly to the technique used, and the *theory* that constitutes a deeper level of justification of practice. Thus, for instance, we can explain the calculation of the limit of a function referring to different technological ingredients, such as 'infinitesimals of equivalent order' or the '$\varepsilon - \delta$ definition' or 'elimination of indeterminations'. These different technological ingredients can make sense and be justified in turn by a discourse of a second level whose aim is to provide a framework of notions, properties and relations to locate, establish and generate technologies, techniques and problems.

Types of problems, techniques, technologies and theories are the basic elements of the anthropological model of mathematical activity. They are also used to describe the mathematical knowledge that is at the same time a means and a product of this activity. Types of problems, techniques, technologies and theories form what is called mathematical *praxeological organisations* or, in short, *mathematical organisations* or *mathematical praxeologies.* The word 'praxeology' indicates that practice (*praxis*) and the discourse about practice (*logos*) always go together, even if it is sometimes possible to find local know-how which is (still) not described and systematised, or knowledge 'in a vacuum' because one does not know (or one has forgotten) what kinds of problems it can help to solve.

The more elementary praxeologies or mathematical organisations are said to be *punctual* if they are based around what is considered a unique type of problems in a given institution. Thus, at high school level, 'to calculate the limit of rational functions at infinity' or 'to demonstrate the existence of the limit of a function using a numerical sequence' can be at the origin of punctual mathematical organisations. When a mathematical organisation (henceforth abbreviated as MO) is obtained by the integration of a certain set of *punctual* MOs in such a way that all of them may be explained using the same technological discourse, it can be said that one has a *local* MO characterised by its technology. For instance, the above mentioned punctual MO can be integrated into a local MO around the calculation of limits of functions, under the technology of the 'algebra of limits', but it can also be integrated into a different local MO depending on the technological discourse used to describe and justify the techniques and also on the different punctual mathematical organisations that are linked together. Going one step ahead, the integration of a number of local MOs accepting the same theoretical discourse gives rise to a *regional* MO. In the same way that a punctual MO can be integrated into different local MOs, a local MO can also be integrated into different regional MOs.

Given this (short) presentation of the general anthropological model of mathematical activity, we can now ask what is needed to create or re-create mathematical organisations? How can one pass from an initial problematic question to the practical and theoretical knowledge structured in a MO? What conditions allow the development of institutionalised mathematical activities? In other words, what are the means available to the mathematician or the mathematics student to carry out a mathematical activity giving an answer to certain problematic questions and crystallising in a MO?

2.2. *Didactic organisations and the moments of the didactic process*

In the Anthropological Theory of Didactics, the process of creation or re-creation of a mathematical organisation is modelled by the notion of *process of study* or *didactic process*. This process presents a non-homogeneous structure and is organised into six distinct *moments*, each of which is characterised depending on the studied mathematical organisation. Each moment has a specific function to fulfill which is essential for a successful completion of the didactic process. These six moments are: the moment of the *first encounter*, the *exploratory* moment, the *technical* moment, the *technological–theoretical* moment, the *institutionalisation* moment, and the *evaluation* moment. According to Chevallard (1999, pp. 250–255, our translation):

> The *first moment* of study is that of the *first encounter* with the organisation O at stake. Such an encounter can take place in several ways, although one kind of encounter or 're-encounter', that is inevitable unless one remains on the surface of O, consists of meeting O through at least one of the types of tasks T_i that constitutes it.[...] The *second moment* concerns the *exploration* of the type of tasks T_i and *elaboration of a technique* τ_i relative to this type of tasks.[...] The *third moment* of the study consists of the *constitution of the technological–theoretical environment* [...] relative to τ_i. In a general way, this moment is closely interrelated to *each* of the other moments. [...] The *fourth* moment concerns the *technical work*, which has at the same time to improve the technique making it more powerful and reliable (a process which generally involves a refinement of the previously elaborated

technique), and develop the mastery of its use. [...] The *fifth moment* involves the *institutionalisation*, the aim of which is to identify what the elaborate mathematical organisation 'exactly' is. [...] The *sixth moment* entails the *evaluation*, which is linked to the institutionalisation moment [...]. In practice, there is always a moment when a balance has to be struck, since this moment of reflection when one examines the *value* of what is done, is by no means an invention of the school, but is in fact on a par with the 'breathing space' intrinsic to every human activity.

It is clear that a 'complete' realisation of the six moments of the didactic process must give rise to the creation of a MO that goes beyond the simple resolution of a single mathematical task. It leads to the creation (or re-creation) of at least the first main elements of a *local* MO, structured around a technological discourse.

The Anthropological Theory of Didactics considers that the notion of praxeological organisation can be applied to any form of human activity, and not only to mathematics. In particular, it can be used to describe the teacher's and the student's practice in terms of *didactic praxeologies* or *didactic organisations*. A didactic praxeology is used when a person or group of persons want to have an appropriate MO available (the *mathematician*'s or *student*'s didactic praxeology) or to help others to do it (the *teacher*'s didactic praxeology). As any praxeology, it has a *practical block* composed of types of *didactic problematic tasks* and *didactic techniques*, and a *knowledge block* formed by a *didactic technological–theoretical* environment. Given the growing interest and necessity to conduct research on teachers and their role in the didactic relationship, the analysis of teachers' didactic praxeologies appears to be a relevant and productive field of investigation for today's didactics of mathematics.

The work presented here began with an observation of two teaching processes about limits of functions.[2] Its main goal was to study how *institutional restrictions* could affect the spontaneous practice of the observed teachers. We are presenting here only one of the observed didactic processes, which will show, not only the kind of analysis we can provide using the Anthropological Theory of Didactics, but also how this analysis allows us to highlight the *didactic restrictions* that affect teachers' practices. Two kinds of didactic restrictions are identified here:

(1) *Specific didactic restrictions* arising from the precise nature of the knowledge to be *taught*. In this study – those related to the content of the *limits of functions* as proposed by official syllabi and textbooks in Spanish secondary schools.
(2) *Generic didactic restrictions* the mathematics teacher encounters when facing the problem of how to teach any proposed mathematical topic in a school institution.

We will show that the conjunction of the two kinds of restrictions determine to a large extent the knowledge related to limits of functions that can be *actually taught* in the classroom. This will provide a first delimitation of the field of possible didactic organisations that can be set up in the considered school institution.

3. The Problem of Teaching 'Limits of Functions'

The problem of teaching 'limits of functions' in secondary schools constitutes a particular case of the *teacher's praxeological problem*. According to the Anthropological Theory of Didactics, this problem consists, essentially, in creating, through a *didactic process*, a specific mathematical organisation in a particular educational institution (Chevallard, 2002a and 2002b; Bosch and Gascón, 2002). To solve this problem, the teacher has some 'given data', such as curricular documentation, textbooks, assessment tasks, national tests, etc., where some components of a mathematical organisation, as well as some pedagogical elements and indications on how to conduct the study can be found. This is how the educational institution 'informs' the teacher about what mathematics to teach and how to do so. Nonetheless, it is clear that an important part of the teacher's problem lies in decoding the information provided by curricular documentation in order to elaborate, in collaboration with the students, a mathematical organisation complete enough to allow the development of a quite coherent mathematical study process.

When considering the 'teaching of limits of functions' as a *research* problem in mathematics education, we need to understand also the choices made by teachers and the institutional restrictions acting upon them. Given that teaching and learning are not isolated but take place in a complex process of didactic transposition (Chevallard, 1985), we need to adopt a broader point of view to make a distinction between: (1) the 'scholarly' mathematical knowledge; (2) the mathematical knowledge 'to be taught' and (3) the mathematical knowledge as it is actually taught by teachers in their classrooms. Figure 1 illustrates these three steps of the didactic transposition process and it includes the 'reference' mathematical knowledge (Bosch and Gascón, 2004) that constitutes the basic theoretical model for the researcher and is elaborated from the empirical data of the three corresponding institutions: the mathematical community, the educational system and the classroom.

3.1. *The reference mathematical organisation*

Spanish official programs and textbooks propose a set of mathematical elements (types of problems, techniques, notions, properties, results, etc.) that

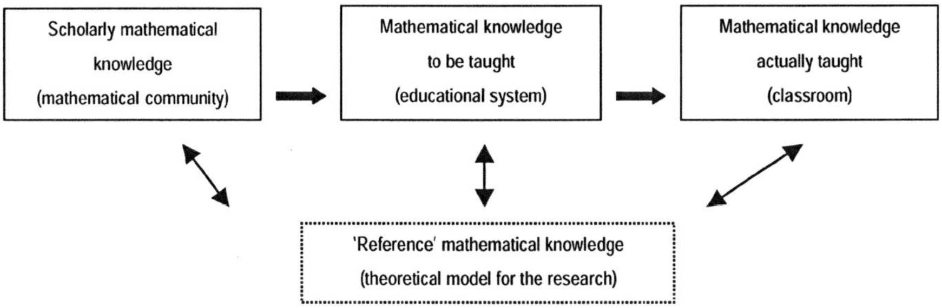

Figure 1. The process of didactic transposition

consititutes the knowledge *to be taught* about the limits of functions. As researchers, we need to interpret these as components of a MO which we will call the *reference mathematical organisation*. This MO constitutes our epistemological model of the 'scholarly knowledge' that legitimates the knowledge to be taught. It is the broader map with reference to which we can interpret the mathematical contents that are proposed to be studied at school. The reference mathematical organisation we are considering here about limits of functions includes and integrates in a regional organisation *two different local* mathematical organisations MO_1 and MO_2 that will assume different roles.

The first mathematical organisation, MO_1, can be named the *algebra of limits*. It starts from the supposition of the existence of the limit of a function and poses the problem of how to determine its value – how to calculate it – for a given family of functions. The two main types of problems or problematic tasks T_i of MO_1 are as follows:

T1: Calculate the limit of a function $f(x)$ as $x \to a$, where a is a real number.

T2: Calculate the limit of a function $f(x)$ as $x \to \pm\infty$.

In both cases the function $f(x)$ is supposed to be given by its algebraic expression and the techniques used to calculate the limits are based on certain algebraic manipulations of this expression (factoring, simplifying, substituting x by a, etc.). For instance:

$$\lim_{x\to 2} \frac{x^2 - x - 2}{x - 2} = \lim_{x\to 2} \frac{(x+1)(x-2)}{x-2} = \lim_{x\to 2}(x+1) = 3$$

$$\lim_{x\to +\infty} \frac{x^2 + 3x + 2}{x^2 + 1} = \lim_{x\to +\infty} \frac{(x^2 + 3x + 2)/x^2}{(x^2+1)/x^2} = \lim_{x\to +\infty} \frac{1 + \frac{3}{x} + \frac{2}{x^2}}{1 + \frac{1}{x^2}} = 1$$

There is also a third type of tasks, much less important, that links the calculation of the limit to the graph of the function:

T3: Determine the limit of a function given its Cartesian graph $y = f(x)$.

Although neither this problem nor the corresponding techniques (based on the reading and interpretation of the graph) are of a proper algebraic nature, in practice this third type of tasks always appears closely related – and even subordinated – to the first ones.

What is, in MO_1, the *minimum technological discourse* needed to generate, explain and justify the properties of limits of functions that are used to calculate them[3], and what is the theoretical foundation of this discourse? A good illustration of the knowledge block of this 'algebra of limits' can be found in the work of Serge Lang (1986) who, for instance, proposes a small axiomatic system to introduce the properties of the notion of limit that will constitute the 'primary resource' of the techniques used to calculate them. This technological ingredient can be informally stated using the following terms:

(1) The limit of the sum of two functions equals the sum of their limits.
(2) The limit of the product of two functions equals the product of their limits.
(3) The limit of the quotient of two functions equals the quotient of their limits.
(4) Inequalities between functions are preserved in the 'passage to limits'.
(5) The limit of a function comprised between two other functions with the same limit equals the value of this limit.

The knowledge (technology and theory) and the know-how (problems and techniques) of MO_1 do not exhaust the mathematical contents that are supposed to be taught in Spanish high schools. Therefore, we need to consider a second component of the reference model, MO_2, which can be designated as the *topology of limits*. This mathematical organisation emerges from the question of the nature of the mathematical object 'limit of a function' and aims to address the problem of the *existence of limit* with respect to different kinds of functions. Some types of problematic tasks T_i that constitute MO_2 are as follows:

T_1: Show the existence (or non-existence) of the limit of a function $f(x)$ as $x \to a$, where a can be a real number, or $x \to +\infty$.
T_2: Show the existence (or non-existence) or one-sided limits for certain kinds of functions (such as monotonic functions).
T_3: Show the properties (1)–(5) used above to justify the way certain limits of functions are calculated.

The *mathematical techniques* usually brought into play in proving these rules are based on the '$\varepsilon - \delta$ inequalities' or on the consideration of special kinds of convergent sequences. A single example can indicate the 'nature' of this work and its difference from the one done in MO_1:

Show that the function $f(x) = \sin(x)$

does not have a limit when $x \to +\infty$.

Let us consider the sequence $x_n = (\pi/2 + 2\pi n)$.

We know that $\lim_{n\to+\infty} x_n = +\infty$

We have: $f(x_n) = \sin(\pi/2 + 2\pi n) = 1$

for all n, which implies $\lim_{n\to+\infty} f(x_n) = 1$

Let us consider the sequence $x'_n = (-\pi/2 + 2\pi n)$.

We know that $\lim_{n\to+\infty} x'_n = +\infty$.

We have: $f(x'_n) = \sin(-\pi/2 + 2\pi n) = -1$ for all n,

which implies $\lim_{n\to+\infty} f(x'_n) = -1$.

We have two sequences that tend to infinity and whose images through f converge to different points. Thus the limit of $f(x) = \sin(x)$ for $x \to +\infty$ does not exist.

The technological discourse of MO_2 is centred on the properties of limits of sequences and the classic $\varepsilon - \delta$ definition of limit. It provides the technical resources needed to solve the problems of the existence of limits. This technology is based on a theory of real numbers structured as a metric space with its different properties: density, completeness, existence of the supremum of every bounded non-empty subset of **R**, Cauchy sequences, etc.

We have, in short, a *reference* MO that integrates, at least, two *local* mathematical organisations, MO_1 and MO_2, which have the following relationships:

(a) Far from being distinct, MO_1 and MO_2 appear to be closely related. As shown, the proof of the rules that support the calculation techniques of MO_1 (that is the *technology* of MO_1) can be considered as a mathematical technique in MO_2 (that is, a part of the *practical block* of MO_2).[4] In fact, it can be stated that MO_1 is partly contained in MO_2.

(b) MO_1 and MO_2 share the same theory of real numbers. Thus, it is possible to state that they can be integrated into the same *reference regional* MO that includes both MO_1 and MO_2 and other MOs. This *regional* MO can, for instance, be the organisation that deals with the question of *differentiability* of certain kinds of functions.

3.2. *The mathematical knowledge to be taught*

The above description of the *reference* MO is now used to describe the mathematical knowledge to be taught about the limits of functions as it appears in the official curriculum of Spanish high schools. The types of problematic tasks most frequently presented in curricular materials and textbooks are as follows:

T_1: Determine the limit of a function $f(x)$ as $x \to a$, with a real and $f(x) = \frac{p(x)}{q(x)}$, where $p(x)$ and $q(x)$ are polynomials or simple irrational functions.
T_2: Determine the limit of a function $f(x)$ as $x \to \pm\infty$, and $f(x) = \frac{p(x)}{q(x)}$, where $p(x)$ and $q(x)$ are polynomials or simple irrational functions.
T_3: Determine the limit of a function at a point from its graph $y = f(x)$.
T_4: Study the continuity of $f(x)$.

The first three types of tasks are particular cases of the constitutive tasks of MO_1. T_4 tasks do not correspond directly to the determination of a limit but are totally subordinate to it. The common techniques introduced to calculate these kinds of limits, for the most part, are based on some algebraic manipulations of the expression of $f(x)$ or on a direct reading of its graph $y = f(x)$. These 'curricular' tasks and techniques make up the practical block of the knowledge *to be taught* and correspond to the practical block of MO_1, mainly. In the following map (Figure 2) $MO'_1 = [T/\tau//]$ is used to indicate the trace left by MO_1 in the textbooks. The letters 'T' and 'τ' indicate the types of problems and techniques of MO_1, while the blanks indicate that the technology and the theory corresponding to this practice are practically *absent* from the curriculum, in the sense that, if they appear, they are not supposed to be used by the students but only presented by the teacher. As such, the reconstruction of MO_1 can be accomplished in the curriculum only in part.

The technological-theoretical discourse proposed by syllabi and textbooks to present, explain and justify this practice clearly comes from MO_2 and, as previously indicated, focuses on the problem of the *existence of the limit* of a function. It uses the standard mathematical discourse but is not accompanied by any mathematical practice within the students' reach. Following the notation proposed by Chevallard (1999), we will use the letters 'θ' and 'Θ' to indicate, respectively, the technology and the theory of a given MO. In our case, the trace left by MO_2 in the textbooks is indicated by $MO'_2 = [//\theta/\Theta]$ in the map and is weaker than the one left by MO_1. It contains only a few technological elements (some definitions and supposedly meaningful comments) whose function is mainly ornamental.

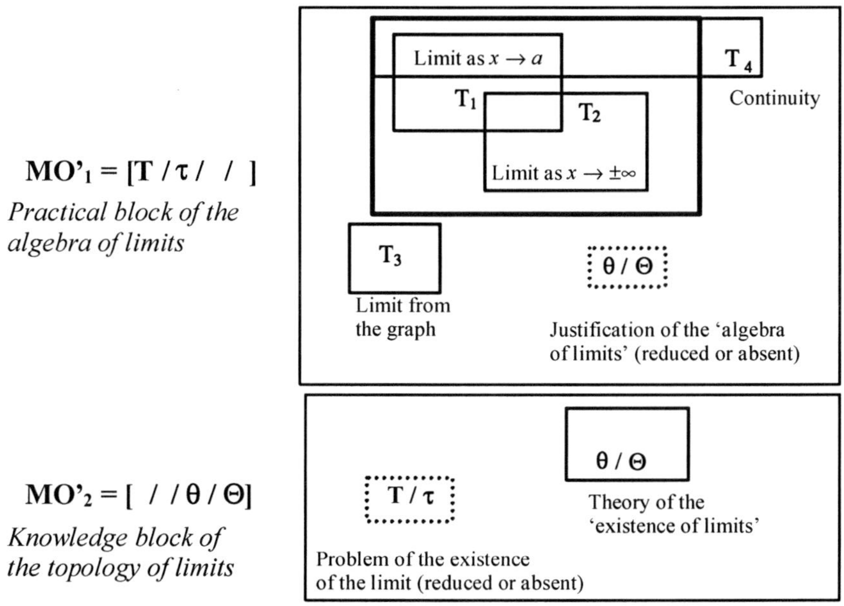

Figure 2. Map of the knowledge to be taught

The blanks, again, indicate an absence. In this case, what is lacking is the practical block of MO_2.

In summary, the considered mathematical knowledge to be taught is composed of the *disjoint union* of the traces left by MO_1 and MO_2 (see Figure 2).

The fact that MO_1 and MO_2 appear *completely disconnected* in the curriculum is mainly due to the absence of both the technological-theoretical block of MO_1 and the practical block of MO_2. The curriculum does not propose the creation of a technological discourse appropriate for the practical block of MO_1, the computation of limits effectively developed by students. Neither does it allow a practice that could be related to the standard mathematical theory about limits of functions (the 'scholarly knowledge') that is proposed instead and which is the technological–theoretical block of *another* mathematical organisation, MO_2. We are not considering here the origin of this *phenomenon of curricular 'two-sidedness' about the limits of functions*, that has to be found in a complex historical process that constitutes the first step of the didactic transposition (see Figure 1).[5] But we want to mention two of its didactic consequences.

The first has already been identified and concerns the major difficulties a teacher will certainly have to face when choosing the concrete mathematical components to teach. In other words, what types of problems must be proposed, what types of techniques can be used to solve them and what kind of explanations and justifications are necessary. The most likely scenario is taking MO_1 (where the existence of limits is not a problem in itself) as the knowledge *to be taught,* for it is the local MO closer (in terms of mathematical components) to the set of tasks, techniques, technologies and theories proposed by the curriculum. But this choice will not remove the difficulties and even contradictions due to the absence of the proper technology of MO_1 and to the presence of technological elements 'external' to MO_1.

There is, however, another phenomenon related to the above-mentioned 'two-sidedness' of the mathematical knowledge to be taught. When looking at the problem of the 'meaning' of limits of functions at secondary school, one can notice that it is precisely the missing technological–theoretical block of MO_1 – how to explain and justify the existence of the limit of a function and the algebraic properties used to determine it – that constitutes the *raison d'être* or the *rationale* of MO_2. In these circumstances, the teacher will encounter the difficulty of motivating the definition of the limits of functions as they are proposed by the curriculum, since this motivation has to be found in a broader MO that includes MO_1 and MO_2 as closely linked components. The same kind of difficulty will appear later at the university level: usually, the knowledge to be taught is mainly based on MO_2 but the practice that motivate this knowledge – the technology of MO_1 – has not been sufficiently developed before.

4. THE MATHEMATICAL KNOWLEDGE ACTUALLY TAUGHT

4.1. *Description of the didactic process*

As indicated before, our research included the observation of a class of Spanish secondary school students (15 to 16-year-olds) during the study of the topic 'limits of functions'. The observation started with the teacher's preparation of the subject and finished with the last session of revising and preparing for the final exam. The main steps of the experimental process are summarised below:

(a) *Data collected about the teacher's performance*

- Videorecording of all the sessions.
- Notes collected during classroom observation.
- Transcript of an interview with the teacher at the end of the process.
- Teachers' didactic materials (books, textbooks, personal notes).

(b) *Data collected about the students' performance*

- Students' notes related to the study process (6 students randomly chosen out of 34).
- Students' solutions to the initial and final exams on the studied topic.
- Students' answers to a questionnaire proposed by the researchers at the end of the teaching process.

(c) *First instrument for the analysis of the global didactic process*
We elaborated two different tables to organise and analyse the collected information of the observed didactic process. Table I included the full transcript of the teaching process according to the following general headings:

TABLE I
Transcript of the teaching process

Episode	Didactic moment	Main player	Mathematical objects	Observed didactic activities

The first column, '*episode*', contained a first intuitive breakdown of the teaching process. The second column, '*didactic moment*', shows the dominant category of moment of study as a summary information, which can help the observer to understand the development of the teaching process. The '*main player*' is the person (teacher or student) who has the responsibility of the specific mathematical task developed (even if it is through interaction with others). '*Mathematical objects*' are those that explicitly appear in the teacher's or students' public discourse (oral or written) in the considered episode. The '*observed didactic activities*' column contains the details of the transcribed and observed public activities in the classroom. This table offers details of the sequence of lessons and a first sequence of the didactic process organised into episodes and *moments*, including the essential components of the *created* MO. As an illustration, we present a small part of the table in the Appendix (it is based on our analysis of the first class).

In fact, this first table only presents 'raw material' which can appear in a non-structured and non-analysed form at the beginning. It does, however, provide the empirical foundation for the second step of the analysis (Table II), the aim of which is to specify the framework of the didactic process in terms of moments.

(d) *Second instrument for the analysis of the global didactic process*
Table II presented a more detailed analysis of the didactic process, and contained the following six columns:

TABLE II
Analysis of the teaching process

Session	Type of mathematic problem	Mathematical techniques	Technological theoretical elements	Dominant moment and sub-moments	Elements of the local didactic techniques

The purpose of this table was to describe the didactic process in terms of the *reference* MO (that is, of the components of MO1 and MO2), taking into account the mathematical elements that have been more or less explicitly present in the class. The table indicates how the didactic process developed, how the different moments were linked, which mathematical objects (types of problems, techniques, technological ingredients, theoretical principles, etc.) appeared, what was their function at every didactic moment, etc. This gives a description of the *practical block of the didactic praxeology*. In order to approach the way the teacher describes and justifies the observed practices – the *knowledge block of the didactic praxeology* – a type of *reference didactic organisation* is also required (Bosch and Gascón, 2002). Given that there are no didactic theoretical models, which are sufficiently well developed to describe the didactic technologies, our study of the teacher's spontaneous didactic technology is only exploratory and preliminary in nature. Empirical data used for the interpretation of the knowledge block of the didactic organisation come from a semi-structured interview with the observed teacher at the end of the didactic process. Themes addressed during this interview were the preparation, planning and management of the didactic process, general matters about the taught MO (components and criteria for their selection), students' difficulties and links with other topics of the syllabus. The interviews had a common structure and specific script and the dialogue with each teacher was freely conducted.

4.2. *The mathematical practice developed in the classroom*

While the knowledge *to be taught* can be reproduced from textbook elements and curriculum documents, the MO *actually taught* appears in students' notes and in the specific teaching practices carried out by the teacher in the classroom. It is clear that the latter heavily depends on the former. They do not however necessarily coincide because the knowledge *to be taught* is not always clearly fixed in curriculum documents and also because of the strong restrictions on the day-to-day teaching praxeologies.

In the observed didactic process, nine types of problems appeared, with different subtypes of problems that are indicated with a quotation mark and the corresponding subindex:[6]

Π_0: Calculate the slope of the straight line $y = ax + b$.
Π'_0: Calculate the slope of straight lines that intersect a curve $y = f(x)$ at a given fixed point x_0 and at a point near this one.
Π_1: Calculate the slope of the straight line tangent to the curve $y = f(x)$ at a given point a.
*Π_2: Calculate the limit of a function $f(x)$ when $x \to a$ (a real).
*Π'_2: Determine the points where a function is not defined and calculate the limit of the function at these points.
*Π_3: Study functions defined piecewise with rational and irrational 'pieces'.
*Π'_3: Study the limits of functions with rational and irrational 'pieces'.
*Π_4: Given the graph of a function, determine the limits of the function at certain determined points.
*Π'_4: Given the graph of a function, determine the points where the function does not have a limit.
*Π_5: Study the continuity of a function at a point.
*Π_6: Study the type of discontinuity of a function at a point.
*Π'_6: Given the graph of a function, identify the points where it is not continuous and determine the type of discontinuity.
*Π''_6: Given a function, find its points of discontinuity.
*Π_7: Study of the conditions under which a function has a limit at a given point.
*Π'_7: Study the conditions for $f(x)$ to be continuous at a given point.
*Π_8: Calculate the limit of an irrational function $f(x)$ when $x \to \infty$ with an indetermination $\infty - \infty$.
*Π_9: Calculate the limit of a sum, difference, product, division or composition of elementary functions.

Most of the activity carried out during the didactic process is aimed at preparing for the emergence of what has to be the *constitutive* tasks of the *actually taught* MO. Thus, most of the types of problems and techniques appear for technological reasons only, for instance, to explain the functioning of a particular technique or to give meaning to a theoretical question. As such, these elements are intended to disappear from the didactic process. Due to the particular circumstances of the specific 'history' of the didactic process, there are also some elements that will be impossible to introduce and which will end up being simple added elements to the created MO.

The types of problems that finally make up the actually created MO in the first observed class are marked by an asterisk. The correspondence between the types of problems Π_k that appeared in the class and the curricular tasks

T_i analysed in Section 3.1 is as follows:

– $\Pi_0$and $\Pi_1$do not correspond to any of the types of tasks T_i
– $^*\Pi_2$, Π_3 and $^*\Pi_9$ correspond to T_1
– $^*\Pi_4$ corresponds to T_3
– $^*\Pi_5$, $^*\Pi_6$ and Π_7 correspond to T_4
– $^*\Pi_8$ corresponds to T_2

The mathematical techniques used in the class to solve these types of problems are not detailed here, because we consider that they can be easily inferred from the delimitation of the types of problems presented above. Consequently, only an example of one of these with its specifications is presented:

τ_2: Replace x by a in the expression of $f(x)$ and manipulate it arithmetically to obtain the final numerical result.
τ_2': Calculate the table of values of $f(x)$ taking values of x close to a $(x > a)$ and deduce the value of the right-hand limit.
τ_2'': Calculate the table of values of $f(x)$ taking values of x close to a $(x < a)$ and deduce the value of the left-hand limit.
τ_2''': Factor the expression of $f(x)$, simplify it and write it distinguishing two cases: $f(x)$ equals the simplified function when $x \neq a$ and $f(x)$ is not defined when $x = a$. Graph the simplified function and determine its limit using τ_2.

In short, one can say that the mathematical organisation developed in the observed class corresponds to the *practical block* of MO_1. However, two types of problems that do not strictly belong to MO_1 – Π_5 and Π_6 that correspond to T_4 – can be identified. Nevertheless, the associated techniques are 'low level' variations of mathematical techniques pertaining to MO_1. Thus, it can be said that the mathematical organisation actually developed does not go much beyond the practical block of MO_1.

4.3. *'Raison d'être' and technology of the studied mathematical organisation*

In the observed didactic process, the *raison d'être* of the mathematical knowledge actually taught responds to the single question of the *calculation of the limit of a function at a point or at infinity*, under the assumption that these limits (at least the one-sided ones) exist or are infinite. Because mathematical technology corresponding to this practice is absent from the curricular documents and textbooks, the observed teacher chooses to *describe and institutionalise* the used techniques as *rather transparent rules*, as if they did not need any justification. This approach of denying the problem is a common strategy in institutionalised human activities when the means of solving it – or approaching it successfully – are not available. Thus, the limit of the quotient of two functions, for instance, is never used to compare their asymptotical behaviour. On the other hand, the technological elements introduced by the teacher are not integrated in the students' practice. They belong to MO_2, the mathematical answer to the question of the existence of the limit, and this question cannot be really raised in the classroom.

5. DYNAMICS OF THE DIDACTIC PROCESS

The didactic process followed in the observed class to recreate the knowledge *to be taught* will be examined now. Our aim is to compare the relations that must be established between the different moments of the theoretical process with the empirical didactic process as it was *actually lived* in the classroom. After using the reference MO to describe both the knowledge *to be taught* (from curricular documents) and the MO *actually taught* (from the students' notes and the teaching practices observed in the classroom), we intend to show how the aforementioned restrictions can affect the possible ways of organising the study of the limits of functions. Our analysis will suggest that any intent to recreate the knowledge *to be taught* in the Spanish high schools will result in a MO that is very close to the MO *actually taught* in the observed class.

5.1. *The moment of the first encounter and the confinement at the thematic level*

As mentioned in section 1.2, following the anthropological model, any didactic process requires a *first encounter* with the MO in question. In the case of the teaching of limits of functions, the ambiguity begins here because neither curriculum documents nor textbooks are explicit enough about this MO and do not answer questions such as: What mathematical knowledge should I teach? Which are its main components? Why is it important? Why is it useful? In this situation, the observed teacher initially proposes a number of type Π_0 problems ('find the slope of a straight line') in preparation for type Π_1 problems ('find the slope of the tangent to a curve in a given point', considering that Π_0 and Π_1 can provide a good first encounter with the MO (see Appendix). This assumptions seems *possible* because all the mathematical objects involved in Π_0 and Π_1 are expected to be unproblematic for the students. Nonetheless, as stated above, neither Π_0 nor Π_1 forms part of the curricular mathematical tasks defining the knowledge *to be taught* around the 'algebra of limits'. What role do those problems play in this case in the *taught* MO? We consider that, for the teacher, Π_0 and Π_1 are an attempt to 'give sense' to the study of the main types of problems that will constitute the *taught* MO. This can explain why he does not propose a first encounter through the 'algebraic' task of calculating the value of a rational function when the value of its denominator is 'very close' to zero (without being equal to zero), which constitutes the 'main mathematical task' of the teaching process (what students should learn to do). He chooses instead a 'geometrical' task which consists in finding the 'slope of a curve at a given point' through calculating the successive slopes of the secants to the curve that tend to the tangent line at this point. And, after this first encounter, the students will not meet this kind of geometrical task again.

5.2. *The exploratory moment and the elaboration of a technique*

Even if the teacher intends to manage the *exploration* of the type of problems Π_1 through the *elaboration and functioning of a technique* τ_1, his attempts rapidly fail almost certainly due to the following factors:

(a) The complexity of Π_1, which seems to be more appropriate for a first encounter with other MOs, for instance involving the 'derivative of a function at a point'.
(b) The lack of the necessary technological elements (related to curves in the plane and the foundations of real numbers) to give more stability and robustness to this mathematical activity.

From the beginning of the third session, the teacher proposes the *exploration* of a new type of problems Π_2 related to the calculation of the limit of a function as $x \to a$ (a real). Thus, the problem of the 'slope of a curve' at a given point completely disappears and the question 'what is the meaning of the limit of a function at a point' becomes secondary. What remains is the *problem of the calculation of the limit* considered as the value that a function 'approaches' when its argument 'approaches closer and closer to' a given real number a. It is this type of problems that will finally form the 'core' of the *actually taught* MO.

The *mathematical technique* τ_2 that is initially elaborated to begin the *exploration* of the first specimen of Π_2 is, in fact, a sub-technique of Π_1. It consists in carrying out algebraic manipulations with the expression of $f(x)$ (factor and simplify the common factors) and in substituting a for x in the simplified expression of $f(x)$. The use of τ_2 increases the need for constituting a technological–theoretical environment to provide a description, interpretation and justification of τ_2 and its numerous variations. In particular, the dynamics of the didactic process requires providing the answer to the following questions (among many others):

1. Why can we replace x by a in the simplified expression of $f(x)$ if we could not do it before?
2. Why, and in which cases, do we need the table of values of $f(x)$ to calculate the limit?
3. Why is it necessary to make a distinction between 'right-hand limit' and 'left-hand limit'?
4. When do we say that $f(x) \to \pm\infty$ as $x \to a$: only if $f(x)$ has a numerator that tends to a real number different from zero and a denominator that tends to zero?

5.3. *Trying to constitute a theoretical–technological environment*

In view of these technological necessities, the restrictions that strongly delimit the teacher's praxeology are clear. Given the structure of the mathematical objects proposed by curricular documents and textbooks around the limits of functions – what we refer to as the 'double-headed' MO = $[T/\tau//] \cup [//\theta/\Theta]$ – the teacher has no technological elements at his disposal that are 'coherent' and 'reliable' enough to be integrated into the mathematical milieu of the student and provide some answers to the questions generated by the activity. This is why the teacher has to find answers to such questions among the previously studied mathematical objects that are, presumably, unproblematic for the students.

The teacher, therefore, decides to consider functions of a particular form:

$$f(x) = (x - a)g(x)/(x - a),$$

a special subtype of the study of functions defined piecewise (Π_3). This option facilitates the justification of the technique τ_2''' used to compute the limit of $f(x)$ at $x = a$ based on the assumed obvious relationship between the graph of functions defined piecewise and their algebraic expressions. Nonetheless, this technique fails because functions defined piecewise have not been mastered by the students. As such, the mathematical activity around the block $[\Pi_3/\tau_3]$, that the teacher tried to situate at the technological level (towards τ_2''') has to be started over again, with a first encounter, an exploration of Π_3 and the carrying out of the technique τ_3 by the students.

There is a deeper reason, however, not related to students' knowledge, for the failure of the teacher's strategy. It is clear that, in the last analysis, reading a graph cannot justify the fact that x can be replaced by a to compute the limit of $f(x)$ as $x \to a$ (central point of τ_2'''). Thus, the teacher is forced to explicitly propose 'reading off the graph' as a justification (and, implicitly, as a new technique τ_4) when computing the limit of any kind of function given by its graph, and not only those defined piecewise.

It is, again, the internal dynamics of the didactic process that determines the presence of a new practical activity around the block $[\Pi_4/\tau_4]$, which is now extended to include a new punctual organisation $[\Pi'_4/\tau'_4]$. It consists in computing limits from the visual interpretation of the graph of a function. Thus, τ_4 and τ'_4 are completely *naturalised*, in the sense that they are invisible and always taken for granted. Because they cannot be really considered as techniques, it becomes impossible to carry out technical work to improve them and, consequently, it is also very difficult to *institutionalise* them.

It follows that, in short, the mathematical activity around $[\Pi_4/\tau_4]$ and $[\Pi'_4/\tau'_4]$ cannot be developed into a relatively 'complete' didactic process, in other words, a process that can integrate all the different didactic moments. What is more, this activity cannot be considered as a justification of the calculation of the limit of a function at a given point because it also gives rise to new technological needs.

A third reason also exists, equally related to the internal dynamics of the didactic process, that can explain why this activity is abandoned and there is a return to type Π_2 problems. It concerns the important technical and technological difficulties that would appear if the problems of Π_4 (where functions are given by their graphs) were connected to those of Π_2 (where functions are given by their analytical expression).

These restrictions may explain why, at the end of the fourth session, the teacher decided to return to the exploration of Π_2 problems with the intention of developing τ_2 (which generates several variations of this technique). This work had been interrupted in view of the need for a technological–theoretical environment. Nonetheless, even if this goal is not attained, the work with τ_2 could not be delayed any further. During the next three sessions, the teacher managed the technical work trying to present successive variations of τ_2 as if they were completely 'natural' and did not require any justification. Thus, for instance, faced with the disorientation of the students, the teacher concluded by saying that, to compute the limit of $f(x)$ as $x \to a$: 'Sometimes we can replace [x by a] and sometimes we can not.'

5.4. *Concentrating the didactic process on the exploration and 'routinisation' of the techniques*

The last sessions were dedicated to what the teacher presented as a new and important type of problems: the study of the continuity of a function, Π_5. The first encounter is conducted through an oral discourse supported by graphs, in a very fast and natural manner. The continuity of a function f at a point a is not presented as a problematic question, but as an application of the computation of the limit of $f(x)$ as $x \to a$ and its comparison with $f(a)$. Thus, τ_5 (calculate the limits of a function at the boundary of its definition domain) and its variations τ_5^i therefore arise as simple 'applications' of previously explored techniques, and the *institutionalisation* of the produced work need only affect the new technological elements that are reduced to the identification of 'continuity' with 'regular behaviour of a function graph'. For instance:

θ_1 ='A function f is continuous at $x = a$ if the two lateral limits as $x \to a$ exists and are equal to $f(a)$'.
θ_2 = 'If a function is not defined at a point, it cannot be continuous at this point'.

Continuity problems are studied in a rather hasty way, as if they only consisted of a change of language in relation to the previously studied ones. In turn, the notion of 'elementary function' suggested by θ_2 assumes an unexpected technological role at the end of the process. Indeed, the teacher improperly extends θ_2 using an implicit definition of 'elementary function': if a function f does not exist at a point a, it cannot be continuous at a; but if f exists at a and is an elementary function, then the limits can be computed by replacing x by a. Subsequently f is continuous at a. As such, it is assumed that the students are already aware which functions are 'elementary' (for instance rational functions) and which are not (for instance defined piecewise). During the whole didactic process, the teacher attempts to justify, unsuccessfully, the mathematical work carried out around $[\Pi_2/\tau_2]$. But, this effort is useless without deeply modifying the knowledge to be taught.

In conclusion, in spite of the major effort made by the teacher to carry out a relatively 'complete' didactic process, the final result is disappointing. Whenever the teacher attempts to go beyond the *exploratory* moment and extend the merely routine use of a technique, he is led to building up the technological discourse using materials extracted from mathematical praxeologies that, in spite of the students' and the teacher's efforts, need to be further explored and 'routinised' by the students. In this way, it is not possible to reach a real technological questioning about the techniques proposed by the teacher. These are only superficially explored and weakly elaborated, thus becoming more rigid in the students' hands. This restricted dynamic of the didactic process involves the construction of a very incomplete and unstructured MO which, moreover, lacks a clear raison d'être, impeding the possibility of becoming globally institutionalised.

6. Didactic restrictions on the possible ways of teaching limits of functions

This last section considers different levels of didactic restrictions, from the most specific ones, arising at the thematic level of the limits of functions, to the most generic ones that arise at the level of school mathematics taken as a whole and even beyond. At every level, these restrictions affect the characteristics of the possible 'ways of teaching' limits in secondary schools, that is, the *possible didactic organisations* that a secondary school teacher can use. Consequently, these restrictions also affect the specific MOs that can actually be taught.

6.1. *Effect of 'thematic confinement' on the spontaneous didactic organisation*

The Anthropological Theory of Didactics suggests a *hierarchy of levels of co-determination* (Chevallard, 2002b) between the different possible levels of MOs that can be considered (punctual, local, regional, etc.) and the way its study is organised at school. This hierarchy is structured in a sequence of levels of MOs and didactic organisations which runs from the most generic level, the society, to the most specific one, a simple mathematical question to be studied, as shown below:

Society → School → Pedagogy → Discipline → Area
→ Sector → Theme → Question

Each level of codetermination introduces particular restrictions showing the mutual determination between mathematical organisations and didactic organisations. The structure of a MO at each level of the hierarchy determines the possible ways of organising its study and, reciprocally, the nature and the functions of a didactic organisation at each level determine, to a large extent, the kind of MOs that can be created (studied) in the considered institution.

This sequence of levels is relative not only to the considered question or group of questions, but also to the corresponding historical period and teaching institution. Let us consider the following mathematical question, for example:

How to compute the limit of a concrete function $f(x)$ as $x \to a$ or as $x \to \pm\infty$?

This question must satisfy certain conditions for it to be studied in a specific teaching institution. One condition is that it arises from a primary question in one of the higher levels of the hierarchy, which is higher than the discipline level, and that its study 'leads somewhere'. In other words, it must not be an 'isolated' question and therefore a 'dead-ended' one (Chevallard et al., 1997, p. 118).

Which institution can assume the responsibility for the questions proposed to be studied in schools (mathematical ones, for example) to satisfy these conditions? Can the teacher be responsible for making such a decision? In general terms, it can be noted that *this problem is out of the teacher's control.* According to Chevallard (2002b), the common situation is that the teacher 'neglects' the higher levels, both those concerning the society and the school, and even the level of the *sector* (in the case of the limits of functions, the sector may be the study of differentiability, for instance). When teachers prepare to teach the limits of functions, they do not decide to which sector the theme belongs. This decision is already taken. They only have to decide how to organise the study of the strict MO around the limits of functions. This 'confinement into themes' constitutes a didactic phenomenon labelled by Chevallard as the '*thematic autism*' of the teacher. It is related to the 'low level' status of the teaching profession.[7] The neglect of the higher levels is not absolute: sometimes mathematics teachers can pay some attention to the *discipline* level and even the school and social ones. However, they rarely express their concerns and opinions as teachers but simply as individuals or, at most, members of political groups or trade unions.

In short, the teacher is destined to go no further than the thematic level, and this situation has important consequences. In particular, the most significant outcome is the disappearance of the reasons of being of the studied MO (Chevallard, 2002b). Indeed, the majority of the mathematical questions that are proposed for study at schools are formed and disappear at the thematic level. Consequently, these questions are only vaguely connected to the higher levels of the organisation (sectors, areas and discipline), which are usually considered as transparent and unquestionable. Furthermore, school mathematical themes are not flexible enough to form local MOs and, therefore, cannot even be united in a functional way in either regional or global MOs. As a result, school mathematical questions are not only very weakly related to the higher levels of determination, but also appear in isolation from each other.

The teacher's responsibility is thus confined to the selection, in each theme, of the mathematical questions to be studied by the students and, although teachers can group the questions in themes, they cannot have an influence on the higher levels of the hierarchy in any relevant way. Thus, a deep gap appears between, on the one hand, the levels of themes and questions (within the teacher's reach) and, on the other hand, those of discipline, areas and sectors (within the limits of the mathematical scope in contrast to the pedagogical environment):

SOCIETY → SCHOOL → PEDAGOGY → DISCIPLINE → AREA → SECTOR
Limits of the mathematical scope

THEME → QUESTION
Field of the teacher's activity

This gap affects the teaching of all mathematical areas.[8] In the case considered here, the teacher's confinement at the thematic level causes the disappearance of *the motivation for studying limits of functions* in secondary schools. Indeed, it becomes impossible to justify why this theme must be studied without going beyond the thematic level. As has already been stated, curricular techniques for computing limits (and, in particular, 'for solving indeterminations') make 'sense' only in a larger regional MO that includes both MO_1 (answering the problem of how to compute limits of functions) and MO_2 (answering the problem of the existence of the limit of functions). Nonetheless, the study of a regional MO would introduce more restrictions, for instance, the two following conditions:

a) To take into consideration a *much more extensive family of functions* than the one usually considered in high schools.
b) To carry out a new 'linkage' (in sectors and in themes into each sector) of the punctual and local mathematical organisations constituting the differential calculus studied at school, so that every new theme contains enough elements to carry out a relatively complete study (in terms of the moments or dimensions of the mathematical activity) of a local MO.
c) Among other outcomes, this new linkage can lead to the disappearance of the 'limits of functions' as a theme in the curriculum of secondary schools and may also result in a *global change of the curricular contents of elementary differential calculus*.

To go beyond the mathematical level, teachers need some means and legitimacy to rebuild the themes of the curricular sector (and, furthermore, the sectors of the corresponding area, for example, all school differential calculus). It is also very difficult for them to extend the family of functions commonly studied in high schools to include those functions that could 'give sense' to the problem of their differentiability – a possible way to bring together the problem of the existence and the computation of the limit of a function at a point. Once the limits of functions in secondary schools have lost their reason of being, the teachers' field of activity is restricted to the level of the specific mathematical questions and the isolated mathematical techniques which can then only appear in a quite opportunistic form.

The outcome is clear. The impossibility of 'giving sense' to the calculation of limits without going beyond the theme, together with a lack of technology to interpret and justify the mathematical techniques used by the students, strongly restrict the field of didactic tasks and techniques that the teacher can use, that is, his spontaneous *didactic practical block*. And, once confined to the thematic level, the teacher is at the mercy of the *common pedagogical ideology* that, concerning the teaching of limits of functions, can be expressed for instance in the following terms: 'the most important thing is that students "understand" the limit concept'; 'algebraic manipulations do not have enough sense by themselves'; 'comprehension of limits requires some kind of geometric or graphical interpretation'; etc. The confluence of both the 'thematic confinement' and the structure of the MO to be taught (the 'two-sided' organisation) can explain the failed attempts of the observed teacher to take back or to give new sense to the didactic process.

6.2. *Restrictions coming from the higher levels of determination*

The study of any kind of mathematical question is restricted by the higher levels of co-determination. In fact, the chain of levels needed to allow any specific mathematical question to be studied starts at the most general levels of the hierarchy: the school and society. Restrictions at these levels are related to the way the question is considered by Society and the kind of educational role conferred on the School. The outcomes and scope of these restrictions are not analysed here. We will only describe some restrictions commencing at the pedagogical and discipline levels, with some incursions into the intermediate ones, and it will be shown that these restrictions are very much related to the spontaneous didactic technology that generates the possible ways of organising the study of limits at school.

The *pedagogical level* is the origin of restrictions affecting the study of any kind of questions and, thus, the different disciplines that are taught in order to answer these questions. Most of these restrictions come from certain conceptions of teaching and learning that are taken for granted. One of them is the belief that there exists a pedagogical domain which is independent of the mathematical one, in the sense that decisions taken in the former domain (grouping students, time organisation, distinctions and links between disciplines, learning assessment, etc.) would not affect the nature of the concrete mathematical tasks that are being carried out in the class (Chevallard, 2000). This distinction would permit discussions about teaching and learning independently of the nature (mathematical, economical or linguistic) of what is being studied. Consequently, the responsibility of this common organisation cannot be left to the mathematics teacher, who is subsequently even more confined to his or her 'thematic level'.

This way of thinking is so generalised and accepted that there is no discussion about whether or not it constitutes an important element of the spontaneous didactic technology. Evidence of this was obtained from the interview we held with the observed teacher (Espinoza, 1998) who used 'generic technological elements' to describe, interpret and justify his practice. He referred, for instance, to the necessity of developing the link between curricular contents and daily life as a tool to increasing students' motivation, and to the use of technology for improving both learning capacities and motivation. It is clear that the pressure of these principles can affect the lowest levels of the mathematical activity carried out in the classroom to compute the limit of the function at a point: the types of tasks that are chosen, the corresponding techniques, the way we describe and justify the performed work, etc.

The *discipline* level – here, mathematics – introduces some restrictions related to the way mathematics is interpreted in educational institutions. In particular, according to Brousseau (1997), 'common teaching models' can be considered as supported by 'naive epistemological models'. In secondary schools, the prevailing epistemological model of mathematics is very eclectic. Most of its main characteristics come from what is called the 'quasi-empirical' epistemology or 'quasi-empiricism' (Lakatos, 1978) identifying mathematical activity with the exploration of open problems. The other characteristics can be related to the 'constructive epistemology' that considers mathematical concepts as the result of human actions and operations (Gascón, 2001).

In the case presented here, the strategy of the observed teacher can be explained by the first kind of characteristics. What has been observed is an attempt from the teacher to regularly guide his students to the exploratory moment, attaching great importance to the exploration of the different problems in a free and creative way. It can be noted that this kind of didactic organisation assigns a very reduced amount of responsibility to the students who merely have to carry out the exploration of the specific problems proposed by the teacher.

It is clear that the possible ways of teaching about questions related to the computation of the limits of functions are also affected by restrictions starting in the *area* and in the *sector* to which those mathematical questions are referred to by the curriculum. For example, the *epistemological model of calculus* prevailing in secondary schools has an important role in specifying the possible ways of organising the study of the limits of functions. Some authors are beginning to produce empirical evidence to support this thesis and are starting to suggest the interdependence between the epistemological model specific to a given school mathematical area and the didactic organisations that can be used to study it (Artigue, 1998; Bloch, 1999; Schneider, 2001).

6.3. *Conclusions*

Considering the dimensions or moments of the mathematical activity that prevail in the observed spontaneous teaching process, we can state that the didactic organisation carried out by the considered teacher is relatively incomplete and biased as it concentrates the didactic process on the *exploratory moment* and the first steps of the *technical work*. In the research presented herein, a second teaching process was also considered and was used as a contrast to the first one. The didactic organisation carried out by this second teacher, even if it appeared very different to the former, was also only centred around two didactic moments. In fact, this second didactic strategy can be considered as a more 'classical' one, based on exhibiting the main technological elements (definition of limit, properties, etc.) of MO_2 and on 'presenting' the principal techniques to compute limits from MO_1. This didactic organisation left room for the technological–theoretical moment only and the moment of the technical work, in which the students 'applied' and 'practised' the techniques the teacher had just showed them through some typical examples on the blackboard (Espinoza, 1998).

In general terms, we can postulate that if the knowledge to be taught is made of a collection of punctual mathematical organisations that are not linked to each other through an operative technological-theoretical discourse, then the possible corresponding spontaneous didactic organisations that the teacher can use will not be able to really integrate the six different moments of the didactic process. Reciprocally, when the didactic technologies available in the teaching institutions are based on naive epistemological models (Euclideanism, naive constructivism or 'quasi-empiricism') and on general pedagogical slogans, then the possible didactic organisations tend to favour only a few of the didactic moments to the detriment of the others. It can be foreseen – but this requires, of course, further empirical research – that these spontaneous didactic organisations will have difficulties in overcoming the problem of the atomisation of the curriculum: it will not favour the integration of mathematical contents previously learned into the new ones, nor the links between different types of problems of the same mathematical organisation, nor, even less, the connections between different mathematical areas (algebra and analysis, for instance).

The concrete teaching process presented here highlights some didactic restrictions coming from different levels of specificity and affecting to a different extent both the mathematical knowledge actually taught and the possible ways of teaching it. These restrictions come from the different institutions involved in the teaching and learning process (society, mathematical community, educational system, school and classroom) and cannot be explained without taking into account the global process of didactic transposition. Problems of great social concern such as the loss of motivation towards scientific activities, the absence of meaning of school mathematical problems, the imposition of 'scholarly' mathematical contents that loses its rationale as it is brought into schools, the atomisation of mathematical curricula, etc., need a deeper understanding of the set of institutional restrictions that regulate teaching and learning processes. Without knowing their functioning and their extent, we will not be able to act on them in a controlled and well-founded manner to ensure progress in mathematics education.

Acknowledgements

We would like to thank Christer Bergsten and Yves Chevallard for their helpful reviews of this English version. This work was supported by DGI BSO2003-04000 (Spain), FONDECYT 1020342 and DICYT 9933ES (Chile).

Notes

1. Brousseau (1997) presents a compilation of his works published between 1970 and 1990.
2. A more complete version of this work can be found in Bosch et al. (2003) and Espinoza (1998).
3. Usually, since limits of functions are studied before their continuity, the 'regularity' of some functions (polynomials, for instance) is used to justify that the limit of a function at a point of its domain equals the value of the function at this point. This kind of argument is clearly a circular one and constitutes, as we will see, one of the weaknesses of the 'mathematical knowledge to be taught'.
4. Here is an example of an *institutional relativity* of the functions that mathematical objects can assume in a MO. While in MO1 the 'rules' play the role of the *technological discourse*, in MO2 they are an integral part of the *mathematical tasks.*
5. Artigue (2003) presents an analysis of the evolution of the teaching of calculus that helps to explain the current situation.
6. The functions $f(x)$ used in the process were essentially rational or simple irrational ones.
7. Didactic phenomena, like social, economic or linguistic ones, are independent of the will, the formation and the capacity of the individual subjects of the institution. As a result, taking the thematic confinement as such a didactic phenomenon is to consider it as a phenomenon the teacher does not create voluntarily and can only influence locally and to a relatively insignificant level.

8. Gascón (2003a) proposes an analysis of the effect of the thematic confinement on the teaching of geometry in Spanish secondary schools.

APPENDIX

TABLE AI
First description of the didactic process.

EPISODE	*Didactic Moment*	*Main Player*	*Present mathematical objects (N : New)*	*Observed didactic activities*
Introduction to the problem of calculating the slope of the tangent to the curve $y = f(x)$ at a given point.	FIRST ENCOUNTER with Π_1	Teacher	*Function* *Point* *Slope (N)* *Limit of a function at a given point (N)* *Curve*	**The teacher (T) starts the lesson saying to the students (S):** **T:** Limit of a function at a given point. Our purpose now is to find the straight of a curve in a point. **The teacher draws on the blackboard:** 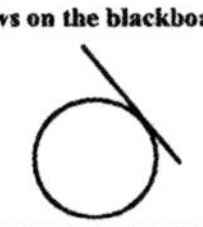
	Short Institution-nalisation	Student → T T → S	*Curve*	**S:** ¿What is a curve? **T:** A curve is a line. **The teacher continues his introduction.**
		T	*Tangent* *Circumference* *Point of a circumference* *Parabola*	**T:** Do you remember when we look for the tangent of a circumference at one of its points? **T:** Now we want find the slope of any curve, for instance a parabola, at a given point. **T draws:** 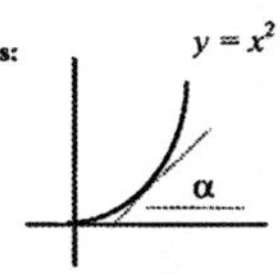
Definition of tan α and of the slope of a curve at a given point.	Institution-nalisation To remember an old problem and an old technique	T	*Angle α* *Tangent of an angle* *tan α = b/a* *slope = tangent* (N) *Tangent of a curve* (N) *Inclination angle α* (N)	**T:** The definition of the tangent is: **The teacher draws on the blackboard:** 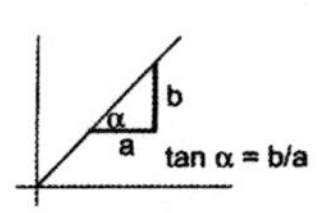 **T:** Thus the slope of a parabola at any point will be the tangent of the inclination angle α. **The teacher writes on the blackboard:** slope = tan α **T:** Now our goal is to calculate tan α.

(*Continued on next page*)

TABLE AI
(*Continued*)

Calculate the slope of the straight line $y = 2x$		T→S S	*Straight line* *Slope of a straight line*	**The teacher proposes a problem to the students.** **T:** Calculate the slope of the straight line $y = 2x$. Come on, do it! **After 3-5 minutes, he asks a student to solve the problem at the blackboard** **S *(writes, without speaking)*:** $\tan\alpha = \frac{4-2}{2-1} = 2$ 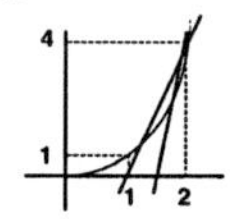
Calculate the tangent of the angle of the secant line passing through two given points of the curve.	Technical Work of Π_0	T→S S	*Slope* *Straight line* *tan α*	**The teacher proposes a problem that cannot be solved using only the mathematical objects known by the students.** **T:** Find the slope of the straight line tangent to the parabola $y = x^2$ at the point $x = 2$. 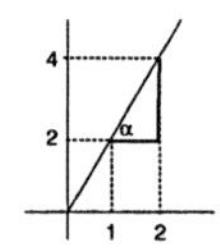 x | y 1 | 2 2 | 4 **T:** We have difficulties to calculate the tangent line; We need another point of the curve to find it.
The limit object is needed to solve the problem.		T	*Slope* (N) $y = x^2$ (N) $\tan\alpha$ (N)	**The teacher proposes another problem: to calculate the slope of the parabola at a point, drawing a sequence of secant straight lines** **T:** Determinate the slope of the parabola $y = x^2$ at a point $x = 1$. **The teacher solves the problem on the blackboard. At the same time he asks some questions to the students:** **T:** For x = 0'5 we calculated it, and we know that the result is....1'5. Now, if x value is 0'75, how much is the slope? Come on, do it. 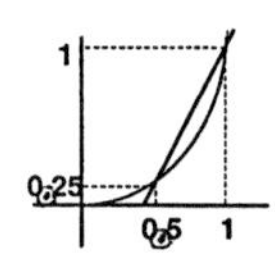 $\tan\alpha = \frac{1-0{,}25}{1-0{,}5} = \frac{0{,}75}{0{,}5} = 1{,}5$ **The teacher gives two minutes for the students to calculate it.**

(*Continued on next page*)

TABLE AI
(*Continued*)

	First Encounter with Π_1	T		**T:** What is the slope value? **S:** 1. **T:** Impossible. How much is it? **S:** 1'75. **T:** 1'75. O.K. **The teacher writes a table on the blackboard.** **T:** 0'5 → 1'5 0'75 → 1'75 **T:** How much is the image for the x value? **S:** 1'x **T:** What are you saying! **The teacher is surprised with this answer. He answers the question writing in the table:**
		T	*Limit* **(N)** *lim*$(1+x)$ = **(N)** $x \to 1$	0'5 → 1'5 0'75 → 1'75 $x \to 1+x$ **T:** The slope for the x value is $1+x$, and the slope of the curve at $x = 1$ is the limit of $1+x$ as x comes near 1. **The teacher writes on the blackboard** **T:** $\lim_{x \to 1}(1+x)$
End of the lesson				

TABLE A2
Analysis of the first session as reported in Table A1.

SESSION	*Type of Mathematical Problems*	*Mathematical Techniques*	*Explicit Technological-Theoretical Elements*	*Dominant Moment & Sub-moments*	*Elements of the Didactical Techniques*
1	**Π_1: Calculate the slope of the straight line tangent tangent to the curve $y = f(x)$ at a given point *a*.**		ϑ_1: Informal definition of a curve like a 'line' in the plane. ϑ_2: Definition of the tangent of an angle: $\tan\alpha = b/a$.  ϑ_3: Definition of the 'slope of a curve at a given point' as a generalisation of the slope of a straight line.	**First Encounter with Π_1**	• Considering the initial problem 'Calculate the slope of a curve at a given point' as the generalisation of some kind of previous solved problems: 'Calculate the slope of straight lines'. • Identify two mathematical objects (the slope of a straight line and the slope of a curve in a given point). • Put in action at the same time four mathematical objects without a clear differentiation between them (straight line slope, curve slope at a given point, straight tangent line and tangent of an angle). • Remember an old MO around Π_0 and modify it progressively to get the new type of problems Π_1. After that, introduce successive variations of the used technique in order to solve the new considered problems. That allows to considering the new problem as a variation of an old one.
	Π_0: Calculate the slope of the straight line $y = ax + b$ π_{01}: Calculate the slope of $y = 2x$	τ_0: Choose two points of the straight line, (x_0, y_0) and (x_1, y_1), draw the characteristic triangle and calculate: $\tan\alpha = \frac{y_1 - y_0}{x_1 - x_0}$ τ_0': Calculate the slope of a general straight line using the formula: $\tan\alpha = \frac{y_1 - y_0}{x_1 - x_0}$	ϑ_4: A straight line is determined by two points or by one point and its slope.	Institution-nalisation (remembering) of the previous work done with the tangent of angles	
	Π_0': Calculate the slope of straight lines that intersect a curve $y = f(x)$ at a given fixed point x_0 and a point close to it. π_{01}': Calculate the slope of the secant straight line that intersects the curve $y = x^2$ at the points (1, 1) and (0'5, 0'25). π_{02}': Calculate the slope of the secant straight line that intersects the curve $y = x^2$ at the points (1, 1) and (0'75, 0'5625). π_{03}': : Calculate the slope of the secant straight line that intersects $y = x^2$ at the points (1,1) and $(x, f(x))$.	τ_0'': Generalise the results obtained in a sequence of secant straight lines at a given curve in order to calculate the slope of the secant straight line that passes through a given point and a general point *x* of the curve.	Θ_1: Generalise a result obtained by an iterative sequence using the induction. ϑ_5: Use of the verbal expression 'limit as *x* comes near x_0' and of the notation $\lim_{x \to 1} (x+1)$		*In this session appears a specific management of the old/new dialectic, supported by a technological ambiguity (the identification of the slope of an angle with the slope of a straight line and the slope of a curve at a given point).* *It is an institutionalisation sub-moment that redefines old knowledge and builds the new mathematical environment of the lesson very quickly.*
	π_{11}: Calculate the slope of the straight line tangent to the curve $y = x^2$ at $x = 2$. π_{12}: Calculate the slope of the straight line tangent to the curve $y = x^2$ at $x = 1$.			**First Encounter with Π_1**	

REFERENCES

Artigue, M.: 1998, 'L'évolution des problématiques en didactique de l'analyse', *Recherches en Didactique des Mathématiques* 18(2), 231–262.

Artigue, M.: 2003, 'Learning and teaching analysis: What can we learn from the past in order to think about the future?', in D. Coray, F. Furinghetti, H. Gispert, B.R. Hodgson, G. Schubring (eds.), *One Hundred Years of* L'Enseignement Mathématique. *Moments of Mathematics Education in the Twentieth Century,* L'Enseignement Mathématique, Genève, pp 213–223.

Bloch, I.: 1999, 'L'articulation du travail mathématique du professeur et de l'élève dans l'enseignement de l'analyse en première scientifique', *Recherches en Didactique des Mathématiques,* 19(2), 135–194.

Bosch, M. and Gascón, J.: 2002, 'Organiser l'étude. 2. Théories et empiries', in Dorier J.-L. et al. (eds.), *Actes de la 11e École d'Été de didactique des mathématiques - Corps - 21-30 Août 2001,* La Pensée Sauvage, Grenoble, pp. 23–40.

Bosch, M. and Gascón, J.: 2004, 'La praxéologie comme unité d'analyse des processus didactiques', in A. Mercier (ed.), *Balises pour la didactique. Actes de la 12e École d'Été de didactique des mathématiques,* La Pensée Sauvage, Grenoble. (in press).

Bosch, M., Espinoza, L. and Gascón, J.: 2003, 'El profesor como director de procesos de estudio: análisis de organizaciones didácticas espontáneas', *Recherches en Didactique des Mathématiques* 23(1), 79–136.

Brousseau, G.: 1997, *Theory of Didactical Situations in Mathematics. Didactique des mathématiques, 1970–1990,* in N. Balacheff, M. Cooper, R. Sutherland, V. Warfield (eds.), Kluwer Academic Publishers, Dordrecht.

Chevallard, Y.: 1985, *La Transporition Didactique: Du Savoir Savant au Savoir Enseigné,* La Pensée Sauvage, Grenoble.

Chevallard, Y.: 1997, 'Familière et problématique, la figure du professeur', *Recherches en Didactique des Mathématiques* 17(3), 17–54.

Chevallard, Y.: 1999, 'L'analyse des pratiques enseignantes en théorie anthropologique du didactique', *Recherches en Didactique des Mathématiques,* 19(2), 221–266.

Chevallard, Y.: 2000, 'La recherche en didactique et la formation des professeurs : problématiques, concepts, problèmes', in Bailleul M. (ed.) *Actes de la xe École d'été de didactique des mathématiques (Houlgate, 18-25 août 1999),* ARDM et IUFM de Caen, Caen, pp. 98–112.

Chevallard, Y.: 2002a, 'Organiser l'étude 1. Structures et fonctions', in J.-L. Dorier *et al.* (eds.), *Actes de la 11e École d'Été de didactique des mathématiques – Corps 21–30 Août 2001,* La Pensée Sauvage, Grenoble, pp. 3–22.

Chevallard, Y.: 2002b, 'Organiser l'étude. 3. Écologie & régulation', in J.-L. Dorier *et al.* (eds.), *Actes de la 11e École d'Été de didactique des mathématiques – Corps 21-30 Août 2001,* La Pensée Sauvage, Grenoble, pp. 41–56.

Chevallard, Y., Bosch, M., and Gascón, J.: 1997, *Estudiar matemáticas. El eslabón perdido entre la enseñanza y el aprendizaje,* ICE/Horsori, Barcelona.

Espinoza, L.: 1998, *Organizaciones matemáticas y didácticas en torno al objeto 'límite de función'. Del 'pensamiento del profesor' a la gestión de los momentos del estudio,* Doctoral Thesis, Universitat Autònoma de Barcelona, Barcelona.

Ferrini-Mundi, J. and Graham, K.: 1994, 'Research in calculus learning: Understanding of limits, derivatives and integrals', in J. Kaput and E. Dubinsky (eds.), *Reserach Issues in Undergraduate Mathematics Learning,* MAA Notes 33, Washington, pp. 31–45.

Gascón, J.: 1998, 'Evolución de la didáctica de las matemáticas como disciplina científica', *Recherches en Didactique des Mathématiques* 18(1), 7–34.

Gascón, J.: 2001, 'Incidencia del modelo epistemológico de las matemáticas sobre las prácticas docentes', *Revista Latinoamericana de Investigación en Matemática Educativa (RELIME)* 4(2), 129–159.

Gascón, J.: 2003a, 'Incidencia del '*autismo temático*' sobre el estudio de la Geometría en Secundaria', in E. Palacián (ed.), *Aspectos didácticos de matemáticas*, Instituto de Ciencias de la Educación de la Universidad de Zaragoza, Zaragoza pp. 81–124.

Gascón, J.: 2003b, 'From the Cognitive Program to the Epistemological Program in didactics of mathematics. Two incommensurable scientific research programs?', *For the Learning of Mathematics* 23(2), 44–55.

Lakatos, I.: 1978, *Philosophical Papers*, Vol.2, Cambrige University Press, Cambridge.

Lang, S.: 1986, *Cálculo,* Addison-Wesley Iberoamericana, New York.

Schneider, M.: 2001, 'Praxéologies didactiques et praxéologies mathématiques. A propos d'un enseignement des limites au secondaire', *Recherches en Didactique des Mathématiques* 21(1.2), 7–56.

Williams, S.: 1991, 'Models of limits held by college calculus students', *Journal for Research in Mathematics Education* 22/3, 219–236.

ALINE ROBERT and JANINE ROGALSKI

A CROSS-ANALYSIS OF THE MATHEMATICS TEACHER'S ACTIVITY. AN EXAMPLE IN A FRENCH 10TH-GRADE CLASS

ABSTRACT. The purpose of this paper is to contribute to the debate about how to tackle the issue of 'the teacher in the teaching/learning process', and to propose a methodology for analysing the teacher's activity in the classroom, based on concepts used in the fields of the didactics of mathematics as well as in cognitive ergonomics. This methodology studies the mathematical activity the teacher organises for students during classroom sessions and the way he manages[1] the relationship between students and mathematical tasks in two approaches: a didactical one [Robert, A., Recherches en Didactique des Mathématiques 21(1/2), 2001, 7–56] and a psychological one [Rogalski, J., Recherches en Didactique des Mathématiques 23(3), 2003, 343–388]. Articulating the two perspectives permits a twofold analysis of the classroom session dynamics: the "cognitive route" students are engaged in—through teacher's decisions—and the mediation of the teacher for controlling students' involvement in the process of acquiring the mathematical concepts being taught. The authors present an example of this cross-analysis of mathematics teachers' activity, based on the observation of a lesson composed of exercises given to 10th grade students in a French 'ordinary' classroom. Each author made an analysis from her viewpoint, the results are confronted and two types of inferences are made: one on potential students' learning and another on the freedom of action the teacher may have to modify his activity. The paper also places this study in the context of previous contributions made by others in the same field.

KEY WORDS: teacher's activity, teacher's discourse, students' activity in the classroom, mathematical tasks, students' enlistment

1. Articulation of Didactical and Psychological Approaches to Mathematics Teacher's Activity

1.1. *Teacher's practices: A complex system, with individual, social and institutional determinants*

In the last few years teachers' practices have been studied from different theoretical viewpoints. Three main questions began to be elucidated: what links can be established between teachers' practices and students' acquisition of knowledge, what determines teachers' and students' activities, and how these results could contribute to improve the pre- and in-service training of the teaching staff? In our work, we are concerned with the first two questions.

Here we present the method we applied to the study of an exercise-based lesson on absolute value in a 10th-grade class.[2] Our purpose was to determine the mathematical contents the teacher brought into play during

Educational Studies in Mathematics (2005) 59: 269–298
DOI: 10.1007/s10649-005-5890-6

the lesson, in relation with the acquisition of knowledge, as well as to try to infer the factors which determined his approach. The latter allows us to assess the 'space of freedom' he may enjoy within the multiple constraints imposed on him. This method proposes a twofold approach: on the one hand – in a didactics-centred approach – we developed a general framework for analyzing teachers' practices taking into account two elements that are very closely linked, students' activities and the teacher's management of the class, (Robert, 2001); and on the other hand – in a cognitive ergonomics approach – we have considered the teacher as a professional who is performing a specific job (Rogalski, 2003).

Articulating these two approaches allows us to see teachers' practices as a complex and coherent system, which is the result of a combination of each teacher's personal history, knowledge and beliefs about mathematics and teaching, and experience and professional history in a given activity (Robert and Rogalski, 2002a). This is reflected in the scenarios the teacher chooses to present to a class, the way he expects them to unfold, how he adapts to students' reactions and in his evaluations at different moments during the process.

1.2. *A twofold approach engaging a didactical and a psychological perspective*

The double approach we propose was developed to allow us to analyze the different determinants of the teacher's activity as well as the activity of students prompted by the teacher in the class.

The psychological analysis of the teacher's classroom practices is based on activity theory (Leontiev, 1975; Leplat, 1997). The notion of activity is also used in the didactic approach from the point of view of students' activity in the sense of the activity we suppose they will develop for performing the teacher proposed tasks. The didactic approach and the psychological approach are used to tackle different issues.

In the didactic approach, our aim is to analyze the results of the teacher's activity in terms of the tasks that he had set for the students, without looking at the reasons for the choices he made, the existence of professional habits, and the nature of the decision making process itself. We are interested in the possible effects of the tasks on the students' mathematical activity during the lesson, according to the possible consequences in students' mathematical learning. We do not study these consequences directly but we analyze the teacher's practices in relation to the potential impact of the students' activities on their learning, insofar as the students engage in these activities.

This first approach takes into account the situation the teacher sets for students, the tools and the aids proposed to them, the use of the blackboard, the routines and regulations observed as the lesson progresses, and as far as possible the implicit didactical contract that the teacher establishes, how he fulfils it or adapts it in class, and his intentions.

In this first perspective, special attention is given to the 'mathematical universes' within which teachers make the students act (Hache and Robert, 1997), and their 'potential widening', that is, how the mathematical content is displayed and opened to students' activity (Hache, 2001) (both Piagetian and Vygotskian perspectives about learning underlie this approach).

To sum up, the issue we tackle is to specify teacher's practices according to students' activity in relation to mathematics learning.

In the psychological approach, we want to identify the functions which are fulfilled by the teacher's activity, with regard to the students. These functions are not limited to the definition of students' tasks and to the progress of the lesson. They are also concerned with how the teacher makes the students engage with the tasks, maintains their mathematical involvement, links individual students' answers to the whole class activity (which we will call "students' *enlistment*"[3]), how he assesses if students follow the lesson, understand the mathematical notions and, what are their difficulties, in order to maintain control in the class while adapting the lesson (which we call *situation assessment* or *diagnosis*). We search for some "internal economy," or "logic" in the teacher's activity, the reasons for his actions, and for the nature of his decisions.

This second approach considers the teacher as a professional, subject to a professional contract, with particular goals, repertories of action, representations of mathematical objects and their learning, and, more generally, personal competencies which determine his activity.

The teacher must define a learning environment with a dynamic organisation of tasks; this is analysed mostly through the first approach. At the same time he seeks to *win* the students *over*, or 'enlist' them for these tasks and 'enlisting' is a key component in the second approach.

To sum up, the issue we tackle is to specify the teacher's activities and to explain his choices relative to his own point of view: doing his work successfully.

These approaches are not in conflict, but rather in a relation of complementarity. We take into account both the fact that there are two main types of means used in classroom management: the organization of tasks for the students (the cognitive – epistemological dimension), and the direct interactions through verbal communication[4] (the mediation – interaction dimension).

Furthermore, what the teacher is doing in terms of organizing students' mathematical activity, through the presentation and management of the mathematical tasks, also has an impact on how he will succeed in maintaining the students' engagement with the tasks and staying in control of the progress of the lesson. Reciprocally, the teacher's actions which are oriented towards students' enlistment or their individual errors or difficulties will put constraints on possible students' mathematical tasks and activity.

As we will see later, our analyses of the same lesson from two different perspectives, each focusing on specific issues, are actually overlapping, even if the same observations are identified differently and different aspects are being stressed.

For instance, in the lesson used for presenting our twofold approach, we will show how a process of "fragmentation" of mathematical tasks may be seen as a means for keeping them within students' reach. Since this is also an effective way of keeping the students on task, and willingly so, i.e. classroom enlistment, such process of fragmentation might be reinforced during the lesson, perhaps against the conscious will of the teacher. Reciprocally, taking into account an individual misunderstanding or a quite unusual solution proposed by a student might result in a loss of control of the classroom mathematical involvement: such risk may lead the teacher to offer a rather "superficial" answer, for example only reminding the whole class of the right notion or a taught procedure.

1.3. *The lesson*

The lesson analyzed here is the second and last lesson about the absolute value of real numbers. It belongs to a chapter about order and approximation, which began in the previous lesson by defining the distance d between two real numbers: $d(a, b) = \mathrm{AB}$ where A and B are points on the real line. The definition of $|x|$ is also given at the beginning of the course: it is OM where M is a point with abscissa x on the real line with origin O. Then it follows that the absolute value is always positive. It is either x or $-x$ depending on which of the two is positive.

Then "c is an approximation of x with precision r" was defined as $d(x,c) \leq r$. The absolute value $|a - b|$ was defined as equal to $d(a, b)$.

After the definitions, a series of equivalent characterizations was presented and justified:

> "Saying $|x - c| \leq r$ is equivalent to saying that x belongs to the interval $[c - r; c + r]$; it is equivalent to saying that $c - r \leq x \leq c + r$, and this is equivalent to saying that x is equal to c with precision r (or c is an approximation of x with precision r)".
>
> Or in a formulaic expression:

(E) $|x - c| \leq r \Leftrightarrow x \in [c - r, c + r] \Leftrightarrow x = c$ with precision $r \Leftrightarrow c - r \leq x \leq c + r$

These equivalences were presented on the blackboard, in a table, with numerical examples. In the following we will refer to this set of equivalent characterizations as (E).

We have to stress that, from this point on, the teacher was not using the graphical representations anymore.

Students began to solve an exercise; they had to finish it as homework.

The lesson we are about to analyze begins with a recall of (E) and the correction of homework: in three tasks (T1, T2, T3) students were required to give equivalent expressions for $|x + 2| \leq 0.5$; three other tasks asked for equivalent expressions of "$x = 2$ with precision 0.5". (The expected answers are given below for each task).

T1: $x = -2$ with precision 0.5
T2: x belongs to $[-2.5, -1.5]$
T3: $-2.5 \leq x \leq -1.5$
T4: $|x - 2| \leq 0.5$
T5: x belongs to $[1.5, 2.5]$
T6: $1.5 \leq x \leq 2.5$

Then the teacher announces that the following exercises, consisting of solving equations and inequalities, will be the last ones in the course about absolute value.

First exercise

T7: $|x| = 11$
T8: $|x| = -1$
T9: $|x - 1.5| = 3$
T10: generalisation: $|x - c| = r$ is equivalent to $x = c + r$ or $x = c - r$

Second exercise

T11: $|x| \leq 4.5$
T12: $|x - 2| \leq 7$
T13: $|x - 2| \leq -5$
T14: generalisation: $|x - c| \leq r$ is equivalent to $c - r \leq x \leq c + r$
T15: $|6 - x| = 1$
T16: $|x| > 5$

Third exercise

T17: $|x - 5| = 1$
T18: $|3 - x| \leq 7$
T19: $|x + 5| \geq 4$
T20: generalisation: $|x - c| \geq r$ is equivalent to $x \geq c + r$ or $x \leq c - r$

Finally, students were asked to copy on their exercise book all that was done in relation with solving equations and inequalities with absolute value, and they were given a last exercise from their textbook as homework for practicing for the next evaluation.

2. An Analysis of Classroom Practices Focused on the Students' 'Cognitive Route' Organized by the Teacher

We will use some categories developed in the field of didactics of mathematics in analyzing classroom practices observed during the lesson described above. The aim is to connect the analysis of the teacher's practice with an *a priori* analysis of students' tasks and further with their activities. This can then lead to analyzing students' potential learning, that is, learning which can be inferred from the type of students' activity triggered by the teacher's decisions. Learning is not directly tackled here but this perspective influences the subsequent analysis.

The mathematical content processed during the lesson – mathematical concepts, properties of examples proposed, types of tools used (for representing or computing), types of tasks given to students – can be considered as a *cognitive route* organized for students in this conceptual field, 'absolute values', in this case. *Potential learning* can then be inferred from the activities students perform while following this type of cognitive route according to teacher's management (Hache and Robert, 1997).

The fundamental unit we consider finally is the couple: {assigned task, lesson in progress}. The series of assigned tasks is linked to the teacher's intentions, while the actual lesson in progress reflects how the teacher adapts his actions to students' behaviour. Our units of analysis are 'episodes' identified in the transcript of the session. Each episode is related to a task or sub-task assigned by the teacher; for the purposes of this paper, episodes are related to each of the T1–T20 tasks. The analysis of each episode looks at the task from several points of view: the mathematical point of view, the teacher's management point of view, and students' activities point of view.

Our didactical approach can then be divided into two steps: the first step is concerned with the mathematical contents and tasks of the episodes; the second step is devoted to actual teacher's management during each task.

In a lesson such as the studied one, the first step leads us to determine the mathematical content of each task, its place in the sequence of lessons, its relation to the broader concept to be taught and the level of knowledge actually needed for the task: is it a direct application, or is there a need for an adaptation? It may be a partial recognition of the knowledge to be used, or recognition of the modalities of this use, or use of intermediaries – notation, unknowns, elements, change of the setting or a combination of different concepts to be applied. A task is considered to be *isolated* when it does not require the use of different "objects of learning" (Robert and Rogalski, 2002b; Robert, 2003); it does not mean that it is unconnected with previous tasks. These categories are brought forth because of the variety of students' activities involved in tasks.

We also reflect on the openness or not of the question in the task, on the guidance provided or not and on the steps that may or not be introduced. Finally we try to determine the means of internal control that students could use. In fact this is an *a priori* characterization of students' tasks, exactly what they are supposed to do in terms of use of knowledge acquired during the course.

These expectations are then, in the second step, compared with what *actually* happens in the classroom. The way students work, what they are exactly asked to produce, the time allowed for each task, the teacher's guidance, oral or written on the blackboard (hints, questions or other interventions) allow us to reconstruct what the students 'have to do' according to the teacher's actual proposals. We must also take into account the moment at which the teacher intervenes, for example, before or after the students' response, the type of help given, direct or indirect, through prompts or answers. It also allows us to infer the customs that have been established and the implicit contract underlying classroom activity. Thus we can finally define the students' activity expected during the lesson and specify the actual use of knowledge, autonomy and initiative. One can recognize here some factors linked to subsequent learning.

2.1. *An initial a priori analysis of the tasks in our lesson*

We worked on a video made during an 'ordinary' lesson in a 10th-grade class and its transcript. As said above, the lesson was planned as a series of exercises (homework and three exercises). Students worked at their desks except when they were asked to come to the blackboard.

The aim of each of the three classroom exercises was to establish the method to be applied in a general case.

Almost all possible combinations in the application of the (E) formulas, in the direction of inequalities and the signs for c and r, are present. But the students have no (independent) means at their disposal to assess the validity of their results – such as a graphic representation; they have to rely on the teacher's or other students' (direct or indirect) evaluation of their work.

2.1.1. *Homework correction*

The first six questions, which were given as homework, only required a direct application of (E). From an expression of the series of equivalences students had to find the other three by placing them in a pre-established table with four columns, drawn on the blackboard. This was done twice, each time for an initial expression given by the teacher: $|x + 2| \leq 0.5$ (T1, T2, T3), and $x = 2$ with precision 0.5 (T4, T5, T6). Students could also read on the blackboard the formula to be used. Given the way that the exercise is presented one could think that their only task was to replace the variables (c and r) by the specific numerical values in the given expression. This is what we call an isolated task since only one formula of those presented in this lesson has to be used in answering each question. Here, "isolated" means that it does not require the use of different new "objects of learning", even if other (old) objects occur, as it is explained above. The task may be more or less simple depending on whether the replacements require the use of their knowledge of algebra. For example, writing $|x+ 2|$ as $|x- \mathrm{c}|$ requires transforming 2 into $-(-2)$ and is not simple.

2.1.2. *The classroom exercises*

The next three exercises were all built using the same format, albeit different from the one used in the homework assignment: students must solve three or four tasks of the same type and then find and express the general solution for this type of problem.

In the first exercise (tasks T7–T10) the student has to solve equations of the $|x - c| = r$ type.

The first two tasks can be solved directly by using the property of the absolute value of an algebraic quantity x mentioned above: '$|x|$ is x or $-x$ depending on which of the two is positive. The third task requires an adaptation of this property to the absolute value of $(x - c)$. The fourth task is to find a 'ready made' formula such as (E) after having discussed the existence of a solution. In fact, we are dealing with isolated tasks, but using the relevant knowledge, just as mentioned above, even if the student does not have to search for it nor to make it explicit, is far from simple. What is involved here is the way in which it is used, since to solve an equation with x is not only to find a few particular values of x, some of which could be obvious, but all values of x satisfying the equation.

In the second exercise students have to solve inequalities of the type $|x - c| \leq r$. The first four tasks of the second exercise are using the same equivalences as in the homework but in the context of solving inequalities. Students have to realize that solving such inequalities can be done using one of the equivalences given in (E). One exception, task T16 (($|x| > 5$), is a transition to the third and last set of tasks. Here again we have six isolated tasks, but none of them is simple. In T11, T13 and T16, students have to adapt their initial property of absolute value, with special work required in the last task, because it does not directly follow from the previous lesson. In T12, T14 and T15 they must adapt the (E) formula.

In T15, identifying the variables (x, c and r) and the given numerical values requires 'arranging' the inequalities by moving expressions from one term to another and using $|6 - x| = |x - 6|$, working both with the opposites and with the absolute value.

Moreover, students have to consider (E) from a different angle. What they initially saw as statements about the order between concrete numbers must now be considered as inequalities where x is an unknown whose values must be determined, even if the expressions look the same.

The last exercise includes one question of each of the three types already seen (**T17**: $|x - c| = r$; **T18**: $|c - x| \leq r$; **T19**: $|x + c| \geq r$) and completes it by a generalisation in T20 of inequalities $|x - c| \geq r$, of which an example was dealt with in T19. For the first two tasks, the simplest procedure, where fewer steps are required, is to use the formulas obtained in T10 and T14, which leads to adaptations similar to those already applied before. In T19 students can adapt what was obtained in T16 by using $-(-5) = 5$, which may be a real adaptation for some students even if it was already used. T20 requires a generalization of what was seen before in this exercise and the previous one.

2.2. *Study of the sequence of events in the classroom: The teacher's control of students' activities during the lesson, the use of "models" and the time allowed for students to do their work*

A priori it seems that the teacher is proposing 20 isolated tasks which require the application of different formulas learnt in the course with some necessary adaptations. Yet we shall see that the management of the lesson can tell a different story. We will examine the activities expected from the students after each intervention by the teacher. According to the way in which the teacher introduces a question, the time allotted to students and the teacher's information, we determine the tasks the students actually have to perform as the lesson proceeds.

Actually, we observed different ways the teacher initiated students' work, but we have to stress first that all the tasks Ti were almost immediately followed by interventions from the teacher proposing a series of sub-tasks. This simplified the tasks for the students, and it forced them to use the formulas given in the lessons on absolute value (initial property or the series of equivalences), in some cases while it was still in the process of being learnt. We also noticed that these formulas were not always explicitly mentioned.

We present two ways this teacher would start a task, taking control immediately or soon after, with some specific examples showing the teacher's decisions following the initial choice. We then give a brief conclusion on students' activities during this lesson.

2.2.1. *One way of starting the task: The teacher immediately takes control of the work done by the students*

In this configuration the teacher begins by indicating an initial sub-task and proceeds with a rapid succession of questions (corresponding to new sub-tasks) to which students reply, usually giving incomplete answers that the teacher sometimes completes. Let us take two very revealing examples.

2.2.1.1. *First example of taking immediate control of students' work.* We consider T9: solution of $|x - 1.5| = 3$, in the first exercise after homework.

The teacher asks a student to read it out and immediately says: "then, the first thing we may have to decide is whether it's like T7, where it's possible, or like T8, where it's impossible."

After the briefest of answers by students, which he validates, it is he who says "why it is possible" (an absolute value can be equal to a positive number) and ads without a pause: "what can we do now to continue to solve this equation?"

After four seconds he takes the first answer given by a student and completes it: the student said "$x - 1.5$", he says "$x - 1.5 = 3$". He does not correct an error in the student's formulation, when the student says "*when the absolute value is positive*" instead of "*when the quantity of which we take the absolute value is positive*".

Without giving the students the time to solve the first equation obtained, he immediately asks: "is that all?" which indicates that the task is not finished. Once again he takes a very vague response from students ("*equal* -3") giving them a 'clean' version, "*then* ... $x - 1.5$*, well, it's going to be equal to* -3".

Finally he asks them to solve it and leaves them 15 seconds before verifying the results.

Thus in these exercises the activities of the students have been completely and immediately organized by the teacher in the series of steps he indicates. Students are not involved in deciding about these sub-divisions: they simply answer the brief questions asked: "*is it possible or not?*", find a first solution, completed by the teacher as an equation, find a second equation, given by the teacher in its correct form, and then solve both equations. The longest time allowed to students is given to make both calculations. Not much is asked of them to justify their answers.

Four other tasks (T5, T7, T11 and T16) are guided in the same manner. In the last three the initial response of the students corresponds to the case in which the expression in the absolute value is positive. The teacher accepts it once more without comment and immediately asks "*is that all?*" He then takes the first adequate response given, concludes: "*so...*" and gives the complete answer.

2.2.1.2 *Second example of the same teacher's activity.* We consider T17: solve $|x - 5| = 1$, in the last exercise.

As soon as the student who has been asked to come to the blackboard arrives the teacher asks: "*then, if we look at our model, are we in the first case or are we already in the second?*" Recall that the teacher uses "model" to indicate the series of equivalences (E), written on the blackboard.

Here the first sub-task is to recognize which of the modalities of the pre-established model (formula) is appropriate. After validating the correct answer, justified with only one word (equation), the teacher imposes a second sub-task: "*are there any solutions?*" The student gives the correct answer and justifies it by referring to the model. The teacher repeats it and asks the student to use the model which has been 'contextualized', that is presented in a specific mathematical task: "*Well then, you go ahead*" (13 seconds for this calculation).

Unfortunately there is an erroneous application of the model: the student uses the (E) formula corresponding to an inequality; and the teacher quickly rectifies it. Another error appears at the end in the solution of $x - 5 = 1$, which could stem from an insufficient command of algebraic calculations. Yet the teacher attributes it to a simple error in calculation and does not explore the previous knowledge involved.

In this case the work done by students is basically centered on identifying the modalities of the formula to be applied and on the final calculations. This way of starting an exercise is almost always used in tasks whose objective is to lead them to use the contextualized model or a formula. Students are guided by the immediate fragmentation into sub-tasks indicated by the teacher, which necessarily leads to the use of formulas: it concerns T2, T3, T4, T12, T13, T17 and T18.

2.2.2. *Another way of starting the task: The teacher takes control after beginning with an "open" search for a solution*

We now study the case where the teacher allows the students to propose a way to find the solution. He quickly rejects the proposals if they are wrong and then hints to the correct solution. This is a slightly different approach and it is mostly used in questions which involve using properties linked to the definition of absolute value and not to (E).

Let us take T15 (solve $|6 - x| = 1$, in the second exercise) as an example. The teacher immediately indicates that an adaptation is necessary ("*it's not exactly the same*"). But this adaptation turns out to be difficult. The teacher corrects each false start but does not always refute it: in the choice of the unknown value/variable (6 or x), the inverse instead of the opposite, without specifying the variable concerned and the possible change in the direction of the inequality. Furthermore, as soon as he can, he manages to propose another sub-task during the discussion: calculate the absolute value of $-a$. The students give the wrong answer. He corrects them and continues the exercise.

We may wonder what produced such errors in students. Is it momentary distraction, a mistake in the calculation or does it reveal a lack of command of algebraic calculations: maybe confusion between the handling of equalities, inequalities and the multiplication by -1. But this possible reorganization of previous knowledge applied to new calculations is not part of the present task. The exercise continues and, as usual, the final calculation is left to the students. And yet this final step that was so carefully prepared again reveals misunderstandings, but the teacher continues to lead students to a 'mechanical' identification of c and r in the model of this inequality.

In T8 (solve $|x| = -1$ in the first exercise) we have another example of the teacher completing the suggestions he has allowed the students to propose. He merely transforms an unsatisfactory expression given by a student ("*the absolute value is always positive*") into "*so, it's never negative... it cannot be smaller, a positive number can never be smaller than a negative number*".

Two of the three questions leading to generalisation (T10, T14, T20) are also strongly guided in this second manner.

In conclusion, in every task we observed an early or immediate intervention of the teacher in the work of the students which simplified these already isolated tasks. Besides, as soon as possible, he proposed himself the application of a formula (called model) or a reference to previous exercises: students are invited to use "models" instead of recognizing or restoring a proof, which of course would be longer and would not lead directly to memorize new formulas. It becomes impossible to restart a proof for a part of the result in one specific problem. Thus the teacher does not allow mixed procedures' to emerge, where there would be, a combination of the general formulations learnt in class and those which are specific to the problem.

This does not mean that he does not adapt to students' reactions, but these adaptations stay within the frame of the tasks he has planned, which do not include an exploration of the conceptual field linked to the concept of absolute value.

There is also no possibility for his detecting an error due to an incorrect learning of previous procedures and not simply to an error in the calculation.

This leads further to a greater fragmentation of the work proposed and students only need to use the partial knowledge required. If the students still cannot handle it, it is further simplified. This is particularly true when it is possible to use a formula (model). It is not up to the students to determine the necessary steps. At best they only need to answer, as a class, the question of how to solve each sub-task, and in particular how to use the model. All they have to do on their own is the calculation that was thoroughly prepared collectively.

The lesson proceeds as if the teacher, because of the constraints under which he works, had to make sure that the time given to the activity of students is used on working on a new acquisition – (E) formula in this case – and that he can trust it will be 'well done'. The calculation is carefully prepared so that students apply what they must learn that day and in particular the application of models. It is worth noting that most of the time allotted to student activity is spent on calculations, more or less the same as needed for the students to settle down to work. Among the institutional conditions limiting the teacher's space of freedom, the additional time constraint imposed (in 2002) by the reduction of the time allotted to 10th-grade mathematics courses in France has had strong effects.

3. TEACHER'S ACTIVITY CONSIDERED AS MANAGEMENT OF A DYNAMIC ENVIRONMENT

We now analyze the work of teachers from a psychological perspective as a particular case of dynamic environment management: their action is concerned with the relation between students and mathematical knowledge (Rogalski, 2003). This relationship has its own dynamics: it is not only determined by the teacher's interventions but is also evolving through processes external to the work proposed in class: beside the individual activity students develop during a lesson, the work done outside the classroom and the processes regulating cognitive acquisitions (maturation, reorganization and forgetting) are determining the dynamics of learning. The management of this environment has various components: the elaboration of scenarios corresponding to teacher's establishing a didactical 'process', its real time development in the classroom while he 'guides' the class and the evaluation of results which lead him to modify his initial project.

The psychological approach aims at identifying the whys and the hows of the teachers' actions; that is, the functions fulfilled by their interventions and the modalities by which they realize these functions. General functions such as diagnosis/prognosis and decision making are permanent functions in any dynamic environment management and they are a natural focus of interest. Beside these general features, the teachers' work has a very important distinctive trait: the 'objects of action' are human beings – the students. Teachers' interventions consist in setting tasks for the students, acting as a mediator while the students carry them out, triggering and controlling their mathematical activity as shown through the didactical approach. Insofar as the activity of other psychological subjects is a goal in the teacher's own activity, he is working on *enlisting* the students for the proposed tasks, which requires specific acts. Moreover, while learning concerns each individual student, teaching is addressed not only toward individuals but also – or mainly – toward the classroom as a whole and we suggest it is particularly important in general secondary schools. One issue in the analysis of the teachers' activity is then to identify the focus of their didactical interventions.

In the analysis of the transcript of the lesson, already analyzed from a didactical perspective, we will look for signs indicating the role of his actions aimed at winning the class over for the proposed tasks and we will assess the hypothesis that the central concern of his activity was the class as a whole, and not individual students. Our analysis was based mainly on verbal indicators, especially on discourse markers.

The discourse markers used in speech give coherence to verbal exchanges; they signal the existence of a link between what a student says and what the teacher answers, or between what the teacher says and the response expected from the student, maybe an oral response or action. They are grammatically optional and do not alter the truth value of what is said (Schourup, 1999). They have a double function: (a) they mark the structure of the verbalized content and play a role in the coherence of the teacher's discourse to the class: words like *then, that's it, there it is,* or *so* – when it is not used as a causal connective; (b) they punctuate the progress of the activity and may mark the role of the speaker: words like *you know, I mean,* etc.

There were many such discourse markers in the lesson (180); they fell in two categories: (a) markers introducing statements that place students in their role as students by the use of the imperative mode or of instructions given in the present tense; these markers played a key role in teacher's actions aimed at enlisting the students for the tasks; (b) markers addressed to the class as a whole, punctuating the progress of the activity in class and guaranteeing that all the students were working towards the same goal at the same time.

3.1. *How is the action of enlistment of students performed?*

Our first hypothesis is that the enlistment of students is a dominant goal. It was Bruner who proposed the term of 'enlistment' in an operational use of Vygotsky's concept of the adult as a mediator in the development of children (Wood et al., 1976; Vygotsky, 1985). It is applied to the action of an adult whose goal is to involve a child in a given task. Vannier-Benmostapha (2002) used the term in a study of the teacher's mediation while teaching the same mathematical concept in three different institutional settings to pupils with various degrees of learning difficulties.

In the teaching process, enlistment is a means to make children or adolescents play the role the educational institution assigns to them as students. It also aims to devolve to them the task the teacher has indicated. If we look at it from the viewpoint of the teacher's intervention in students' activities, we can see that an early enlistment maintained throughout the duration of the tasks is a necessary condition for the teacher to be able to act on the student/knowledge relation through students' accomplishment of the assigned tasks.

Enlisting may be performed in various ways, with two contrasting poles: motivating students for and through an autonomous work, with the support of collective work; or, on the contrary, strongly directing the students' activity, triggering and orienting it for well defined – and small – tasks performing, and punctuating the succession of tasks through explicit markers.

3.1.1. *Triggering the activity of students through discourse markers*
Markers used to 'trigger' the activity of the student who is interrogated, or of one or more student at their desks, are very frequent. The observed teacher mostly used the words "*then*" (*alors*, in French) and sometimes "*so*" (*donc*) with the same meaning. The wording can be explicitly imperative: "*Well then, write down what you just said*', "*So, now you're going to copy it*" or phrased as a question the student must answer: "*Now, who can explain it to him*?," "*Well then, Alice, can you come and do it?*". In both cases the imperative nature of the question is clear since the students immediately respond by an answer or by an action. They obviously have not been perceived as rhetorical questions marking the continuation of the teacher's presentation which is immediately followed by his own answer, such as: "*Then, what are c and r here? c is equal to 6 and r is 1*". These activity triggers are numerous. The teacher we observed used them in almost half of his interventions (58 out of 110), mainly the word '*then*' (51 occurrences).

3.1.2. *Punctuating the succession of tasks*
There are two main types: closing markers, linked to a repetition of the result obtained, "*that's it!, so, do you agree?, ok?*," and opening markers: "*then, so*," explicit temporal indications of the next step, "*and then,*" "*now*" *we're going to...*", "*finally*" etc., or an identification of the next step, "*first question in exercise 2*", "*let's continue*", "*the last one is* $|x| > 5$". A significant number follow the pattern S below, where the closing marker may appear before or after the reminder of the result obtained, or else the closing markers may be absent and teacher simply repeats the last result obtained.

(S): = <closing marker> <reminder of result> / <opening marker> <temporal marker and/or identification of the next task>

In the teacher's 110 oral interventions, and the 20 tasks of the lesson, there were 24 S patterns. Thus we can conclude that there is a strong temporal articulation of the students' activity structured by discourse markers, with a predominance of <*that's it / then*>. These patterns of interaction, which we may call schemes, used by the teacher, offer the students an essential component for the control of their activity by indicating when one task is finished and the other begins.

3.1.3. *Orienting students' task performing*

Another possible means for keeping the students on task is to orient their task performing through proposed or imposed ways of acting. In the lesson observed, there was a forced use of the formula which plays this controlling role.

We have seen examples where the teacher introduces, at a very early stage, a sub-task which forces the students to apply the formulas previously studied in class. Recall that teacher says "models" when he wants students use their formulas. This mediation by the teacher can be seen as a means for keeping the students enlisted through a strong orientation to what has to be done for task performing. It was marked by the use of verbal indicators which link the work being done in the lesson to the model. Most often, after an explicit reference to the model, the teacher would say "*here*" or occasionally "*there*", used as an equivalent of "*here*", as a reference to work in progress.

In some cases he refers to a general model:

- $|x - c| \leq r$ is equivalent to of x equals c with precision r, all right, so **here**, if we want to have a minus when we have $|x + 2| \ldots$ we have to write $2 = -(-2)$. (T1)
- how are A and B defined? $x \leq c - r \ldots$ and $x \geq c + r$, now you do it **here**.

Sometimes it can be a specific exercise treated as a generic case, or as an illustration of the general model, for example:

- back **there** we saw a case where it was sometimes possible and sometimes not; **here** we could ask the same question. (T9)
- before, when we had $|x|$ greater than 5, we said that... **there** it's not x but $x + 5$, but it's done in exactly the same way.

During the lesson, out of the 17 tasks which were not generalisations, there were seven referrals to the 'general model' and four to a 'generic case'.

3.2. *Classroom as the focus of the teacher's activity*

The processes of students' enlistment are concerned with both individual students and the classroom as a whole. The use of the pronouns '*you*' or '*we*' in punctuating the succession of tasks or in orienting students' task performing can be seen as an indicator of a focus being on the classroom as a whole, and not on individual students. In the last case, the teacher would use the pronoun '*tu*', rather than '*vous*'. There are other significant indicators of the fact that the teacher was focusing on the whole class: (1) the way the teacher reacts to students' oral contributions, from their seats or at the blackboard, is an opportunity to share with the class the activity of the student who speaks, thus having the whole class follow the development of the task; (2) from this perspective, the fragmentation of tasks is also a way of constantly keeping *all the students* working *on the same task*; (3) the type of decision made when faced with a student's error or unexpected answer may also give insight into this point.

3.2.1. *The response to students' oral contributions*

There are several possible reactions to students' contributions: letting them develop a proposal, letting other students react, or directly interacting. In the last case, reaction may be a direct repetition; we distinguish three categories: without any explicit evaluation, repetition emphasized by a discourse marker, and repetition expressing a positive conclusion of an item. The teacher's reaction may also involve some correction; we also identified three categories: repetition with corrected wordings; response with an implicit correction – generally as a question about the others' agreement, and response which explicitly corrects an error. The teacher's oral response to students' answers is a way to make his activity public. It is addressed to the class and indicates the point reached in the development of the task, and it is not necessarily linked to any previous or current evaluation.

Table I presents examples of the six categories of teacher's responses to students' oral contributions.

Almost two thirds of the interventions made by students during the lesson (71 out of 109) are repeated by the teacher for the class. This confirms how students' activity is closely managed and that it is publicly done. Half of the teachers' reactions involves an element of correction, almost equally divided among the three categories; half of them are direct repetitions, which means that the teacher is not only often correcting wrong answers as soon as they are produced, but that he is making students' contribution 'public' with regards to the classroom.

TABLE I
Categories of teacher's responses

Type of responses	Example	Context
Direct repetition		
• repetition of the student's wording	T: which gives an absolute value of ...	– teacher to class
	s: $x + 2$	– student at the blackboard
	T: $x + 2$	– teacher to class
• repetition with a simple marker	s: there's +2 and −2	– student from desk
	T: so, +2 and −2	– teacher, with the discourse marker *so*
• repetition with markers, expressed agreement and conclusion of an item in the exercise	s: x is between −2.5 and −1.5	– student at the blackboard during the correction of an exercise
	T: that's it, so now we can conclude that x is between −2.5 and −1.5	– teacher, with the indication that the student's answer is valid
Response with an element of correction	T: explain more clearly what you mean, what did we see in class?	– teacher, after an ambiguous formulation
	s: so there's $x - c$ in the absolute value	– student from the class (same formulation used to say "there's $x - c$ in brackets")
	T: the absolute value of $x - c$ is ...	– teacher: formal oral formulation of the expression $\|x - c\|$
Response with an implicit correction	s: less than or equal to 0.5	– student writes $x + 2 = 0.5$
	T: to 0.5,	– (simple repetition)
	So, what about the rest of you, do you agree with what André wrote?	– teacher gives an implicit indication on the quality of the response
Response with an explicit correction of an error	T: then we have $x - c = r$...	– teacher is generalising
	ss.: and	– students in the class
	T: no, not *and*, *or* $x - c = -r$	– teacher: correction of an error often made by students ('and' instead of 'or' for union of sets)

3.2.2. *Incidents as "decision points"*

The 'public' nature of the teacher's interactions with students allows him to know almost all the time what the students are doing: during some longer phases of students' calculating at their desks, he would walk around in the classroom and look over their shoulders at their written work. In oral contributions he would not wait to see where their activity will lead them, and he would evaluate them on the spot. Sometimes, students were leaving the mathematical route he decided for them, either in their errors or in their ways of tackling a task: his reactions are indicative of him focusing on the whole classroom.

In the analysis of the teacher's activity we identified points at which he had to make a decision. This is particularly clear when we see how he manages 'incidents', that is, errors or unexpected interventions by students (Roditi, 2003). Errors are almost always corrected as soon as they are identified: the right formulation or answer is directly given by the teacher; the deep nature of the error is not questioned. There are probably several reasons. One is the possible difficulty of making a deep diagnosis on the spot. Another is linked to the focus on the class as a whole: questioning a particular student on an erroneous conception could engage a diagnosis process out of reach for the students, and which might create, therefore, a disturbance in the cognitive activity of the class, even if it could be of interest for the 'wandering' student.

In several episodes students proposed a solution to the exercise which did not include the application of one of the 'models' presented in class. Some proposals tried to relate the problem to one they were familiar with. For solving equations with absolute value, one student said "*we take out the absolute value and we solve the rest*". Another suggested, for solving $|6 - x| = 1$, "*we can work with a new variable:* $-x$" (this could allow students to use the model with the new variable and $c = -6$). In such cases, the teacher was first trying to understand the proposal, and as soon as he identified a reason – for himself – he stopped interacting with the student and came back to shared ways of working with the 'new' knowledge being learned ('*So, I see.., but what about* $|6 - x| and |x - 6|$?').

4. CROSSING THE TWO ANALYSES OF THE SAME PROTOCOL, REPRESENTATIVENESS OF THE LESSON, AND CONSEQUENCES FOR TEACHING AND LEARNING

The two analyses of the same protocol (analysis of the mathematical route proposed to students and that of the teacher's discourse during the lesson) were done independently. As we have seen, the results obtained in the two complementary analyses converge and reinforce one another. Both show that the students' activity is fragmented and also reveal the central role the 'models' (mathematical formulas) play in the teaching/learning process in the case under scrutiny (concerning algebra: it certainly depends on the content). Furthermore, both support the hypothesis that the fragmentation of tasks into small and precise sub-tasks enables the teacher to lead students through a predetermined cognitive route as well as to enlist them and to maintain a close control of the class. We elaborate on that below. We will then give some comments on the representativeness of this lesson for this teacher, and infer implications on students' learning from the twofold perspective and on the degree of freedom the teacher enjoys.

4.1. *Fragmentation of tasks into small units*

The fragmentation of tasks (exercises) into "small" units has already been described in this paper and analyzed from the didactical point of view. It can also be seen as a means of ensuring that, as the lesson proceeds, the class is working on the same task: if, on the contrary, a whole task, which could be approached in various ways, was proposed and the students were allowed to explore different possibilities, they would probably follow different paths. Then their questions or their proposals would only be of interest to those who were trying to elucidate the same problem at a given time, and the interventions of the teacher would be of interest only for them. In that case the teacher may no longer obtain the enlistment of all students. Some may be sufficiently involved in their exploration to benefit from the interactions, but others, maybe many others, would lose interest because they would consider that the teacher's interventions were 'beside the point'.

4.2. *How representative is the lesson studied*

Obviously the observation of only one lesson is not enough to characterize the practice of this teacher, not even of the whole of his activity on the subject of absolute value.

The analysis of the interview with the teacher gave us some indications as to the representative character of the lesson, in which the teacher himself had chosen to be observed:

"I tried to choose a lesson as normal – in quotes – as possible in relation to others [. . .] It was a lesson given to a half-class as any other [. . .]. There are other activities in which the students work in groups but then I arrange the tables in a different way, so they aren't facing the blackboard; but when they are facing the blackboard it's always more or less the same".

His comments on the reason why he used the blackboard as he did also confirm the importance he attaches to "*not having the class lose interest*": "*when we go to the blackboard it's because the whole class is working together*".

The same concern is expressed when he says that when he validates a correct initiative coming from a student it is "*to keep the class involved*"; the teacher wants them "*to be involved in the lesson as it develops*". This confirms the importance he attributes to enlisting the students and to the fact that the class as a whole 'sticks' to the ongoing lesson.

Finally we have an indication which confirms the role of models (mathematical formulas to be used) identified from a didactical perspective as well as in the organization of the teacher's discourse (such as "*you remember..., so here...*"). When asked if he could have done 'otherwise' he answers: "*yes I could have given some explanations differently, but the general frame of the lesson would have been the same because, after all, there aren't umpteen ways of working with equations with absolute values*".

Furthermore, previous studies in secondary schools essentially based on the first approach (Hache and Robert, 1997; Robert, 2001; Robert and Vandebrouck, 2003; Roditi, 2003) have always shown a great coherence in the practices of the same teacher. The organization of lessons of the same type has few variations. Fragmentation often occurs, especially at the beginning of the presentation of new notions. The use of the blackboard hardly varies. Difficulties, common to many, arise in class when during the same class there is an alternation between the presentation of knowledge by the teacher and the time devoted to work by the students.

4.3. *Conclusions regarding students' learning*

Our inferences concerning students' potential acquisition of knowledge are based on a socio-constructivist theoretical framework integrating concepts developed by Piaget and Vygotsky and their discussion of each other's work.[5] Our hypothesis is that the potential activity that the teacher proposes to the students, through the mathematics they work on, will, at least partially, determine their knowledge of the underlying mathematical object. From the data obtained we infer which tasks may be appropriated by students and what are the object of their mathematical activity. We also deduce which mathematical concepts the teacher is leading – mediating through his interventions – during the lesson.

In the lesson observed, the subdivision of tasks in intermediate questions and the constant prompting of students while they are working seem to allow the students to appropriate them: both devolution and enlistment succeed. Students really participate in a mathematical activity which can contribute to an acquisition of a concept, or at least initiate it. Yet the analysis shows that the tasks of students are *a priori* isolated: they only deal with the chapter being studied and require few adaptations of what is learned. They are often further simplified *a posteriori* by the subdivisions presented in the management of the lesson. There are reasons to fear that the cognitive component, 'organization of knowledge', may well be the victim of this fragmentation and of the fact that there is little dynamic interplay between the mathematical content presented and its use in exercises.

If we examine this from the point of view of the mediation by the teacher between students and mathematical knowledge we see that the management of students' activities leaves no room for them to wonder how to tackle the solution of a problem. The question of 'what has to be done' is immediately given by the teacher. The same applies to the procedures used for the solution of problems, even when the students are asked to provide the answers. Furthermore the time allotted to students' responses only allows brief answers by some students to 'well formulated questions'; only the time for the students, all of them, if possible, to make the final and already well defined calculations, is less limited. The interventions provide a framework for what the students can do on their own. Nothing is left undefined, students never face uncertainties: there is little room for autonomy.

We also observed that activities referring to the same notion are presented sequentially, often at different times: students use their cognitive tools (definitions, or formulas, solving procedures) separately, one after the other. They only need to have knowledge of the tools needed for a particular lesson and their use is prompted by the subdivisions organized by the teacher. Under these conditions there is no need to devolve to students the means to control their activity, which is one of the main factors of acquisition in a constructivist view of learning. They need not structure their knowledge to act; the teacher does it for them. Neither do they need to mobilize 'old' knowledge and assess their applicability to new situations. Finally, the forced use of 'models' strongly favors reasoning which goes from the 'decontextualized' to the 'contextualized': it gives even less weight to all that can actively develop the capacity of students to establish relationships, to explore different possibilities and to organise their knowledge.

There is no certainty that this management of the class results in the fragmentation of knowledge in students since students often do learn what has not been explicitly taught, knowledge being more or less implicitly devolved. During the lesson we did observe some participation of students in a discussion that didn't follow the path proposed. For instance, in the example referred to above, a student proposed to "*take out the absolute value and solve the rest*"; even when the teacher disqualifies the proposal with irony ("*fine, we get rid of the absolute value and then it's easy!*"), the student refuses to be discouraged and repeats his proposal. There were other similar episodes during the lesson: so we see that some students at least do not lose all their autonomy in spite of a very strong enlistment.

Besides, the students do 'follow' the fulfilment of tasks closely controlled by the teacher both in the sense of the cognitive route and of a participatory behaviour, albeit simply by listening. This may later allow them to reproduce these steps by themselves. It is difficult to infer if most of them will be able to handle more complex activities on their own.

4.4. *Teacher's activity and alternatives open to the teacher*

The hypotheses on the degree of freedom enjoyed by the teacher are based on the analysis of the various determinants which impinge on his activity.

One of the important external determinants is the official syllabus for this course. It is very restrictive in its reference to absolute values. The comment indicates that "*the absolute value of a number makes it easy to speak of the distance between two numbers*". We have to stress that the syllabus, the comments and the textbooks are not completely coherent on this notion. The textbooks which propose an implementation of the program are organized in a sequence of small units. The program restricts the inter-relation between the various new concepts learned, and with those previously acquired, because of its content and because of the limited time allotted to it. There are more constraints than maneuvre space.

Furthermore, the teacher's conceptions of the object to be taught and of the relation students/knowledge taught are subjective determinants of his professional activity. They condition the 'didactical process' he wants his students to follow (planned cognitive route) as well as the management of the processes which develop during the lesson. Our hypothesis, backed by the convergence of both analyses, is that this teacher believes that, in the case of absolute values, his job is to teach students *how to apply models* (*general mathematical formulas*).

Finally, when the wish to enlist the whole class leads to decisions analogous to those based on cognitive conceptions, these determinants strongly reinforce one another. This makes it improbable that the teacher will look for other alternatives.

Only the observation of very poor results in his students' acquisition of the knowledge taught could lead him to make changes in his practice. But there is little chance of his being destabilized since the present system hardly allows for the emergence of situations in which students need to use anything other than locally procedural knowledge.

5. Discussion and Perspectives

Our twofold approach comes from an imperious necessity to connect results on the organization of learning in the classroom with studies about professional practices. Indeed, in French research, as we recall below, there have been many studies allowing a fine-grained diagnosis of classroom functioning according to students' learning; but these studies gave few insights into why teachers would not change their practices even when they knew and agreed with these studies. Furthermore, in most of the various studies of the teaching/learning issue, the main focus was on students' learning; teachers' activity as a professional was not really taken into account. Although two approaches used two theoretical perspectives, one centered on students and the other on teachers, we think that their connection of the two differs from ours. After addressing relations between our approach and others, we conclude by articulating some research perspectives for the future.

5.1. *French research on teachers in the didactical institution*

Many French studies on the didactics of mathematics insist on the place of the teacher within the didactical system.[6] They have developed within Brousseau's theory of didactic situations (Brousseau, 1996; 1997; 1998) or in the frame of Chevallard's anthropological approach (Chevallard, 1999). In both, the models proposed take into account the place of the subject matter in the educational system (Coulange, 2001; Margolinas, 2002). They underline the institutional and epistemological constraints that limit the (re)production of new situations (Arsac et al., 1992), determine the regulations affecting the proposed didactical situation (Comiti and Grenier, 1997; Mercier, 1998; Hersant, 2001), or, more generally, the constraints which limit the activity of the mathematics teacher (Perrin-Glorian, 1999). These studies focus their attention simultaneously on the teacher, seen as the 'generic one', and on the student seen as the 'didactical subject'. Mercier, for instance, stresses that the theory of situations "*describes the organisation of the didactical space as a space of action for the students and the teacher, but it does not describe the actual actions of the teachers, of the students or their interaction*" (Mercier, 1998, p. 282).

Considerably less attention is given in these works to the individual exercise of the profession, and this is a great difference with ours. It is clear that their main aim is not to explain teachers' choices but to model the global system. Inferences from both types of research are different: our work may lead to building hypotheses about teacher training and other works may lead to a better understanding of the general constraints of the system.

5.2. *Research centred on teachers in the class context*

Schoenfeld (1998) proposed a theory where teaching is seen in context. He emphasised that the way teachers exercise their activity is mainly determined by their personal conception of mathematical knowledge, their own learning history and their beliefs about the ways students learn mathematics. He explains how, once these determinants are identified, it is possible to predict the decisions the teacher will make in case there is a (didactical) incident. He does not establish an explicit link between the teacher's decisions and the learning of students, nor does he refer to the 'exogenous' constraints. We believe that, on the contrary, these constraints also have a direct influence on the teacher's activity: the determinants of his activity do indeed have a social and institutional component.

5.3. *Research centred on communication in the mathematics classroom*

Various studies focused on communication in the mathematics classroom stressed the existence of (invariant) '*patterns of interaction*' (Voigt, 1985); Voigt used the term '*funnelling*', coined by Heinrich Bauersfeld, to describe those prevalent in the so-called *traditional classrooms* (the case of the lesson we analyzed). More generally, the Sinclair-Coulthard's approach to discourse analysis (not studied in mathematics classrooms) (Sinclair and Coulthard, 1975) "*found in the language of traditional native-speaker school classrooms a rigid pattern, where teachers and pupils spoke accordingly to very fixed perceptions of their roles and where the talk would be seen to conform to highly structured sentences*" (McCarthy, 1997).

Our focus on the interactive dimension, is also found in studies on communication in mathematics classes (Steinbring et al., 1998), initiated by the German school of the didactics of mathematics (Voigt, 1985; Krummheuer, 1988, among others), and developed in analyses of significant events for students when they interact – with the teacher or other students – in class. These studies are mainly focused on the students' point of view: the teacher is seen as an organizer of "*learning opportunities*" through the different types of interactions in which the children participate (Cobb and Whitenack, 1996, p. 215); the teacher plays a key role in "*fitting out routines and ways of interacting*" (Wood, 1996, p. 102).

Steinbring's studies (1997; 2000) on the interactive development of mathematical meaning strongly stress the role of mathematical knowledge. He analyses "*what kinds of ideas and meanings regarding mathematical knowledge are constituted during the course of this process [of classroom interaction], and how do the communicative patterns and the epistemological constraints of the mathematical knowledge influence each other*" (Steinbring, 1997, p. 79). His epistemological perspective and the twofold approach we propose could be considered as perspectives from two sides of a looking glass: ours is oriented towards the teacher's activity in mathematics' teaching, Steinbring's towards the student's activity in mathematics learning. There is however an important difference: He considers "*mathematics class as an autonomous culture in which the understanding and growth of mathematical knowledge develops in a self-referential way*" (Steinbring, 2000, p. 146), we consider it as a place where the teacher does a job whose aim is to act on the relation between students and knowledge to be learned, within the framework of the constraints imposed on him and the resources he can use.

5.4. *Research bringing together different theoretical perspectives*

Although we tackle issues similar to those stressed by Jaworski: the '*teaching triad*': '*Mathematical challenge*' (MC), '*management of learning*' (ML), and '*sensitivity to students*' (SS) for analyzing teachers' practices (Jaworski, 1998, 2003), the structure of our approach differs from hers. MC and ML are both considered in the first approach, but *establishing norms* and *fostering ways of working* (in ML) are mainly analyzed through our second approach. SS is partly considered in its cognitive dimension in the first—it concerns the adaptation of tasks to observed students' activity—and partly in the second: interaction with individual students or with the class to keep students enlisted.

The framework we propose shares important features with Even and Schwarz (2003), who studied a high-school mathematics class. These authors also apply two different theoretical perspectives in the analysis of the same mathematical lesson, and one of them is the activity theory on which our second approach is based. The results of their 'classic cognitive approach' confirmed that it is possible to 'consider the whole group of students as an entity' and recognize 'the central role played by the teacher'. They interpret the results obtained in the two approaches as conflicting. This would merit an in-depth discussion.

5.5. *Conclusion and perspectives*

Are our findings of a general nature? The method we propose – a cross-analysis based on bringing together two theoretical approaches – show results that converge. It allowed us to determine the properties of teaching activity and its relation with the activity of students. It could also at a later stage be extended to include the relation between the activities of teachers and the learning of students and thus go beyond the description of regularities.

This method permits systematic comparisons between the different dimensions of teachers' activities in the classroom: moment in course development, content, type of class. It could also be useful in comparing the activity of different teachers and thus examine the impact of social and institutional determinants by identifying what is invariant in their practice.[7]

Interpreting classroom research in primary mathematics education (Krummheuer, 2000) or analyzing mathematics teachers' practices at the end of compulsory schooling might call for somewhat different approaches. Indeed, the status of each child and of the whole class might be not the same in primary education and in more advanced levels.

Moreover, high on the research agenda should be the issue of how much can be validly inferred about the learning of students by studying the activities of teachers, since this is the main motivation for these studies on the teaching of mathematics and most particularly for those studies which aim at contributing to improve teacher training. Admittedly, it is reasonable to expect that students' knowledge would be related to the mathematical cognitive route the teacher organizes for them. But to expect, even on the most reasonable grounds, is not the same as to prove.

NOTES

1. To alleviate the text, the masculine pronoun 'he', rather than the compound 'he or she' will be used throughout the text.
2. The first level of the three in the French 'lycée'. All students (15 or 16 years old) at this level follow the same curriculum in mathematics.
3. This term is used by Bruner and others (Wood, Bruner, Ross): *"This means that vis-à-vis the 3-yr-old the tutor has the initial task of enlisting the child as tutoring partner"* p. 95 and we decided to take it in spite of the military connotation. It means that the teacher tries to keep the students in the class, with him, even before they start working on mathematical tasks.
4. In our approach, when we analyse the verbal exchanges between the teacher and the students, we are basically concerned with the teacher's interventions and their purpose. An approach such as Steinbring's (1997, 2000), on the contrary, sees these exchanges from the point of view of the individual student who wants to use them as a help in learning. Others, like Voigt or Krummheuer study the interaction of the teacher/class "couple". The teacher may also use his "verbal actions" in class as a means of reconsidering his understanding (representation) of what the students are doing, in other words, as an "on-line" or on-the-spot diagnosis; we have not studied it here.
5. Vygotsky's critique of Piaget's first two books on thought and reasoning in children was published in 1932. Piaget's response to Vygotsky's comments appeared in the first English edition of Thought and Language (Vygotsky, 1962; Piaget, 1962/2000). (In the first Vygotsky's French edition: Vygotsky, 1985, pp. 45–100 et pp. 387–399.)
6. Teachers' practices were examined in contributions presented in the last four *Écoles d'Été de Didactique des Mathématiques'*. Different studies focus on specific components of the teachers' practices; they appear in the following proceedings:

Noirfalise, R. and Perrin-Glorian, M.-J. (ed.): 1995, *Actes de la VIIIème École d'Eté de didactique des mathématiques*, IREM, Clermont-Ferrand.
Bailleul, M., Comiti, C., Dorier J.-L., Lagrange, J.-B., Parzysz, B. and Salin, M.-H. (eds.): 1997, *Actes de la IX° École d'été de didactique des mathématiques*, ARDM & CA Bruz.
Bailleul, M. (ed.): 1999, *Actes de la X° École d'été de didactique des mathématiques*, IUFM & ARDM, Caen.
Dorier, J.-L., Artaud, M., Artigue, M., Berthelot, R., and Floris, R. (eds.): 2002, *Actes de la XIème École d'été de didactique des mathématiques*, La Pensée Sauvage, Grenoble.

7. Studies which compare the same teacher in different situations or two different teachers tend to confirm the existence of invariants. They are present in the same teacher, even when he teaches very different contents. Maurice and Allègre (2002) observed it in the time given to students to find the answer. Other studies show the role played by the contents being taught. For example Zaragosa (2000), in a study done in primary school, found that the mode of devolving a problem to children depended both on the experience of the teacher and on the type of situation: modeling versus application of procedures, arithmetic versus geometry.

REFERENCES

Arsac, G., Balacheff, N. and Mante, M.: 1992, 'Teacher's role and reproducibility of didactical situations', *Educational Studies in Mathematics* 23, 5–29.

Brousseau, G.: 1996, 'L'enseignant dans la théorie des situations didactiques', in R. Noirfalise and M.-J. Perrin-Glorian (eds.), *Actes de la 8ème École d'été de didactique des mathématiques*, IREM de Clermont–Ferrand, pp. 3–16.

Brousseau, G.: 1997, *Theory of didactical situations in mathematics: Didactique des mathématiques 1970–1990*, (trans.) N. Balacheff, M. Cooper, R. Sutherland and V. Warfield (eds.), Kluwer Academic Publishers, Dordrecht.

Brousseau, G.: 1998, *Théorie des situations didactiques*, Grenoble: La Pensée Sauvage.

Chevallard, Y.: 1999, 'L'analyse des pratiques enseignantes en théorie anthropologique du didactique', *Recherches en Didactique des Mathématiques* 19(2), 221–265.

Cobb, P. and Whitenack, J.W.: 1996, 'A method for conducting longitudinal analysis of classroom videorecordings and transcripts', *Educational Studies in Mathematics* 30, 213–228.

Comiti, C. and Grenier, D.: 1997, 'Régulations didactiques et changements de contrat', *Recherches en Didactique des Mathématiques* 17(3), 81–102.

Coulange, L.: 2001, 'Enseigner les systèmes d'équation en Troisième. Une étude économique et écologique', *Recherches en Didactique des Mathématiques* 21(3), 305–353.

Even, R. and Schwarz, B.B.: 2003, 'Implications of competing interpretations of practice for research and theory in mathematics education', *Educational Studies in Mathematics* 54, 283–313.

Hache, C.: 2001, 'L'univers mathématique proposé par le professeur en classe', *Recherches en Didactique des Mathématiques* 21(1.2), 81–98.

Hache, C. and Robert, A.: 1997, 'Un essai d'analyse de pratiques effectives en classe de seconde, ou comment un enseignant fait fréquenter les mathématiques à ses élèves pendant la classe', *Recherches en Didactique des Mathématiques* 17(3), 103–150.

Hersant, M.: 2001, *Interactions didactiques et pratiques d'enseignement – le cas de la proportionnalité au collège*, Thèse de doctorat, Université Paris7.

Jaworski, B.: 1998, 'Mathematics teacher research: Process, practice and the development of teaching', *Journal of Mathematics Teacher Education* 1, 3–31.

Jaworski, B.: 2003, 'Research practice into influencing mathematics teaching and learning development: Towards a theoretical framework based on co-learning partnerships', *Educational Studies in Mathematics* 54, 283–313.

Krummheuer, G.: 1988, 'Structures microscopiques des situations d'enseignement des mathématiques', in C. Laborde (ed.), *Actes du premier colloque franco-allemand de didactique des mathématiques*, La Pensée Sauvage, Grenoble, pp. 41–51.

Krummheuer, G.: 2000, 'Interpretative classroom research in primary mathematics education. Some preliminary remarks', *Zentralblatt für Didaktik der Mathematik* 5, 124–125.

Leontiev, A.N.: 1975, *Activité, conscience, personnalité*, Moscou: Edition du progrès.

Leplat, J.: 1997, 'Regards sur l'activité en situation de travail, Paris: PUF.

Margolinas, C.: 2002, 'Situations, milieux, connaissances', in J.-L. Dorier, M. Artaud, M. Artigue, R. Berthelot and R. Floris (eds.), *Actes de la 11e école d'été de didactique des mathématiques*, La Pensée Sauvage, Grenoble, pp. 141–155.

Maurice, J.-J. and Allègre, E.: 2002, 'Invariance temporelle des pratiques enseignantes: Le temps donné aux élèves pour chercher', *Revue Française de Pédagogie* 138, 115–124.

McCarthy, M.: 1997, *Discourse analysis for language teachers*, Cambridge University Press, Cambridge.

Mercier, A.: 1998, 'La participation des élèves à l'enseignement', *Recherches en Didactique des Mathématiques* 18(3), 279–310.

Perrin-Glorian, M.-J.: 1999, 'Problèmes d'articulation de cadres théoriques: l'exemple du concept de milieu', *Recherches en Didactique des Mathématiques* 19(3), 279–322.

Piaget, J.: 1962/2000, 'Commentaries on Vygotsky's criticisms of "Language and thought of the child", and "Judgement and reasoning in the child"', *New Ideas in Psychology* 18, 241–259.

Robert, A.: 2001, 'Recherches sur les pratiques des enseignants de mathématiques du secondaire: imbrication du point de vue de l'apprentissage des élèves et du point de vue de l'exercice du métier d'enseignant', *Recherches en Didactique des Mathématiques* 21(1/2), 7–56.

Robert, A.: 2003 'Tâches mathématiques et activités des élèves: une discussion sur le jeu des adaptations introduites au démarrage des exercices cherchés en classe', *Petit x* 62, 61–71.

Robert, A. and Rogalski, J.: 2002a, 'Le système complexe et cohérent des pratiques des enseignants de mathématiques: une double approche', *Canadian Journal of Science, Mathematics and Technology Education (La Revue Canadienne de l'Enseignement des Sciences des Mathématiques et des Technologies)* 2(4), 505–528.

Robert, A. and Rogalski, M.: 2002b, 'Comment peuvent varier les activités mathématiques des élèves sur des exercices. Le double travail de l'enseignant sur les énoncés et sur la gestion en classe', *Petit x* 60, 6–25.

Robert, A. and Vandebrouck, F.: 2003, 'Recherches sur l'utilisation du tableau par des enseignants de mathématiques de seconde pendant des séances d'exercices', *Recherches en Didactique des Mathématiques* 23(3), 389–424.

Roditi, É.: 2003, 'Régularité et variabilité des pratiques ordinaires d'enseignement. Le cas de la multiplication des nombres décimaux en sixième', *Recherches en Didactique des Mathématiques* 23(2), 183–216.

Rogalski, J.: 2003, 'Y a-t-il un pilote dans la classe ? Une analyse de l'activité de l'enseignant comme gestion d'un environnement dynamique ouvert', *Recherches en Didactique des Mathématiques* 23(3), 343–388.

Schoenfeld, A.: 1998, *Toward a Theory of Teaching-In-Context*, http.//www.berkeley. edu/faculty/aschoenfeld/TeachInContext/tic.html.

Schourup, L.: 1999, 'Discourse markers: Tutorial overview', *Lingua* 107, 640–667.

Sinclair, J.McH. and Coulthard, R.M.: 1975, *Towards an Analysis of Discourse*, Oxford University Press, Oxford.

Steinbring, H.: 1997, 'Epistemological investigation of classroom interaction in elementary mathematics teaching', *Educational Studies in Mathematics* 32, 49–92.

Steinbring, H.: 2000, 'Interaction analysis of mathematical communication in primary teaching: The epistemological perspective', *Zentralblatt für Didaktik der Mathematik* 5, 138–148.

Steinbring, H., Bartolini Bussi, M.G. and Sierpinska, A. (eds.): 1998, *Language and Communication in the Mathematics Classroom*, National Council of Teachers of Mathematics, Reston, VA.

Vannier-Benmostapha, M.-P.M.: 2002, *Dimensions sensibles des situations de tutelle et travail de l'enseignant de mathématiques. Étude de cas dans trois institutions scolaires, en CLIPA, 4ème technologique agricole et CM2*, Thèse de Sciences de l'Éducation, Université Paris5.

Voigt, J.: 1985, 'Patterns and routines in classroom interaction', *Recherches en Didactique des Mathématiques* 6, 69–118.

Vygotsky, L.: 1962, *Thought and Language*, MIT Press, Cambridge, U.S.A.

Vygotsky, L.: 1985, *Pensée et Langage*, Editions Sociales, Paris.

Wood, 1996: 'Events in learning mathematics: Insights from research in classroom', *Educational Studies in Mathematics* 30, 85–105.

Wood, D., Bruner, J.S. and Ross, G.: 1976, 'The role of tutoring in problem solving', *Journal of Child Psycholology and Psychiatry* 17, 89–100.

Zaragosa, S.: 2000, *Interactions verbales dans le processus de dévolution*, Thèse de doctorat d'Université, Sciences de l'Education, Paris5.

ALINE ROBERT
Equipe Didirem, Université Paris7,
2 place Jussieu, 75005, Paris France,
E-mail: robert@math.uvsq.fr

JANINE ROGALSKI
Laboratoire Cognition & Usages, Université Paris8/CNRS,
2 rue de la liberté, 93526 Saint-Denis Cedex 2, France,
E-mail: rogalskij@univ-paris8.fr

MARIA G. BARTOLINI BUSSI

WHEN CLASSROOM SITUATION IS THE UNIT OF ANALYSIS: THE POTENTIAL IMPACT ON RESEARCH IN MATHEMATICS EDUCATION

ABSTRACT. In this commentary, I will critically elaborate on the potential impact of the coordinated papers of this volume on further development of research in mathematics education. The papers, which share common theoretical frameworks, will be categorized into three different classes: 'demolishers of illusions', 'economizers of thought' and 'energizers of practice'. I will analyze the role played by psychology and related sciences as a possible enrichment of the frameworks, especially where technologies are concerned. Finally, I will discuss the possible conflict between the need to consider the phenomena elicited in this kind of studies and the sophistication required by the theoretical constructs, which makes the results of these studies very difficult to communicate to the international community.

KEY WORDS: anthropological theory, demolishers of illusions, didactical contract, didactical engineering, economizers of thought, energizers of practice, theory of didactic situations

1. INTRODUCTION

Research studies presented in this volume offer analyses of teachers' and students' classroom activity in a variety of mathematical contexts and at different school levels. In most cases, the observed classes were ordinary classes, with small or no intended intervention of the researchers in the design of the lessons. The papers are analytical studies aimed at modelling the observed classes, using different theoretical tools, but most often the theory of didactic situations (TDS) and anthropological theory (AT).

My concern in this commentary paper is to critically elaborate on the potential impact of this rich set of coordinated papers on further development of research in mathematics education.

2. A FIRST CATEGORY OF RESULTS: DEMOLISHERS OF ILLUSIONS

In the discussion document of the ICMI study *What is research in mathematics education and what are its results?,* held in Washington, DC, in 1994, some categories of results were listed: demolishers of illusions; economizers of thought; energizers of practice (Sierpinska and Kilpatrick,

Educational Studies in Mathematics (2005) 59: 299–311
DOI: 10.1007/s10649-005-5478-1

1998, p. 8). With respect to this categorization, although not exhaustive, some results reported in this volume belong to the first category: in fact, they aim at explaining why the intentional goal of the teacher has not been (or even could not be) fulfilled. This aim is supported by both approaches, the TDS and AT.

For the former, Brousseau wrote:

> The theory of situations is neither an ideology nor a particular didactic method. In this sense, it has no technical alternative. It does not directly recommend this or that particular didactic procedure. Its theoretical concepts only allow one, for reasons of consistency, to predict the role of certain factors in some circumstances. This way it puts limitations on what it is possible to do or change in teaching, just as thermodynamics discards the possibility of building a *perpetuum mobile* but does not give precise guidelines for the construction of an ideal engine. (from a letter of Guy Brousseau to Anna Sierpinska, Sierpinska, 2000, p. 86)

Consistently with this position, in this volume, Brousseau and Gibel produce a thorough analysis based on the TDS to show that:

> Although the students, faced with a problem situation elaborated and conducted by the teacher, have certainly produced reasonings, they have not made much progress in their practice of reasoning. Indeed, they have not reflected back on their reasonings, on their validity, relevance or adequacy because the teacher was not able to process them. He could not respond to these reasonings by logical arguments based on the objective situation; he was forced to use rhetorical means. Now, it is not the complexity of the students' reasonings that forced the teacher to use this type of means but the fact that the problem situation could not be devolved to the students. This implies that it is not the teacher's management of the whole class presentation and discussion of the students' work that is challenged here, but rather the nature itself of the situation set up by the teacher, which strongly constraints the possibilities of really taking into account the students' reasonings. (Brousseau and Gibel, this volume)

In other words, the application of the TDS has shown that the 'ski-pass situation', although fascinating and appealing, was not (and could not be) productive, for this kind of students, to foster an effective practice of reasoning, that takes into account also metacognitive processes.

Margolinas et al. (this volume) too, pushing the TDS to include the modelling of teacher didactic knowledge, show that (and why) the observed teachers could not learn how to observe the pupils' strategies in classroom situation.

In the anthropological approach a similar aim is assumed. Chevallard, in the foundations of this approach (1991) wrote about didactic transposition:

> In fact, the particular effectiveness [of the concept of didactic transposition] consists in showing a *difference* where the teacher denies it exists, in questioning the spontaneously assumed identity [of the mathematics taught and research

mathematics] in the aim of revealing the inadequacy concealed by its apparent obviousness. (Chevallard, 1991, p. 45, [editor's translation])

Accordingly, in this volume, Barbé et al. writes:

The work presented here began with an observation of two teaching processes about limits of functions. Its main goal was to study how *institutional restrictions* could affect the spontaneous practice of the observed teachers. We are presenting here only one of the observed didactic processes, which will show, not only the kind of analysis we can provide on it using the Anthropological Theory of Didactics, but also how this analysis allows us to highlight the *didactic restrictions* that affect teachers' practices. Two kinds of didactic restrictions are identified here.

I. *Specific didactic restrictions* arising from the precise nature of the knowledge to be *taught*. In this study – those related to the content of the *limits of functions* as proposed by official syllabi and textbooks in Spanish secondary schools.
II. *Generic didactic restrictions* the mathematics teacher encounters when facing the problem of how to teach any proposed mathematical topic in a school institution. We will show that the conjunction of the two kinds of restrictions determine to a large extent the knowledge related to limits of functions that can be *actually taught* in the classroom. This will provide a first delimitation of the field of possible didactic organizations that can be set up in the considered school institution (Barbé et al., this volume).

These papers present results that are 'demolishers of illusion' as the main aim is to study the constraints that make innovation in the classroom very difficult to implement.

3. A SECOND CATEGORY OF RESULTS: ECONOMIZERS OF THOUGHT

The attention to the constraints of the teaching profession is shared by another group of papers of this volume, that, however, seems to be slightly oriented to produce – albeit in the future – models for teacher training. In the same categorization quoted above, they may be shifted to the category of 'economizers of thought'. In these papers, the analysis based on either the TDS or the AT is oriented to produce models of teaching and learning activity that might be, in the future, applicable to the design of new activities in teacher training.

This is the case of Hersant and Perrin-Glorian's paper, where the authors study, in ordinary classes, a practice, which consists of problem solving sessions in small groups of students followed by whole class discussions of the solutions to the problems. This format of organization, where a whole class discussion follows the either individual or small group solution of a problem, has been studied by several researchers (e.g., Bartolini Bussi,

1996; Cobb and Bauersfeld, 1995), but Hersant and Perrin-Glorian claim that their intentions are very different.

The aim of the research is to gain knowledge and understanding of teaching phenomena; it is not to produce immediate action or to improve teaching in a direct way. Moreover, our project is not one of didactic engineering. Indeed, the researcher intervenes neither in the design of teaching nor in its realization. [. . .] We hypothesize that teaching practices are very complex and that researchers generally do not take into account the 'economy' of ordinary practices, that is, the teachers' attempts to balance the various professional constraints under which they work and the degree of freedom they have. Clarification and understanding of ordinary practices is, for us, an essential issue and a first step towards research on teacher training (Hersant and Perrin-Glorian, this volume).

Sensevy et al. (this volume) produce a model to study mathematics teacher's action. In their study they had suggested that the observed teachers adopt a classical situation, used to illustrate TDS, namely the 'race to 20'. However, this situation is not considered a means to improve teaching but rather to foster the emergence of a wide range of teacher's actions to observe. The authors conclude the paper by presenting a possible direction for future research:

[To develop] a teacher training unit about the teacher's action in the so-called 'investigative activities' in mathematics classes. It is a question of documenting the extremely technical nature of the actions a teacher must undertake in an adidactical situation. Our aim is thus to create the conditions to transform general teaching techniques (so often remarkably well mastered by teachers) into specific mathematics teaching techniques and, in doing so, to grant them their true scope. (Sensevy et al., this volume)

Also Robert and Rogalski study the teaching practice, yet with a stronger orientation towards application in teacher training. After a short review of other studies developed within the TDS and AT approaches, the authors claim:

Considerably less attention is given in these works to the individual exercise of the profession, and this is a great difference with ours. It is clear that their main aim is not to explain teacher's choices but to model the global system. Inferences from both types of research are different: our work may lead to building hypotheses about teacher training and other works may lead to a better understanding of the general constraints of the system. (Robert and Rogalski, this volume)

In the papers quoted in this section, although with different emphasis, the application to the design of activity in teacher training is considered a possible outcome of the analysis of teaching practices.

4. A THIRD CATEGORY OF RESULTS: ENERGIZERS OF PRACTICE

Some papers of this volume refer explicitly to didactical engineering, described by Artigue (2002) as follows:

> But didactic engineering, in mathematics as well as in other disciplines . . . , appears to be a very efficient methodology in research whose progress requires the support of didactic constructions that cannot be observed through naturalistic observation of the [didactic] system, which is then intentionally disrupted to enable the observation of its common functioning. (Artigue, 2000)

Flückiger (this volume) presents an analysis of the results of a work of didactical engineering in the context of introduction of long division in primary school. The long-term activity is designed with different kinds of sessions (calculation session, journal writing session, debate sessions and so on). The paper provides ideas for teaching, hence it may be considered within the category of 'energizers of practice' (referring as above to Sierpinska and Kilpatrick, 1998, p. 8). However, the author claims that the goal of the paper is theoretical.

> The goal of the study was not to lead the children to invent a division algorithm that would later be instituted in the classroom. Nor was it a question of testing a new teaching method for written division problems, currently learned in fifth grade. The goal was rather to devise a *research methodology* for studying the genesis of numerical knowledge over time, under didactically controlled conditions (Flückiger, this volume)

Sadovsky and Sessa (this volume) present a part of an implementation of a didactical engineering project aimed at exploring the possibilities of making a connection between arithmetical and algebraic practices. They detail the aims of the paper as follows.

1. To specify the conditions that enable the creation, in the classroom, of a *milieu* for generating new questions that are constitutive of the mathematical knowledge required for getting involved in algebraic practices.
2. To analyze the knowledge related with the arithmetic–algebra transition, which emerges when grade 7 students (12-year-old, last year of primary education) discuss the productions of others, after being invited to take a position concerning such productions during an adidactic phase of work (Sadovsky and Sessa, this volume)

The authors are aware of the complexity of the teachers' work. Yet, they are interested in describing the conditions that enable rather than the constraints that inhibit effective teaching.

Ideas for teaching may be taken also from Assude's paper. The paper compares the production and management of didactic time (i.e. the scheduling of the teaching of a piece of knowledge) of teachers before and after the integration of Cabri in their courses. Some strategies to save time are described: for instance, the production of a poster that 'freezes' the dynamic properties of Cabri, condensing the essential information of activity. This and other strategies are produced by the teachers in the second year of the Cabri experiment, on the base of their experience and of the interaction with the research team. Assude's study is realized in a context where the teachers themselves are members of the research team. The team shared the research task as follows:

> The teachers were responsible for choosing and implementing classroom situations, the researchers served as resource providers through the production of a 'task box' and they observed the classes; researchers and teachers analyzed the data together. (Assude, this volume)

The last three papers, without giving up on the theoretical elaboration, offer to readership, with different emphases, examples of effective activities in the mathematics classroom, that may be used as 'energizers of practice'.

5. THE POTENTIAL IMPACT OF THE CONTRIBUTIONS TO THIS VOLUME ON FURTHER DEVELOPMENT OF RESEARCH IN MATHEMATICS EDUCATION

The last category of results has the potential to capture the attention of many researchers all over the world. This is not to deny the importance of the others: one of the main points in didactical research is to understand why an innovation does not work (or even cannot work) in spite of the teacher's best intentions. Hence the thorough analyses of the social conditions of the teaching and learning processes (both at the micro and at the macro levels) developed by the contributors to this volume must not be ignored. However, as studies of concrete classrooms are naturally biased by the local conditions, what could be interesting are not the results per se, but the methods that have allowed to obtain the results. Yet, the production of these results in the papers of this volume seems to require an extraordinary theoretical sophistication, apparent in the large amount of technical terms and deep discussions of the relationships between the theoretical constructs themselves. This aspect is evident already in the length of the theoretical introductions of the papers. For instance, in Brousseau and Gibel's and in Margolinas et al.'s papers the theoretical introductions fill about one-third of each paper (which is a lot considering that the papers are longer than the average in this journal). The authors themselves are aware of this. For instance, Brousseau and Gibel write:

> We will do our best to restrict the specialized terminology of the theory to those terms that are indispensable for understanding our particular study, and we will try to justify its use in each case. We hope that these 'didactic' precautions will

not prevent readers more familiar with our theoretical approach from appreciating our work. (Brousseau and Gibel, this volume)

This theoretical sophistication is unavoidable in the internal debate of studies conducted from the same theoretical perspective, but it is quite hard to be communicated outside and may act as an obstacle to the diffusion of results and methods.

In many groups of mathematics educators the interest for 'energizers of practice' might be larger than for other kinds of results. This is the case for instance, of my research community, where, for historical reasons, we have adopted a model of didactical research (Arzarello and Bartolini Bussi, 1998) which puts the innovation component to the fore, combining it with the epistemological and cognitive components. This may also be the case of other groups of researchers. For example, in research reported in (Cobb and Bauersfeld, 1995; Stein et al., 2000; Lampert, 2001), the study of the classroom practices seemed to be related to the design of specific activities and to the study of the conditions of effective functioning in the classroom.

In spite of their apparent theoretical complexity, some theoretical constructs of TDS and AT are being adopted by researchers worldwide (see a discussion of this issue in Boero, 1994). Lampert (2001) often refers to TDS as presented in Brousseau (1997). In her book, 'Teaching problems and the problems of teaching', Lampert analyzes the activity of a whole school year as a teacher in a fifth grade mathematics classroom from the perspective of the classroom teacher who is also 'a prominent member of the worldwide mathematics education research community' (Seeger, 2001). This case is very interesting.

Lampert's thoughts and research have always set up landmarks in the development of discussion within that community. [...] Being not only a researcher but also a mathematics teacher in elementary grades, she was applauded for proving through personal example the possibility of the fruitful coexistence of the world of educational scholarship and the world of the primary classroom. (Seeger, 2001, p. 169)

Lampert mentions the notion of 'didactical contract' in her notes to a description of an interaction with a student, Varouna:

I wanted to direct her work but in a way that was not a simple 'telling the answer'. (Lampert, 2001, p. 125)

The theoretical sophistication in Lampert's analyses is not comparable with that presented in this volume, if we refer to the complex four-dimensional structure of the didactical contract in Hersant and Perrin-Glorian, for example. The problems of this 'metaphorical' (Boero, 1994, p. 33) use of a construct will be reconsidered in the conclusion.

6. THE ROLE OF PSYCHOLOGY

In this section I will comment on the role of psychology in the analyses of classroom situations presented in this volume.

In her article, Flückiger (this volume) combines the psychological theory of conceptual fields (Vergnaud, 1990) with TDS, as far as the theoretical construct of didactic memory is concerned. The paper aims at showing the power of complementing tools of analysis coming from different research field (didactics and psychology) to enlighten each other in order to understand pupils' processes.

A similar approach is used by Robert and Rogalski (this volume) to analyze the teacher's activity in the classroom. The psychological lens that is used to analyze the teacher's classroom practices is claimed to be based on Leont'ev activity theory. As this approach is adopted by other researchers, a connection between this paper and the international literature is easy to find. Bartolini Bussi (1998) and Mariotti (2000) analyze long-term processes that cannot be reduced to the juxaposition of short-term ones, by using the distinction made after Leont'ev (1978, 1981) among the levels of activity, actions and operations.

> The first is the global level and corresponds to motives (the objects of teaching in the broad sense, including not only the mastery of mathematical concepts and procedures but also the attitude toward mathematics). Activity is actualized by means of actions, defined by conscious goals, which include, for instance, the school tasks, solved individually or collectively; actions are realized by means of operations that directly depend on the conditions of attaining concrete goals. (Bartolini Bussi, 1998, p. 69)

The analysis carried out by Robert and Rogalski (this volume) hints at this distinction, if we interpret the students' 'enlistment' as one of the motives realized by means of actions (the proposed tasks). The teacher's operations are the communicative strategies used during teacher-student interaction, and they are carefully examined by the authors (e.g. discourse markers, punctuation, orientation, different kinds of repetitions or responses, and so on). Hence, the study of Robert and Rogalski enriches the collection of effective (or non effective) communicative strategies studied in situations of classroom interaction, within an activity theoretical approach, and connect them to the sophisticated analysis carried out by means of TDS.

In this volume, however, the impact of psychology on the studies seems very limited. From time to time we feel we are at the doorstep of psychology, with implicit references to it, that are explicitly excluded by the authors a few lines below. For instance, in the study presented by Hersant and Perrin-Glorian:

The problems are chosen so as to partly require what the students already know, but they include questions or can be extended to problems whose solution requires some new knowledge. During the whole class discussion of students' solutions or attempts at them, the teacher extends the problems in the direction of the new knowledge. He helps the students to *synthesize* the solutions obtained by different students using old knowledge and he extends the problems by highlighting or asking those additional questions, which are then solved collectively in class. This way, new knowledge is introduced as a solution to these new questions in the same context and linked to the old knowledge.

This description evokes a rich elaboration of the Vygotskian idea of the zone of proximal development, suggesting a focus on individual processes and requiring psychological tools. However, this interpretation is explicitly rejected by the authors, who claim that in this study they do not consider the effects of teacher's practice on students' learning.

Another example of this intentional distance from cognitive issues is offered by Brousseau and Gibel, who devote a large part of their long theoretical introduction to the discussion of a model of subject's 'reasoning', in particular of the reasoning that 'can be modelled by inferences of the form "If the condition A is satisfied, so is (or will be) the condition B"' (Brousseau and Gibel, this volume). The production of this kind of inferences has been studied extensively, and in a variety of ways, in mathematics education literature about argumentation and proof. Some of these studies demonstrate the fruitfulness of integrating didactic and psychological perspectives. For example, in a doctoral thesis by Pedemonte (2002) titled, 'Étude didactique et cognitive des rapports de l'argumentation et de la demonstration dans l'apprentissage des mathématiques', the didactical perspective underlying the design of the experimental part was combined with a cognitive analysis, which allowed the author to compare, by means of Toulmin's model (1958), the different phases of argumentation and proof.

The scarce attention paid to psychological aspects of mathematics learning in the present volume may discourage some readers. In the international literature on mathematics education, the number of studies of classroom processes using tools borrowed or adapted from psychology and related sciences is growing. The cognitive dimension was one of the main components of the didactical approach presented by Arzarello (in press) in his plenary lecture at ICME10. In his lecture the descriptive level offered by psychology was enriched by the recent studies in the field of cognitive science (e.g. Lakoff and Nuñez, 2000) and of neurosciences. Analytical tools from psychology are extensively used by researchers of the same community to study classroom processes (e.g., Bartolini Bussi et al., 1999; Mariotti, 2000).

Psychology has come to play a greater role also in the domain of studies on technology in mathematics education by way, among others, of psychological concepts related to instrumental activity developed by Rabardel (1995; see, for example, Trouche, 2002a,b).

7. A MISSING ISSUE: TECHNOLOGIES

Only one paper in this volume looks at classrooms where technology is used to teach and learn mathematics (Assude), and even in this paper the focus is not on technology but on time management, which is a universal issue in classroom teaching. It is too bad, because the two main theoretical frameworks presented in the volume, TDS and AT, together with other theories such as Rabardel's theory of instrumentation of tools, mentioned above, have been used in quite sophisticated ways to study the integration of software such as Cabri (e.g. Laborde, 2000) or graphic and symbolic calculators in the mathematics classrooms (e.g., Lagrange, 2002a,b; Trouche, 2002a,b).

Technology is a very intrusive component of the 'objective milieu' (Hersant and Perrin-Glorian, this volume), that starts the process of 'instrumentation' (Rabardel, 1995). This feature seems to force the researchers to take seriously the cognitive perspective. An example may be quoted (Artigue, 2002), concerning a didactical engineering on the use of graphic calculators to approach the concept of derivative (Maschietto, 2002). Because of the amount of metaphors, gestures and analogies used by the students and suggested by the very physical activity performed on the symbolic calculators, it seemed necessary to introduce, in the standard framework of TDS, analytical tools taken from the so-called 'embodied cognition' field of research (Lakoff and Nuñez, 2000). The need of combining didactical engineering with cognitive science is justified by the very nature of the artifacts the students are using:

> If one zooms in on the graphical representation of a function in the neighborhood of a point where it is differentiable, one ends up seeing a straight line, which is a linear object. Could this dynamic process be a basis of a kind of metaphorical thinking that would support the focus on local behavior [of functions] at the secondary school? How can this process be endowed with mathematical meaning? What can be the responsibilities of the students in this construction? What remains the teacher's responsibility? How best to connect the respective contributions of the teacher and the students in advancing knowledge in the classroom? (Artigue, 2002 [editor's translation])

Studies of this kind are appealing to the international readership, because, on the one hand, they address relevant and up-to-date problems, and, on the other, without giving up on theory, they do not bury their results under excessive theoretical sophistication (or, maybe, only a sophisticated terminology).

8. Concluding Remarks

What could be the impact of studies presented in this volume on international mathematics education research? Kilpatrick (1994) commented on the meagre impact of the papers published in France within the frameworks of TDS and AT on the studies produced by the American community of researchers. He suggested three reasons that might be applied to other communities as well: a linguistic barrier, as most studies developed within the quoted frameworks were published in French; a cultural barrier, as the influence of radical constructivism in the USA devalued the importance of the social phenomena of culture transmission that were at the core of TDS and AT; and the theoretical sophistication induced by the need of building a system in the most precise way.

The linguistic barrier is now less important, because of the more and more frequent publication in English of the results (e.g. Brousseau, 1997): this volume, too, is an effort in this direction. The cultural barrier is falling too, as the attention to social phenomena in the mathematics classroom is growing all over the world (e.g., Cobb and Bauersfeld, 1995; Seeger et al., 1998; Lampert, 2001). In my opinion, the third issue still deserves attention. We have seen in Section 5, in the paradigmatic case of didactical contract, that the theoretical sophistication in TDS is enormous and not easy to communicate. Yet, in spite of their apparent theoretical complexity, some theoretical constructs elaborated in TDS and AT have been 'transposed' to research programs that do not share the same theoretical framework. As Boero (1994) wrote, this transposition evidences once more the importance of the phenomena studied in TDS and AT and, hence, the need to make the underlying elaborations available to the community of researchers all over the world. However, extracting theoretical constructs from the system that defines their meaning is risky: a term might be used metaphorically as evocative of existing problems and not as a precise working tool for the analysis of classroom activities. This risk cannot be avoided, but might be reduced, if the authors of the papers of this volume and the researchers who share the same theoretical frameworks were willing to contribute to an international debate around some precise questions: To what extent is this theoretical sophistication necessary for the interpretation and design of classroom activities? Is it possible to distil the essential features of theoretical constructs, to relate them to other theories or approaches and to compare the outcomes of using them in the interpretation and design of classroom activities? It may not be easy to take distance from the internal debate within a restricted scope of the TDS-AT framework, whose structure has become more and more complex over the years. Yet, if the challenge is accepted, this effort would nurture collaboration between researchers of different schools and the fruitful elaboration of shared theoretical paradigms for research studies.

References

Artigue, M.: 2002, 'Ingégnierie didactique: quel rôle dans la recherche didactique aujourd'hui?', *Revue Internationale des Sciences de l'Education* 8, 59–72.

Arzarello, F.: in press, 'Mathematical landscapes and their inhabitants: Perceptions, languages, theories', in *Proceedings of ICME 10*, Plenary Lecture.

Arzarello, F. and Bartolini Bussi, M.G.: 1998, 'Italian trends in research in mathematical education: A national case study from an international perspective', in A. Sierpinska and J. Kilpatrick (eds.), *Mathematics Education as a Research Domain: A Search for Identity*, Kluwer Academic Publishers, Dordrecht, pp. 243–262.

Bartolini Bussi, M.G.: 1996, 'Mathematical discussion and perspective drawing in primary school', *Educational Studies in Mathematics* 31, 11–41.

Bartolini Bussi, M.G.: 1998, 'Verbal interaction in the mathematics classroom: A Vygotskian analysis', in H. Steinbring, M.G. Bartolini Bussi and A. Sierpinska (eds.), *Language and Communication in the Mathematics Classroom*, NCTM, Reston, VA, pp. 65–84.

Bartolini Bussi, M.G., Boni, M., Ferri, F. and Garuti, R.: 1999, 'Early approach to theoretical thinking: Gears in primary school', *Educational Studies in Mathematics* 39, 67–87.

Boero, P.: 1994, 'Situations didactiques et problèmes d'apprentissage: Convergences et divergences dans les perspectives de recherche', in M. Artigue, R. Gras, C. Laborde, and P. Tavignot (eds.), *Vingt ans de didactique des mathématiques en France: Hommage à Guy Brousseau et Gérard Vergnaud)*, La Pensée Sauvage, Editions, pp. 17–50.

Brousseau, G.: 1997, *Theory of Didactical Situations in Mathematics. Didactique des Mathématiques, 1970–1990*, Kluwer Academic Publishers, Dordrecht.

Chevallard, Y.: 1991, *La trasposition didactique du savoir savant au savoir enseigné*, La Pensée Sauvage, Editions, Grenoble.

Cobb, P. and Bauersfeld, H. (eds.): 1995, *The Emergence of Mathematical Meaning: Interaction in Classroom Cultures*, Lawrence Erlbaum Associates, Inc., Mahwah, NJ.

Kilpatrick, J.: 1994, 'Vingt ans de didactique française depuis les USA', in M. Artigue, R. Gras, C. Laborde, and P. Tavignot (eds.), *Vingt ans de didactique des mathématiques en France: Hommage à Guy Brousseau et Gérard Vergnaud)*, La Pensée Sauvage, Editions, Grenoble, pp. 84–96.

Laborde, C.: 2000, 'Dynamic geometry environments as a source of rich learning contexts for the complex activity of proving', *Educational Studies in Mathematics* 44, 151–161.

Lagrange, J.: 2002a, 'Les outils informatiques entre "sciences mathématiques" et enseignement. Une difficile trasposition?', in D. Guin and L. Trouche (eds.), *Calculatrices Symboliques. Tranformer un outil en un instrument du travail mathématique: un problème didactique*, La Pensée Sauvage, Editions, Grenoble, pp. 89–116.

Lagrange, J.: 2002b, 'Etudier les mathématiques avec les calculatrices symboliques. Quelle place pour les techniques?', in D. Guin and L. Trouche (eds.), *Calculatrices Symboliques. Tranformer un outil en un instrument du travail mathématique: un problème didactique*, La Pensée Sauvage, Editions, Grenoble, pp. 151–186.

Lakoff, G. and Nuñez, R.: 2000, *Where Mathematics Comes From: How the Embodied Mind Brings Mathematics into Being*, Basic Books, New York.
Lampert, M.: 2001, *Teaching Problems and the Problems of Teaching*, Yale University Press, New Haven and London.
Leont'ev, A.N.: 1978, *Activity, Consciousness and Personality*, Prentice Hall, Englewood Cliffs, NJ.
Leont'ev, A.N.: 1981, *Problems in the Development of Mind*, Progress, Moscow.
Mariotti, M.A.: 2000, 'Introduction to proof: The mediation of a dynamic software environment', *Educational Studies in Mathematics* 44, 25–53.
Maschietto, M.: 2002, *L'enseignement de l'analyse au lycee: les debuts du jeu global/local dans l'environnement de calculatrices*, Doctoral Dissertation, Université Paris 7 and Università di Torino, IREM Paris 7.
Pedemonte, B.: 2002, *Etude didactique et cognitive des rapports de l'argumentation et de la démonstration dans l'apprentissage des mathématiques,* Doctoral Dissertation, Université Joseph Fourier, Grenoble I, Università di Genova.
Rabardel, P.: 1995, *Les hommes and les technologies. Approche cognitive des instruments contemporains*, Armand Colin, Paris.
Seeger, F.: 2001, 'A large territory: One year in an elementary mathematics classroom', *Mind, Culture and Activity: An International Journal* 10, 168–172.
Seeger, F., Voigt, J. and Waschescio, U. (eds.): 1998, *The Culture of the Mathematics Classroom*, Cambridge University Press, Cambridge.
Sierpinska, A. and Kilpatrick, J. (eds.): 1998, *Mathematics Education as a Research Domain: A Search for Identity*, Kluwer Academic Publishers, Dordrecht.
Sierpinska, A.: 2000, *The 'Theory of Didactic Situations': Lecture Notes for a Graduate Course with Samples of Students' Work*, Master in the Teaching of Mathematics, Concordia University (see also TDS Lecture 8 accessible from http://alcor.concordia.ca/~sierp/).
Stein, M.C., Smith, M.S., Henningsen, M.A. and Silver, E.A.: 2000, *Implementing Standards-Based Mathematical Instruction: A Casebook for Professional Development*, Teachers College Press, New York.
Toulmin, S.E.: 1958, *The Use of Arguments*, Cambridge University Press, Cambridge.
Trouche, L.: 2002a, 'Les calculatrices dans l'enseignement des mathématiques: une évolution rapide des matériels, des effets différenciées', in D. Guin and L. Trouche (eds.), *Calculatrices Symboliques. Tranformer un outil en un instrument du travail mathématique: un problème didactique*, La Pensée Sauvage Editions, Grenoble, pp. 21–54.
Trouche, L.: 2002b, 'Une approche instrumentale de l'apprentissage des mathématiques dans des environments de calculatrice symbolique', in D. Guin and L. Trouche (eds.), *Calculatrices Symboliques. Tranformer un outil en un instrument du travail mathématique: un problème didactique*, La Pensée Sauvage Editions, Grenoble, pp. 21–54.
Vergnaud, G.: 1990, 'La théorie des champs conceptuels', *Recherches en Didactique des Mathématiques* 10(2–3), 133–170.

Dipartimento di Matematica
Università di Modena e Reggio Emilia
Via Campi 213/B 41100 Modena
E-mail: bartolini@unimo.it

HEINZ STEINBRING

ANALYZING MATHEMATICAL TEACHING-LEARNING SITUATIONS — THE INTERPLAY OF COMMUNICATIONAL AND EPISTEMOLOGICAL CONSTRAINTS

ABSTRACT. This is a commentary paper in the volume on "Teachings situations as object of research: empirical studies within theoretical perspectives". An essential object of mathematics education research is the analysis of interactive teaching and learning processes in which mathematical knowledge is mediated and communicated. Such a research perspective on *processes of mathematical interaction* has to take care of the difficult relationship between mathematics education theory and everyday mathematics teaching practice. In this regard, the paper tries to relate the development in mathematics education research within the *theory of didactical situations* to developments in interaction theory and in the *epistemological analysis of mathematical communication*.

KEY WORDS: communication, epistemology, relationship between mental and social constraints

Over the past decades, the elaboration of a distinct scientific identity for mathematics education has been a lengthy and difficult process on both the international and national planes. The development of research in mathematics education can be followed by looking at the various ways this research has focused on the three main elements of the didactic system: (1) the mathematical knowledge, (2) the student and (3) the teacher. In the tradition of didactics of mathematics in Germany, for example, (cf. Steinbring, 1998), these three elements have been represented as vertices of a system of relations, called the "didactic triangle" (see Figure 1).

In the early days of mathematics education, the functioning of the didactic triangle was usually interpreted in a very schematic way. Teaching and learning of mathematics was described in mechanistic terms by, for example, the so-called "sender-receiver model". The sender, in our case the mathematics teacher, passes unequivocal messages, for example, didactically prepared mathematical contents, on to the receiver, that is, the students. Consequently, teaching and learning is a mere handing over of mathematical knowledge from the teacher to the students. This interpretation has been vehemently criticized in mathematics education and replaced by other conceptions about the interplay of mathematics, students and teacher (Bazzini, 1994; Even and Loewenberg Ball, 2003; Steinbring, 1994; Seeger and Steinbring, 1992; Verstappen, 1988). Also, the (radical)

Educational Studies in Mathematics (2005) 59: 313–324
DOI: 10.1007/s10649-005-4819-4

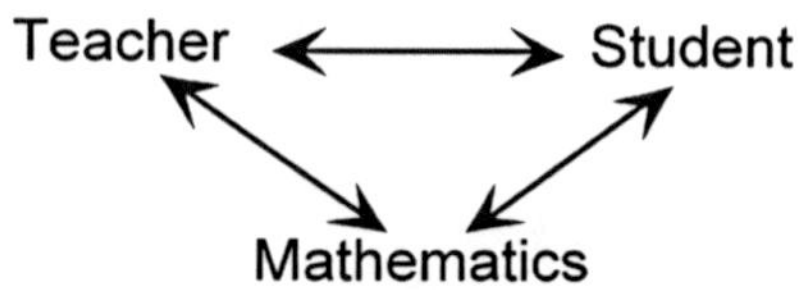

Figure 1. The didactic triangle.

constructivism – especially in mathematics education – fundamentally questioned the assumption that mathematical knowledge could be "handed over" from one person to another, because each person must independently construct mathematical knowledge and his or her own interpretations of this knowledge (see, among others, von Glasersfeld, 1991).

Nevertheless, the didactic triangle (Figure 1) does contain the essential elements of mathematical teaching-learning processes. The long-standing development of mathematics education research, as well as specific features characterizing this research, can be described by how the three vertices of the triangle have been regarded as isolated or related to each other. Thus the main paradigmatic aspects of mathematics education research approaches can be presented briefly in the following way.

PARADIGM 1: FOCUS ON SCHOOL MATHEMATICS

From this perspective, didactical studies and investigations concentrate on school-mathematical knowledge and its didactical elementarization as well as on content-related or substance-related analyses. These researches are essentially based on a view of mathematics as instructional subject matter, which is given and seen as relatively autonomous and independent from learning, understanding and teaching. The autonomy of a specific mathematics education research approach is not really visible here yet.

PARADIGM 2: FOCUS ON THE LEARNER AND HIS OR HER COGNITIVE POTENTIAL

This position in educational research moves the learner with hisorher cognitive aptitude into the center of interest. This is a pedagogical position that has been held for a long time in early childhood and elementary education. Research carried out in this perspective used, among other things, methods borrowed or adapted from related disciplines, such as psychology, pedagogy and sociology. This kind of research has not been simply reducible to mathematical analyses and it substantially contributed to the development of a profile of mathematics education as an independent research domain.

PARADIGM 3: FOCUS ON PROCESSES OF MATHEMATICAL INTERACTION

In the last 25 years or so, research approaches, which took the interplay between mathematical knowledge, the teacher and the students as the object of their scientific investigations, have become more and more prominent in mathematics education. Here, everyday mathematical instruction and learning processes are taken seriously and are analyzed from a social interactionist perspective (e. g. Bauersfeld, 1978, 1988; Cobb and Bauersfeld, 1998; Krummheuer, 1984, 1998; Maier and Voigt, 1991; Voigt, 1994; Steinbring, 2000b, 2004). The everyday mathematics teaching is seen as an independent culture, which is not completely and directly determined by the scientific discipline of "mathematics". In this culture, there emerges a specific type of mathematical knowledge and mathematical language.

As a fundamental research object of mathematics education, mathematical interaction or communication processes are complex structures, in which questions of (school-) mathematical knowledge and cognitive processes of learning and understanding as well as social processes of instructing have to be put in relation to each other. Research on such processes requires methods from different disciplines. But the research object "processes of mathematical interaction" is a research concern, which clearly distinguishes mathematics education from these disciplines and accounts for its scientific identity and independence.

The "theory of didactical situations" (developed in France mainly by Guy Brousseau) represents "Paradigm 3" in an exemplary fashion, by including the relationships between mathematical knowledge, the teacher and the students among its most basic assumptions. The main goals of this theory were to describe the structure and the functioning of mathematical learning-teaching processes as precisely as possible, and furthermore to elaborate theoretical concepts for modeling the different phases of these processes (phases of *action*, *formulation*, *validation* and *institutionalization*) and the mechanisms of regulation of didactical interactions between the teacher and the students (e.g., *didactic contract*).

The mathematics-student-teacher triplet is emphasized as an object of analysis in many contributions to this volume. For example, in the paper by Brousseau and Gibel, this triplet is presented in the following way.

In a didactic analysis of a lesson, it is necessary to distinguish several situations:

- the mathematical situation (the objective situation) the student is faced with and has to act upon;

- the situation of autonomous learning (what it is about, in terms of the didactic objectives of the lesson); and
- the didactic situation (the way the teacher conducts the lesson: his interventions, the arguments he uses) (Brousseau and Gibel, this volume).

In Sensevy et al. (this volume), we find the following description:

> The didactic relationship is a ternary relation between the teacher, the students, and the pieces of knowledge at stake. We assume that the teacher's work consists in initiating, establishing, and monitoring this relationship. Using the concept of didactic relationship is a way to emphasize the communicative nature of teaching techniques, and to focus on the fact that the core of the relationship between the teacher and the students is their sharing of this piece of knowledge. We consider that the didactic relationship is fundamentally threefold: understanding between the teacher and his students thus implies not only analyzing their respective positions but, especially, taking into account the knowledge that will be the focus of the lesson (Sensevy et al., this volume).

Flückiger, on the other hand, notes only briefly:

> In the framework of a didactic system modeled by the teacher-pupil-knowledge triplet, the study focused on the elaboration of knowledge of division by the pupil subsystem, in a research-controlled didactic context. (Flückiger, this volume)

In the theory of didactic situations, as in any research grounded in Paradigm 3, the triplet cannot be ultimately reduced to the individual components; it depends on the whole system including the mutual relations and interactions between these three elements.

But if one intends to emphasize the distinctive features of instructional research conducted on the basis of the "theory of didactical situations" as compared to other research conceptions, one can realize that this approach tended to be prescriptive rather than descriptive in its traditional roots and development, and that, in a way, it was based on a priori theory with a multi-layered conceptual system. In many cases, the (partly ideal) teaching episodes and examples described in publications referring to this theory, have usually served for illustrating and explaining the theory.

By contrast, in other mathematics education approaches in the area of interaction research (cf. Cobb and Bauersfeld, 1995; Krummheuer, 1998; Voigt, 1994), the emphasis was mainly on a highly detailed description of the observed and documented mathematical interaction processes, with a deliberate dismissal of a priori theoretical and conceptual instruments of analyzing the empirical phenomena.

For an "outside" observer, the contributions in this volume reveal, in a concise manner, a remarkable turn in research that still identifies its foundations with the theory of didactic situations. This research is increasingly leaning towards analyzing real, everyday instruction processes, which are, in general, not influenced by the researchers and their theoretical models. Thus descriptive moments of analysis are increasingly used, nevertheless keeping a "balance" with the developed "theory of didactical situations". Regarding this, one can speak of a certain further development to the present conceptions about the role of theoretical concepts and empirical observations in mathematics instruction research.

The problem in the relation between (a priori elaborated) theory of didactics of mathematics and empirical observations of mathematics teaching (to be analyzed as genuinely as possible) cannot be solved in a one-sided way from within the frame of research in mathematics education alone. Whether we want it or not, any observation and description of everyday learning and teaching processes is necessarily determined by points of view evolving during the process of observation, which are pre-given and general, hence theoretical. An entirely "unbiased" and "conceptionless" observation and description of empirical phenomena is not possible. On the other hand, instructional *research* means researching and exploring something that is new and can be, in principle, derived only from the empirical observations and documents, and which cannot be supplied exclusively and directly from a multi-layered didactical theory.

The dialectics between the empirical phenomena of mathematics teaching and the theoretical means, concepts and instruments of its exploration and research cannot be turned off; this makes it necessary to explain, more and more explicitly in the course of the research, the first implicit or hidden assumptions about the functioning of processes of mathematical interaction.

Some such hidden, underlying assumptions concern, in particular, mathematical knowledge. Is it given as a finished product (by the mathematicians who do the research)? How does it develop in processes of interaction? How can it be characterized? Which – different – roles does it have during instruction for the students and for the teacher? There are also hidden assumptions about the communication processes between the participants of the events. Or about how individual learning takes place; or how social and cognitive conditions of teaching and learning mathematics are related to each other.

The elaboration of the first underlying assumptions about the conditions and the functioning of complex processes of mathematical interaction is a demanding task, which one often can only carry out *post factum*, after having developed theoretical analyses of mathematics teaching episodes. Ultimately, there are always new, still unfocused conceptions and assumptions, which enter the research process, no matter how sustained and intentional the explication of such assumptions may be. It is true that studies of processes of mathematical interaction based on the explicit "theory of didactical situations" have developed together with the growth of this theory, which was enriched, over the years, by concepts for analyzing and characterizing the complexity of mathematical teaching and learning processes. But there are still some assumptions for which a further explication following the research would be helpful. A further clarification of the scientific-theoretical status and the role of mathematical knowledge within the teaching events could add to a better description and understanding of the conditions of *interactive* development of mathematical knowledge.

Furthermore, in the contributions to this volume, the conceptions and assumptions about the relations between the individual and the social group, or between cognition and communication are elaborated more explicitly only at some places. One example is the paper by Sadovsky and Sessa, where the authors state:

> The personal processes of learning are embedded in the weave created by the two types of interaction – subject/*milieu* and teacher/student – which are separate only in the theoretical analysis. The perspective... conceives of mathematical learning at school as a process, in which the cognitive and the social adaptations are interwoven. (Sadovsky and Sessa, this volume)

In the following, I would like to make some remarks about the role, in the interaction, of social communication and individual consciousness as well as about the epistemological conditions of mathematical knowledge. These remarks intend to contribute to making more precise and explicit the background assumptions about the conditions of the functioning of processes of mathematical interaction. In his "Theory of Society" (Luhmann, 1997), the sociologist Niklas Luhmann uses the concept of systems, in particular self-referential and autopoietic systems, as a central analytic tool to characterize "communication" as such a system, and to explain the functioning of systems of consciousness such as people, teachers, learning students, etc. With regard to teaching and learning processes, in which teachers and students communicate with each other, the following confrontation of different concepts of "systems" is very revealing.

Referring to Heinz von Foerster (1993, p. 244), Luhmann distinguishes the so-called "trivial machines" from "non-trivial machines".

> ...[In] trivial machines (...) the input impulse is transformed into output according to a certain rule, so that whenever one enters the information..., the machine runs and produces a certain result. If one enters a different input, it operates again and produces, if it has different functions, a different result. (Luhmann, 2002, p. 97–98 [author's translation]).

Trivial machines always give the same output for the same input. Non-trivial machines, however, work differently.

> Non-trivial machines... always reflect on their own state and ask questions such as 'Who am I?', 'What have I just done?', 'What mood am I in?', 'How strong is

my interest still?, and so on every once in a while, in order to only then produce the output. A self-referential loop is built in. (Luhmann, 2002, p. 98 [author's translation]).

Of great interest is Luhmann's relation to pedagogy and instruction.

I earned a massive negative reaction of pedagogues when I explained to them that they want to educate their students like trivial machines that have to give right answers to certain questions. If the answer is wrong, it is wrong, if it is right, it is right. If it is wrong, the machine has an error, if it is right, it is good. The system does not allow the student, for example, to put the question into question, or look for creative ways out, or take an esthetic viewpoint on a mathematical formula, spread on the paper like concrete poetry, or do something which can only be explained if one knows the state he is in at the given moment. (Luhmann, 2002, p. 98–99 [author's translation]).

In one of the papers in this volume, Robert and Rogalski characterize an instructional episode in a way which suggests that, to a certain extent, the participating students had been treated by the teacher as "trivial machines".

If we examine this [the mathematical content and its use in exercises] from the point of view of the mediation by the teacher between students and mathematical knowledge we see that the management of students' activities leaves no room for them to wonder how to tackle the solution of a problem. The question of 'what has to be done' is immediately given by the teacher. The same applies to the procedures used for the solution of problems, even when the students are asked to provide the answers. Furthermore the time allotted to students' responses only allows brief answers by some students to 'well formulated questions'; only the time for the students, all of them, if possible, to make the final and already well defined calculations is less limited. The interventions provide a framework for what the students can do on their own. Nothing is left undefined, students never face uncertainties: there is little room for autonomy. (Robert and Rogalski, this volume).

The above description exemplifies the danger that such tightly guided instruction may indeed lead to educating students like trivial machines. Just like any living being, students also should be understood as autonomous, autopoietic systems in organized and intentional teaching and learning processes. The concept of "autopoietic system" was introduced by Maturana and Varela (see, e.g., 1987). It refers to systems, which exist and develop autonomously by means of their self-referential relation. They consist of components, which are produced permanently in order to keep the system going. Luhmann has further developed this concept and applied it to mental and social systems as well.

In mathematics teaching, teachers and students interact by communicating with each other. Niklas Luhmann characterizes "communication" as the constitutive concept of sociology:

But when communication is to come about, the . . . closed . . . autopoietic system has to become active, as a social system which reproduces communication by means of communication and does nothing but this. (Luhmann 1996, p. 279 [author's translation]).

In communicative situations, such as mathematics instruction, different persons are participating as mental consciousness systems. By means of the conversations and interactions between these individual persons a new autopoietic system develops – according to Luhmann – the communication system. What is the difference between social and mental systems and how do they act upon each other? The mental system is based on consciousness and the social system is based on communication.

A social system cannot think, a mental system cannot communicate. Causally seen, there are still immense, highly complex interdependencies. (Luhmann, 1997, p. 28).

How can these interdependencies be understood?

Communication systems and mental systems (or consciousness) form two clearly separated autopoietic domains. . . . These two kinds of systems are, however, closely connected to each other in a particular tight relation and mutually form a 'portion of a necessary environment': Without the participation of consciousness systems there is no communication, and without the participation in communication, there is no development of the consciousness. (Baraldi et al., 1997, p. 86 [author's translation]).

Thus, it is clearly stated that, basically, one can assume a separation of communication and consciousness and distinguish between the social and the individual. This is an unequivocal position, which cannot be found in statements such as "the cognitive and the social adaptations are interwoven" (Flückiger, this volume). At the same time, however, Luhmann's theory gives ample support to the claim that both systems – communication and consciousness – represent a necessary environment for each other, each system stimulating the other and thus triggering developments and changes in it. Therefore, the question is not whether the social or the individual represents the foundation for the other. Instead, the existence of a special mutual relationship between social and individual is assumed, in which both positions are necessary in a complementary way, while at the same time, there is a fundamental separation between them.

The participating persons are involved in communication and interaction. In the communication system, they repeatedly carry out communicative operations. But it is only by means of a distanced observation, that it is possible to reflect upon the current communication system.

The observation is able to identify objects and can [. . .] distinguish the inner processes of a system from what does not belong to it, can recognize causal relations between the within and the without, can assign an aim to the system. (Baraldi et al., 1997, p. 125 [author's translation]).

Considering the background of this explication of the assumptions about the relation between the social and the mental as two strictly separated systems, which, however, require each other as necessary environments, it is interesting to see that the students now clearly obtain a different theoretical position within the instructional communication.

Think about a class: A teacher observes the students – that is common. The students observe the teacher – they have to. The teacher also observes that the students observe him. But now, in addition, the interaction observes the students, sometimes even the teacher – that is rare, but it may happen: The teacher becomes the topic of discussion about instruction. The social system observes mental systems; the mental systems can observe social systems: 'Why is this asked exactly now, why does he always have to ask the questions I can't answer?' What happens here can be thematized in a psychological or communicative way. (Luhmann, 2002, p. 147–148, [author's translation]).

This fundamental view on the complexity of instructional processes, on the interacting persons, the teacher as well as the students, in which it is clearly distinguished between the individual mental systems, which cannot be accessed directly, and the communicative system of 'mathematical instructional interaction', opens up different orientations and new questions. In spite of the instructions, goals, controls and 'unequivocal hints' planned by the teacher for the course of the mathematics instruction, from this point of view, the students with their consciousness and their interactions have a significant autonomous position, which of course also indirectly affects the teacher's consciousness in the form of possible irritations and stimulations.

Only if students are educated as non-trivial machines, perceived as autopoietic consciousness systems within the instructional communication, and allowed to participate in it, the conditions of a true mathematical instruction communication can be analyzed in a theoretical way and from a mathematics education perspective. If the students are participating in a tight, strictly guided instruction, in which the teacher asks seemingly unequivocal questions and mathematical problems, to which he expects the right and unequivocal answers or solutions from the student, ultimately, it can only be controlled whether the student repeats in a quasi ritual way what the teacher has suggestively said. A deeper-going, individual mathematical understanding of the students cannot be examined.

If mathematical knowledge in learning processes could be reduced to this description [of a ready made curricular product], the interpretation of mathematical communication would become a direct and simple concern. When observing and analyzing mathematical interaction one would only have to diagnose whether a participant in the discussion has used the 'correct' mathematical word, whether he or she has applied a learned rule in the appropriate way, and then has gained the correct result of calculation, etc. (Steinbring, 2000a, p. 138).

If students are understood as autonomous consciousness systems, which express mathematical relations and ideas with their own concepts, words and descriptions, the analysis and interpretation of the children's communicative remarks within mathematics instruction becomes an important and challenging task for the researcher.

. . .when observing children's statements, it is impossible immediately to deduce a construction or a justification of new knowledge from the use of abstract notation, universal definitions or the introduction of variables. The students shall, must and can only try to make attempts to justify by using their own descriptions and covered in the expressions they have used until now. . .. But when the young students interactively describe new relations and new knowledge with the old exemplary, and partially concrete interpretations and references, an epistemological analysis is faced with the problem of finding out the extent to which these documented statements and contributions, with their customary, familiar descriptions, contain justifications of *new* knowledge or whether they are mere repetitions of knowledge already known mediated by the use of familiar expressions. (Steinbring, 2000a, p. 141)

The epistemological analysis of mathematical communication thus becomes a reconstruction process of the interpretations of mathematical knowledge, which have been interactively developed within the instruction. Often, the participants' individual mathematical intentions and interpretations cannot be directly derived and classified based on the documents (videos or instruction protocols and transcripts) and descriptions of the global structures and courses of the instructional communication. The analysis of the particular interpretations of mathematical knowledge, which have emerged within the common interactions, and their specific epistemological status represent a challenge for qualitative, epistemologically oriented instructional research (this theoretical task cannot be further represented here; it has been elaborated in detail as a result of a comprehensive research project (Steinbring, 2004).

For research in didactics of mathematics to obtain an autonomous identity, it is essential that the mathematics-students-teacher triplet be understood as an entity that cannot be further reduced to simpler elements. The still on-going development of the theoretical characterization of the object of research in mathematics education – processes of mathematical interaction and instruction – continues to reveal its immanent complexity and diversity. The reflection and explication of the concealed or perhaps only implicit assumptions about the conditions, the functioning and the theoretical positions of the elements of this triplet in different research approaches is thus important in order to further develop the theory, methods and instruments of research in mathematics education.

REFERENCES

Baraldi, C., Corsi, G. and Esposito, E.: 1997, *GLU. Glossar zu Niklas Luhmanns Theorie sozialer Systeme*. Frankfurt am Main: Suhrkamp.

Bauersfeld, H.: 1978, 'Kommunikationsmuster im Mathematikunterricht – Eine Analyse am Beispiel der Handlungsverengung durch Antworterwartung', in H. Bauersfeld (ed.), *Fallstudien und Analysen zum Mathematikunterricht*, Schroedel, Hannover, pp. 158–170.

Bauersfeld, H.: 1988, 'Interaction, Construction and Knowledge: Alternative Perspectives for Mathematics Education', in D.A. Grouws, T.J. Cooney and D. Jones (eds.), *Effective Mathematics Teaching*, NCTM and Lawrence Erlbaum, Reston, Virginia, pp. 27–46.

Bazzini, L. (Ed.): 1994, Theory and Practice in Mathematics Education, *Proceedings of the Fifth International Conference on Systematic Cooperation Between Theory and Practice in Mathematics Education*, Grado, Italy, Padua, ISDAF.

Cobb, P. and Bauersfeld, H. (Eds.): 1995, *The Emergence of Mathematical Meaning – Interaction in Classroom Cultures,* Vol. 2, Lawrence Erlbaum Hillsdale, New Jersey.

Even, R. and Loewenberg Ball, D. (Eds.): 2003, 'Connecting research, practise and theory in the development and study of mathematics education', *Educational Studies in Mathematics*, Special Issue, 54, 2–3.

Foerster, H. von: 1993, *Wissen und Gewissen: Versuch einer Brücke*, Frankfurt am Main: Suhrkamp.

Glasersfeld, E. von (Ed.): 1991, *Radical Constructivism in Mathematics Education,* Kluwer Academic Publishers, Dordrecht, Boston, London.

Krummheuer, G.: 1984, 'Zur unterrichtsmethodischen Diskussion von Rahmungsprozessen', *Journal für Mathematik Didaktik* 5(4), 285–306.

Krummheuer, G.: 1998, 'Formats of Argumentation in the Mathematics Classroom', in H. Steinbring, M. G. B. Bussi and A. Sierpinska (eds.), *Language and Communication in the Mathematics Classroom,* National Council of Teachers of Mathematics, Reston, Virginia, pp. 223–234.

Luhmann, N.: 1996, 'Takt und Zensur im Erziehungssystem', in N. Luhmann and K.-E. Schorr (eds.), *Zwischen System und Umwelt. Fragen an die Pädagogik,* Suhrkamp, Frankfurt am Main, pp. 279–294.

Luhmann, N.: 1997, *Die Gesellschaft der Gesellschaft*, Suhrkamp, Frankfurt am Main.

Luhmann, N.: 2002, Einführung in die Systemtheorie, Carl-Auer-Systeme Verlag, Heidelberg.

Maier, H. and Voigt, J. (Eds.): 1991, *Interpretative Unterrichtsforschung*, Aulis, Köln.

Maturana, H.R. and Varela, F.J.: 1987, *The Tree of Knowledge: The Biological Roots of Human Understanding*, New Science Library, Boston, London.

Seeger, F. and Steinbring, H. (Eds.): 1992, 'The dialogue between theory and practice in mathematics education: Overcoming the broadcast metaphor', *Proceedings of the Fourth Conference on Systematic Cooperation between Theory and Practice in Mathematics Education (SCTP).* Brakel. (IDM Materialien und Studien 38). Bielefeld: IDM Universität Bielefeld.

Steinbring, H.: 1994, 'Dialogue between theory and practice in mathematics education', in R. Biehler, R.W. Scholz, R. Sträßer and B. Winkelmann (eds.), *Didactics of Mathematics as a Scientific Discipline*, Kluwer Academic Publishers, Dordrecht, pp. 89–102.

Steinbring, H.: 1998, 'From 'Stoffdidaktik' to Social Interactionism: An evolution of approaches to the study of language and communication in German mathematics education research', in H. Steinbring, M.G.B. Bussi and A. Sierpinska (eds.), *Language and*

Communication in the Mathematics Classroom, National Council of Teachers of Mathematics, Reston, Virginia, pp. 102–119.

Steinbring, H.: 2000a, 'Interaction analysis of mathematical communication in primary teaching: The epistemological perspective', *Zentralblatt für Didaktik der Mathematik* 5, 138–148.

Steinbring, H.: 2000b, *Epistemologische und sozial-interaktive Bedingungen der Konstruktion mathematischer Wissensstrukturen (im Unterricht der Grundschule). (Abschlussbericht zu einem DFG–Projekt),* Dortmund: Universität, Dortmund.

Steinbring, H.: 2005, *The Construction of New Mathematical Knowledge in Classroom Interaction – An Epistemological Perspective.* Mathematics Education Library (MELI) Vol. 38, Springer, New York, Heidelberg.

Verstappen, P.F.L. (Ed.): 1988, *Report of the Second Conference on Systematic Cooperation Between Theory and Practice in Mathematics Education,* Enschede, S.L.O., Lochem/Netherlands.

Voigt, J.: 1994, 'Negotiation of mathematical meaning and learning mathematics', *Educational Studies in Mathematics* 26, 275–298.

Fachbereich 6, Mathematik
Universität Duisburg Essen
Campus Essen

9 780387 253534